Micro-Electro Discharge Machining: Principles, Recent Advancements and Applications, Volume II

Micro-Electro Discharge Machining: Principles, Recent Advancements and Applications, Volume II

Editors

Irene Fassi
Francesco Modica

MDPI • Basel • Beijing • Wuhan • Barcelona • Belgrade • Manchester • Tokyo • Cluj • Tianjin

Editors

Irene Fassi
STIIMA
CNR, National Research
Council of Italy
Milano
Italy

Francesco Modica
STIIMA
CNR, National Research
Council of Italy
Bari
Italy

Editorial Office
MDPI
St. Alban-Anlage 66
4052 Basel, Switzerland

This is a reprint of articles from the Special Issue published online in the open access journal *Micromachines* (ISSN 2072-666X) (available at: www.mdpi.com/journal/micromachines/special_issues/Micro_Electro_Discharge_Machining_II).

For citation purposes, cite each article independently as indicated on the article page online and as indicated below:

LastName, A.A.; LastName, B.B.; LastName, C.C. Article Title. *Journal Name* **Year**, *Volume Number*, Page Range.

ISBN 978-3-0365-6987-1 (Hbk)
ISBN 978-3-0365-6986-4 (PDF)

Contents

About the Editors

Irene Fassi

Irene Fassi is currently Research Director at CNR-STIIMA, where she leads the research group MEDIS, performing research activities in micro and meso-scale manufacturing and robotics. She gives courses and seminars in robotics, precision engineering, and advanced manufacturing systems at the University of Brescia and Politecnico di Milano. She is member of several Expert Evaluators Panels: European Commission, Agence Nationale de la Recherche, Norwegian Research Council, Eureka, MiSE, and MIUR. She has been the general or scientific chair of several international Conferences in micromanufacturing, microsystems, and robotics, including the World Congress on Micro and Nano Manufacturing. She is a member of the Executive Board of SIRI (Italian Robotics and Automation Association), the Italian Association of Manufacturing Technologies (AITeM), the ASME Technical Committee on Micro and Nano Manufacturing, and currently serves as President of the International Institution for Micro Manufacturing.

Francesco Modica

Francesco Modica is a full-time researcher at STIIMA-CNR, where he is currently researching mainly micro-manufacturing and micro-engineering at the STIIMA-CNR MicroManufacturing laboratory in Bari. From 2000 to 2009, he conducted research activities in mechatronic analysis and simulation of machine tools. From 2005 to 2009, he was also involved as researcher and project manager in collaboration with Sintesi s.p.a., developing mechatronic solutions (devices and simulations) to improve machine tools' dynamic performance. In 2009, he was visiting researcher at Katholieke Universiteit Leuven, applying to micro-electrical discharge machining. He has successfully led different research activities concerning micro-manufacturing processes (mainly related to micro-EDM, micro injection molding, and Additive Manufacturing for micro applications) and the design and fabrication of micro-components and devices. He is currently a member of the Institute board of STIIMA and Head of the secondary location of STIIMA Bari-Lembo.

Preface to "Micro-Electro Discharge Machining: Principles, Recent Advancements and Applications, Volume II"

Micro electrical discharge machining (micro-EDM) is a thermo-electric and contactless process most suited for micro-manufacturing and high-precision machining, especially when difficult-to-cut materials, such as super alloys, composites, and electro conductive ceramics, are processed.

The second volume of the Special Issue on "Micro-Electro Discharge Machining: Principles, Recent Advancements and Applications" confirms the growing interest in the micro-EDM technology as a suitable and efficient technology for machining novel, multilateral components, with demanding requirements in terms of precision, accuracy, and productivity.

This volume consists of 10 original research papers which involve several approaches to micro-EDM and cover the enhancement of the process performance, such as the material removal rate, surface roughness, or machining accuracy, using advanced optimization methods. Some studies also consider several dielectric fluid additives and investigate the processability of new materials. Others investigate the combination of Reverse-micro-EDM with laser beam micromachining or explore new applications for the micro-EDM for fabricating antimicrobial nanosilver colloid.

Authors who submitted their papers to this Special Issue are from academia, institutions of research and industry from Asia and Europe. We want to thank them all for their valuable work. We would also like to acknowledge all the reviewers for dedicating their time to provide careful and timely reviews to ensure the quality of this Special Issue.

Irene Fassi and Francesco Modica
Editors

micromachines

Editorial

Editorial for the Special Issue on Micro-Electro Discharge Machining: Principles, Recent Advancements and Applications, Volume II

Irene Fassi [1,*] and Francesco Modica [2,*]

1 STIIMA CNR, Institute of Intelligent Industrial Technologies and Systems for Advanced Manufacturing, National Research Council, Via A. Corti 12, 20133 Milan, Italy
2 STIIMA CNR, Institute of Intelligent Industrial Technologies and Systems for Advanced Manufacturing, National Research Council, Via P. Lembo 38/F, 70124 Bari, Italy
* Correspondence: irene.fassi@stiima.cnr.it (I.F.); francesco.modica@stiima.cnr.it (F.M.)

Citation: Fassi, I.; Modica, F. Editorial for the Special Issue on Micro-Electro Discharge Machining: Principles, Recent Advancements and Applications, Volume II. *Micromachines* **2023**, *14*, 29. https://doi.org/10.3390/mi14010029

Received: 15 December 2022
Accepted: 16 December 2022
Published: 23 December 2022

The second volume of the Special Issue on "Micro-Electro Discharge Machining: Principles, Recent Advancements and Applications" confirms the growing interest in the micro-EDM technology as a suitable and efficient technology for machining novel, multi-material components, with demanding requirements in terms of precision, accuracy and productivity.

This volume consists of 10 original research papers which involve several approaches to micro-EDM and cover the enhancement of the process performance, such as the material removal rate, surface roughness, or accuracy of the machining, using advanced optimization methods. Some studies also consider several dielectric fluid additives and investigate the processability of new materials. Others investigate the combination of Reverse-micro-EDM with laser beam micromachining or explore new applications for the micro-EDM for fabricating antimicrobial nanosilver colloid.

In more detail, in order to improve the machining accuracy of detail features in micro-EDM milling, a theoretical model is developed by Jing et al. [1] to simulate the micro-EDM milling process with a straight-line single path. In particular, the model is obtained by the accumulative difference mechanism in time and space. Micro-EDM milling experiments were carried out to verify the simulation model, showing that the maximum mean relative deviation between the microgroove profiles of simulation and experiments is 11.09%, with a good consistency in profile shapes.

The influence of the near-dry WEDM technique to reduce the environmental impact of wet WEDM was investigated by Chaudhari et al. [2]. The study employed a teaching–learning-based optimization (TLBO) algorithm to find the optimal combination of process parameters for material removal rate (MRR) and surface roughness (SR) considering near-dry WEDM of NiTiNol Shape Memory Alloy. Even if near-dry WEDM shows lower MRR in respect of wet-WEDM, it can machine a better surface morphology in terms of reduction in surface defects and better surface quality.

A mathematical–statistical computational (MSC) model for predicting high productivity and quality of the machined area is formulated by Straka et al. [3] by the application of non-linear programming (NLP) methods using MATLAB. The method is applied to maximize the process performance of micro-WEDM on a workpiece made of steel MS1 sintered via direct metal laser sintering (DMLS). Experimentation and results show the model's effectiveness in optimizing process performance.

Wire-cut electro-discharge machining (Wire-EDM) of polymer composite material (PCM) was studied by Ablyaz et al. [4]. Tests were performed on a workpiece made of a laminated fibrous polymer composite with carbon fiber twill as reinforcement/filler and epoxy as a binder material. The machining can be performed thanks to improved conductivity obtained using 1 mm thick titanium plates sandwiched on the PCM. The results demonstrated that voltage and pulse duration and their interaction are the significant factors affecting the cut-width accuracy for machining the PCM workpiece.

Sidhu et al. [5] used the analytic hierarchy process (AHP), a multiple-criteria decision-making technique, to achieve a reliable outcome for different responses in electric discharge machining (EDM) of metal matrix composites (MMCs). They identified the optimal process conditions by considering two materials, 65 vol% SiC/A356.2, and 10 vol% SiC-5 vol% quartz/Al composites, revealing that, in the presence of a suspended particle dielectric medium (PMEDM), a graphite tool electrode and higher pulse-on time coupled with lowest pulse-off time contributed to minimizing the residual stress with the desired MRR.

Ablyaz et al. [6] investigated a hybrid magnetic field assisted powder mixed electrical discharge machining on machining the aluminum-silicon carbide (Al-SiC) metal matrix composite to obtain a higher surface finish and enhanced the material removal rate. The dielectric mediums employ plain EDM oil, SiCp mixed, and graphite powder mixed EDM oil. They find that MRR augmented considerably with increased magnetic field intensity and peak current. At the same time, the quality of the machined surface improved significantly in graphite powder mixed dielectric flushing conditions with an intermediate external magnetic field environment. Moreover, micro-hardness enhancement was quantified as compared to base material due to the transfer of the material (SiCp).

Gattu et al. [7] mixed three different powders at different concentrations in a dielectric fluid: electrically conductive carbon nanofiber (CnF), semiconductive silicon (Si) powder, and insulative alumina powder (Al_2O_3). The study evaluated effects on material removal rate (MRR), relative electrode wear rate (REWR), and surface roughness on machining (EDM) of ultrafine particle type tungsten carbide and observing single discharge crater and hole machining tests. The results showed that adding CnF enhanced the material removal rate under all conditions. In contrast, Si and Al_2O_3 powders only improved the machining performance at a high discharge energy of 110 V. Improvement in surface roughness was observed prominently at high voltages for all the powders. However, alumina improved the surface roughness the most among the three powders.

Esser et al. [8] observed the discharge phenomena in the discharge gap by using a high-speed camera to study the effect on the machining process of tool vibration used to improve flushing conditions. They found that the discharges occurred in periodic intervals, and the intensity increased with the amplitude of tool vibration. Consequently, it was determined that, by adjusting the vibration parameters, it is possible to achieve optimum stability by improving the discharge distribution uniformity, increasing the machining efficiency and reducing the tool wear.

Reverse-μEDM was considered by Kishore et al. [9] with the fabrication of the tool plate realized by Nd:YAG-based laser beam micromachining (LB μM). The Grey relation analysis technique was used for optimizing LBμM parameters for producing tool plates with arrayed micro-holes in elliptical and droplet profiles. A duty cycle of 1.25% and a current of 20% were found to be an optimal setting for the fabrication of burr-free shallow striation microholes with a minimal dimensional error. After that, analogous protrusions were produced by Reverse-μEDM. Since the tool has no apparent cleavage or burrs at the micro-hole cut edges, it allows faster machining by restricting high-order discharges and short-circuiting during reverse-μEDM and obtaining protrusions free from tip damage.

Finally, an alternative study has been performed by Tseng et al. [10] with the implementation of a Micro-EDM Monitoring System to Fabricate Antimicrobial Nanosilver Colloid. The new system can replace the traditional oscillograph observation method, and its advantage consists of instantly observing and controlling discharge conditions. The monitoring system can use the discharge rate to control the energy consumption of the electrodes to standardize the nanosilver colloid. By experimentation, it was found that the nanosilver colloid prepared by EDM is free of any chemical additive that, compared to other preparation methods, is more applicable to biotechnology and the human body.

We thank all the authors who submitted their papers to this Special Issue, "Micro-Electro Discharge Machining: Principles, Recent Advancements and Applications". We would also like to acknowledge all the reviewers for dedicating their time to providing careful and timely reviews to ensure the quality of this Special Issue.

Conflicts of Interest: The authors declare no conflict of interest.

References

1. Jing, Q.; Zhang, Y.; Kong, L.; Xu, M.; Ji, F. An Investigation into Accumulative Difference Mechanism in Time and Space for Material Removal in Micro-EDM Milling. *Micromachines* **2021**, *12*, 711. [CrossRef] [PubMed]
2. Chaudhari, R.; Kevalramani, A.; Vora, J.; Khanna, S.; Patel, V.K.; Pimenov, D.Y.; Giasin, K. Parametric Optimization and Influence of Near-Dry WEDM Variables on Nitinol Shape Memory Alloy. *Micromachines* **2022**, *13*, 1026. [CrossRef] [PubMed]
3. Straka, L'.; Gombár, M.; Vagaská, A.; Kuchta, P. Efficiency Optimization of the Electroerosive Process in μ-WEDM of Steel MS1 Sintered Using DMLS Technology. *Micromachines* **2022**, *13*, 1446. [CrossRef] [PubMed]
4. Ablyaz, T.R.; Shlykov, E.S.; Muratov, K.R.; Sidhu, S.S. Analysis of Wire-Cut Electro Discharge Machining of Polymer Composite Materials. *Micromachines* **2021**, *12*, 571. [CrossRef] [PubMed]
5. Sidhu, S.S.; Ablyaz, T.R.; Bains, P.S.; Muratov, K.R.; Shlykov, E.S.; Shiryaev, V.V. Parametric Optimization of Electric Discharge Machining of Metal Matrix Composites Using Analytic Hierarchy Process. *Micromachines* **2021**, *12*, 1289. [CrossRef] [PubMed]
6. Ablyaz, T.R.; Bains, P.S.; Sidhu, S.S.; Muratov, K.R.; Shlykov, E.S. Impact of Magnetic Field Environment on the EDM Performance of Al-SiC Metal Matrix Composite. *Micromachines* **2021**, *12*, 469. [CrossRef] [PubMed]
7. Gattu, S.D.; Yan, J. Micro Electrical Discharge Machining of Ultrafine Particle Type Tungsten Carbide Using Dielectrics Mixed with Various Powders. *Micromachines* **2022**, *13*, 998. [CrossRef] [PubMed]
8. Esser, G.; Yan, J. Direct Observation of Discharge Phenomena in Vibration-Assisted Micro EDM of Array Structures. *Micromachines* **2022**, *13*, 1286. [CrossRef] [PubMed]
9. Kishore, H.; Nirala, C.K.; Agrawal, A. Laser Micromachining in Fabrication of Reverse-μEDM Tools for Producing Arrayed Protrusions. *Micromachines* **2022**, *13*, 306. [CrossRef] [PubMed]
10. Tseng, K.-H.; Chung, M.-Y.; Chiu, J.-L. Implementation of Micro-EDM Monitoring System to Fabricate Antimicrobial Nanosilver Colloid. *Micromachines* **2022**, *13*, 790. [CrossRef] [PubMed]

micromachines

Article

An Investigation into Accumulative Difference Mechanism in Time and Space for Material Removal in Micro-EDM Milling

Qi Jing [1,2], Yongbin Zhang [2], Lingbao Kong [1,*], Min Xu [1] and Fang Ji [2]

1 Shanghai Engineering Research Center of Ultra-Precision Optical Manufacturing, School of Information Science and Technology, Fudan University, Shanghai 200438, China; qjing17@fudan.edu.cn (Q.J.); minx@fudan.edu.cn (M.X.)

2 Institute of Machinery Manufacturing Technology, China Academy of Engineering Physics, Mianyang 621000, China; zcaep6@163.com (Y.Z.); jfang2013@caep.cn (F.J.)

* Correspondence: LKong@fudan.edu.cn; Tel.: +86-21-3124-2565

Abstract: In micro-electrical discharge machining (micro-EDM) milling, the cross-section of the microgroove machine is frequently not an ideal rectangle. For instance, there are arc shapes on the bottom and corners, and the sidewall is not steep. The theoretical explanation for this phenomenon is still lacking. In addition to the tip discharge effect, the essential reason is that there is an accumulative difference in time and space during the shape change process of a tool electrode and the microstructure formation on a workpiece. The process parameters are critical influencing factors that determine this accumulative difference. Therefore, the accumulative difference mechanism in time and space is investigated in this paper, and then a theoretical model is developed to simulate the micro-EDM milling process with a straight-line single path. The simulation results for a cylindrical electrode at the two rotational speeds of 0 (nonrotating) and 300 rpm are compared, while the results for a cylindrical electrode and a square electrode at a rotation speed of 0 are also compared to verify that different process parameters generate accumulative differences in the time and space of material removal. Finally, micro-EDM milling experiments are carried out to verify the simulation model. The maximum mean relative deviation between the microgroove profiles of simulation results and those of experiments is 11.09%, and the profile shapes of simulations and experiments have a good consistency. A comparative experiment between a cylindrical electrode and a hollow electrode is also performed, which further verifies the mechanism revealed in the study. Furthermore, the cross-section profile of a microgroove can be effectively controlled by adjusting the process parameters when utilising these accumulative differences through fabricating a microgroove with a V-shaped cross-section by a square electrode and a microgroove with a semi-circular cross-section by a cylindrical electrode. This research provides theoretical guidance for solving the problems of the machining accuracy of detail features in micro-EDM milling, for instance, to machine a microgroove with an ideal rectangular cross-section.

Keywords: micro-EDM milling; accumulative difference; material removal; tool electrode wear

Citation: Jing, Q.; Zhang, Y.; Kong, L.; Xu, M.; Ji, F. An Investigation into Accumulative Difference Mechanism in Time and Space for Material Removal in Micro-EDM Milling. *Micromachines* **2021**, *12*, 711. https://doi.org/10.3390/mi12060711

Academic Editors: Irene Fassi and Francesco Modica

Received: 26 May 2021
Accepted: 14 June 2021
Published: 17 June 2021

1. Introduction

Micro-electrical discharge machining (micro-EDM) is a kind of machining technology that removes materials by means of the extremely high temperature generated by spark discharge. The tool electrode and the workpiece retain a certain gap between them and do not come into contact, and there is no macro-cutting force. Microtool electrodes with a scale of several microns can be fabricated; therefore, the size of machined structures can also reach the level of several microns [1]. This process is particularly suitable for machining both hard-to-cut [2–6] and easily deformed conductive materials [7].

As an electrical machining method, the machining localisation and precision of micro-EDM are very high. There is no stray corrosion, which often appears in micro-electrochemical machining (micro-ECM) [8], when the dielectric fluid material is kerosene or air. Furthermore, the corrosion of machine tools and environmental pollution can also be avoided. There is no burr, which is commonly found in micro-milling [9], at the edge of the machining structures. It is hence more suitable for the processing of microstructures on hard metal materials than micro-milling. By controlling the tool electrode with complex three-dimensional movements, complicated micro three-dimensional structures can be fabricated expediently by micro-EDM milling [10], which is generally difficult to achieve in MEMS, LIGA and other semiconductor processes [11,12]. Laser micromachining technology can also be used for machining conductive materials such as metals. Although they have a similar principle, which is to melt materials and vaporize them through a thermal effect, the verticality of the sidewalls and surface quality of the structures machined by micro-EDM are better than those machined by laser micromachining [13]. In addition, micro-EDM offers a better economy, flexibility and convenience than laser micromachining [14]. Some studies have shown that micro-EDM can also be used to machine other materials such as glass, ceramics and semiconductors by adding an auxiliary electrode or changing the working fluid [15,16]. Furthermore, there are many studies focusing on hybrid micro-EDM, and many significant results have been achieved [17]. Therefore, micro-EDM is playing an increasingly important role in the field of micromachining and has many potential applications.

It is typically difficult to fabricate a forming tool electrode for complex micro three-dimensional structures, and the dimensional accuracy of the forming tool is hard to maintain because of the high tool wear. Hence, die-sinking EDM technology is generally not employed to machine complex structures. Wire electrical discharge grinding (WEDG) was developed to address in situ preparation of microtool electrodes in the 1980s, which improved the machining accuracy and promoted the use of micro-EDM [18]. Subsequently, three-dimensional structures, such as microcavities, were fabricated by controlling the microtool electrode for layered milling [10]. However, problems remain in micro-EDM milling, such as tool-electrode wear. Furthermore, the material removal volume of a workpiece is often very large during micro-EDM milling; thus, the axial wear of the tool electrode is considerable due to the relatively small radial dimension, which not only affects the machining efficiency but also seriously influences the machining accuracy. To solve this problem, some studies have focused on the reduction of electrode wear, and others have focused on compensation strategies [19]. Yu et al. [20] proposed the uniform wear method (UWM) for the machining of micro moulds. This method maintains the original electrode shape and converts the three-dimensional electrode wear to a linear process, which greatly simplifies the compensation strategy. Subsequent research put forward various strategies to advance this method [21]. Yu et al. [22] proposed a combined linear uniform (CLU) method that combined the linear compensation method (LCM) and the UWM. The material removal rate, electrode wear ratio and surface roughness were improved using the proposed method compared to those by the uniform wear method while machining a square cavity with slanted surfaces. Li et al. [23] proposed a compensation method based on the scanned area (BSA) in each machining layer. Three-dimensional microcavities were generated using this method. The machining efficiency was improved, and the tool wear ratio was reduced compared with the UWM and CLU methods. Pei et al. [24] introduced a fixed-length compensation (FLC) method where compensation was applied after a fixed machining length. The dimensional precision and shape accuracy of the machined groove were enhanced by this method.

The above studies are all based on the UWM and assume that the wear of the electrode end is uniform through layered thickness control and path planning. Thus, the axial length wear of the electrode and its compensation are the main focus, and whether there is nonuniform wear of the electrode is usually not considered. In addition, the machining efficiency is limited due to the relatively small monolayer thickness. Therefore, Zhang et al. [25] introduced a large monolayer thickness milling method by modifying the conventional FLC model and achieved higher machining efficiency, precision and quality compared with the layer-by-layer machining method. Zhang et al. [26] developed a two-dimensional geometrical simulation model to predict the machined surface of the tool and the workpiece. They provided a better realisation of a cone-shaped tool and indicated that the cone angle of the electrode was formed through the relative motion of the electrode along the workpiece. Pei et al. [27] proposed an improved fixed-length compensation method that was validated experimentally by replacing a cylindrical tool with a tubular-shaped tool, and a truncated conic tool end was observed in experiments. They also developed an analytical model to relate the truncated conic shape to fixed-length compensation parameters. The model was no longer confined to face milling in these studies, but monolayer thickness was increased so that the electrode sidewall is involved in peripheral milling at the same time, which obviously causes radial wear of the electrode and changes the end of the electrode to a conical or truncated conical shape.

However, other studies have found that the shape of the electrode end is also changed during micro-EDM milling by the nonuniform wear of the electrode. This nonuniform wear will lead to an arc shape of the machined groove profile and significantly affect the machining accuracy during the milling of the cavity. Yan et al. [28] presented an electrode wear compensation method based on a machine vision system. Tool electrode wear can be evaluated directly by this vision system. They found that an arc shape at the end of the tool electrode appeared after just one layer of machining. Pham et al. [29] carried out a number of experiments to analyse the shape change of the electrode. They found that the electrode tended to change during machining towards a constant shape and that the milled groove was an arc shape. Karthikeyan et al. [30] found that the shape of the electrode end changed after channel machining during a study of micro-EDM milling. They thought that the tool rotation motion was a key factor. Nguyen et al. [31] presented geometric models to identify and analyse the error components of the 3D micro-EDM milling process. They found that aside from the inherent machining gap and the unavoidable electrode wear, the corner radius of the virtual electrode in the model was also of prime importance in determining the machining accuracy, which indicated that the rounded shape of the tool edge leads to significant geometrical inaccuracies. However, these studies mainly focused on the influence of various process parameters on the machining results under certain specific situations. The mechanism was rarely discussed and analysed. What is more, a generic solution, which may realise the prediction of micro-EDM milling, has not been proposed yet.

At present, in addition to the requirements of dimensional accuracy for many micro three-dimensional structures, important detailed features in the structures are also needed to meet high accuracy requirements. For instance, a folded waveguide with a rectangular cross-section is a common slow-wave structure of the Terahertz travelling wave tube [32], which has strict accuracy requirements on the cross-section, such as the steepness of the sidewall, the flatness of the bottom and the corner radius of the sidewall and bottom. However, it can be seen from previous studies that the nonuniform wear of the tool electrode made the accuracy of the detailed features worse; for instance, an uneven microcavity bottom and a larger sidewall corner radius and cavity bottom. Although a flatter cavity bottom can be achieved by using a smaller microtool electrode through overlapping moving paths, the size of the corner radius is still hard to control. Moreover, it is also difficult to realise the overlap of the moving electrode paths for some narrow grooves with a large aspect ratio, which is usually machined only by the single-path layered milling method. In this case, accuracy control problems will occur easily if the process parameters are inappropriate. Therefore, the effective control of the nonuniform wear of the tool electrode is of great significance in improving the dimensional accuracy of the micro-detailed features, reduce the overlapping rate of the machining paths during the machining process and thereby improve the machining efficiency.

On the other hand, the cross-sectional profile is not necessarily rectangular for some more complex three-dimensional microstructures. Some cross-sections are semi-circular, such as the electron beam tunnels in Terahertz travelling wave tubes [32]. Some are trapezoidal or even V-shaped profiles, such as surface micro-textures [33] and microflow channels [34]. From this perspective, the nonuniform wear characteristics of the electrode in micro-EDM milling can be used to obtain various groove profiles.

Axinte et al. [35] demonstrated that, despite differences in their nature, many energy beam (EB) controlled-depth machining processes could be modelled using the same mathematical framework in the study of freeform surfaces machining. They have developed approaches and algorithms for the inverse problem to generate freeform surfaces using different EB machining processes and various workpiece target materials [36]. Wan et al. [37,38] proposed a time-variant and space-variant tool influence function (TIF) model in the study of precision optical polishing. In order to work out the dwell time, they used a linear transformation to replace the convolution operation. These studies in other fields make us realise that a generic solution can be explored from this perspective to solve the problem of machining accuracy of microstructures caused by the wear of tool electrode.

Micro-EDM is a "profile-copying" machining method, so the shape of the tool electrode and the shape of the machining structure influence and replicate each other. For example, in the process of microgroove milling with a cylindrical electrode, if the cross-section of the microgroove is not an ideal rectangle, it means that the axial cross-section of the electrode is also no longer an ideal rectangle, which means that there is nonuniform wear at the end of the electrode. A literature review indicates that the mechanism and the influence rules of the factors are still unclear for the nonuniform wear phenomenon of tool electrodes and the corresponding profile imprinted onto the workpiece. Hence, further studies should be undertaken to explain this phenomenon and provide theoretical guidance for solving the problems of the machining accuracy of detail features.

To solve this practical problem, the accumulative difference mechanism in time and space of material removal in micro-EDM milling is analysed and investigated combined with the key process parameters. Then, a micro-EDM simulation model is established to verify the mechanism, and machining experiments are also carried out to demonstrate the validities of the simulation model and the revealed mechanism. In addition, grooves with different cross-sections are generated by experiments through the utilisation of this accumulative difference in time and space. This work provides theoretical guidance for solving the problems of the machining accuracy of detail features in micro-EDM milling, for instance, to machine a microgroove with an ideal rectangular cross-section.

2. Accumulative Difference Mechanism in Time and Space for Material Removal

Micro-EDM milling is a kind of EDM method that controls the microtool electrode to move along a certain path, as shown in Figure 1. To keep the radial shape of the electrode unchanged as much as possible, the layered machining method is generally employed. In layered machining, the thickness of one layer is set to be less than the critical discharge gap so that the discharge sparks mainly occur on the bottom end face of the electrode and the opposite face on the workpiece. The sidewall of the electrode hence can be prevented from discharging as much as possible [20]. For each layer of micro-EDM milling, the machining path is shown in Figure 2a. It can be regarded as consisting of many simple paths. The most common single path, named the straight-line path, is shown in Figure 2b. The straight-line path is the basis of micro-EDM milling, which is, therefore, the main study objective in this paper.

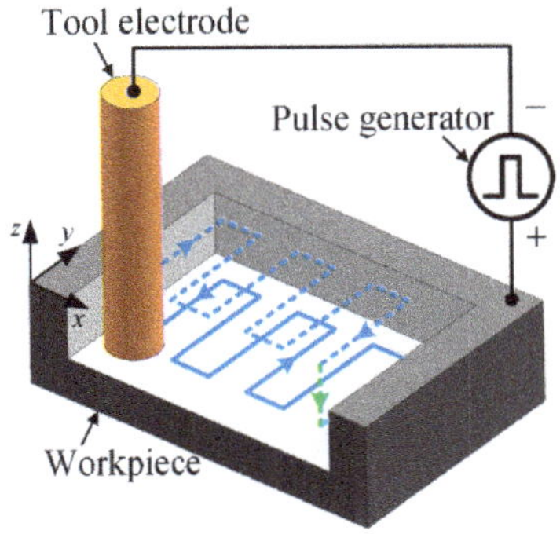

Figure 1. Schematic diagram of micro-EDM milling.

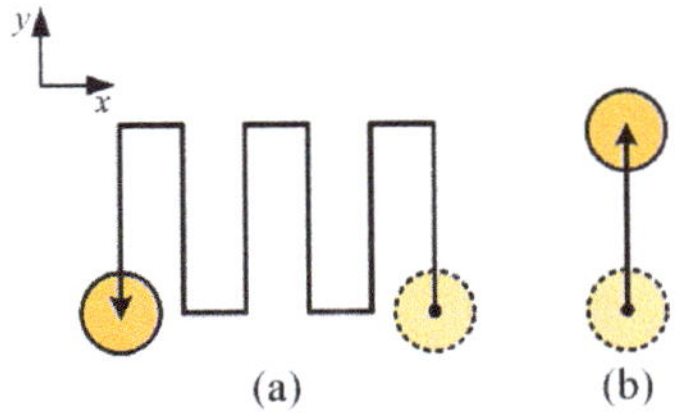

Figure 2. Schematic diagram of (**a**) milling paths in one layer and (**b**) straight-line milling path.

Generally, micro-EDM milling can be simplified as a discharge model between the bottom end surface of the cylindrical electrode and the corresponding surface of the workpiece, as shown in Figure 3. The relative movement between the two surfaces consists of the rotational motion of the cylindrical tool electrode itself and its movement along the machining path. The two surfaces are eroded continuously by the discharging that occurs when a high-frequency pulse breaks down the dielectric fluid.

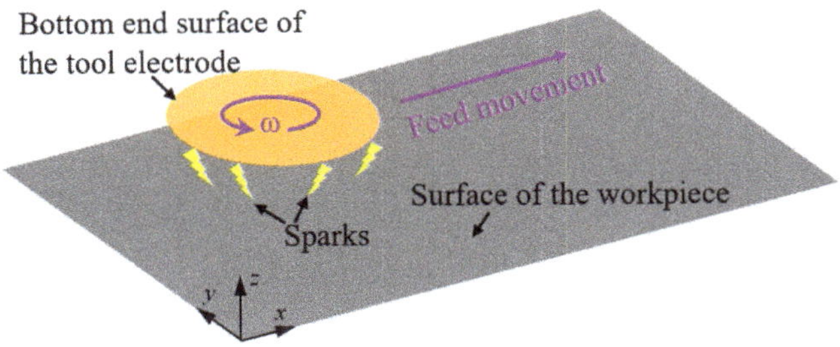

Figure 3. Schematic diagram of the simplified discharge model of micro-EDM milling.

The coordinate of an arbitrary point on the bottom surface of the tool electrode is set as (x_T, y_T), and the coordinate of an arbitrary point on the surface of the workpiece is as (x_W, y_W). After n pulses, the z-coordinate can be employed to represent the change in the depth direction of an arbitrary point on the tool electrode or the workpiece, as expressed in Equation (1).

$$\begin{cases} z_T(x_T, y_T) = z_{T0} + \sum_{j=1}^{n} d_j \\ z_W(x_W, y_W) = z_{W0} - \sum_{j=1}^{n} h_j \end{cases} \quad (1)$$

where $z_T(x_T, y_T)$ represents the z-direction position of point (x_T, y_T) and $z_W(x_W, y_W)$ represents the z-direction position of point (x_W, y_W). z_{T0} and z_{W0} are the initial z-direction positions of the tool electrode bottom surface and the workpiece surface, respectively. d_j and h_j represent the removal depth at this point of the j-th pulse on the electrode and the workpiece, respectively. Both d_j and h_j can be equal to zero.

The tool electrode and the workpiece both undergo electrical discharge erosion as the machining continues, and the materials are removed from both. Thus, the surface morphologies of the tool electrode and the workpiece, which can be represented by z_T and z_W, respectively, are changing continuously. Furthermore, the positional relationship of the spatial point pair that is formed by one point on the electrode and one point on the workpiece also changes with the relative movement between the electrode and workpiece. The spatial position of the pair of erosional craters generated by the effective discharge pulse will hence change over time. Therefore, the formation of the tool electrode and the workpiece profiles is a process in which the erosions of the discharge pulses at different times and spaces accumulate continuously.

During this process, taking the tool electrode as an example, m points can be selected randomly on the bottom end surface of the electrode, and a matrix $\boldsymbol{D}$ can be presented to show their removal depths in time and space after n pulses.

$$\boldsymbol{D}(m \times n) = \begin{bmatrix} d_{11} & d_{12} & d_{13} & \cdots & d_{1n} \\ d_{21} & d_{22} & d_{23} & \cdots & d_{2n} \\ d_{31} & d_{32} & d_{33} & \cdots & d_{3n} \\ \vdots & \vdots & \vdots & \ddots & \vdots \\ d_{m1} & d_{m2} & d_{m3} & \cdots & d_{mn} \end{bmatrix}$$

where the i-th row vector $\boldsymbol{d}_i = \begin{bmatrix} d_{i1}, & d_{i2}, & d_{i3}, & \cdots, & d_{in} \end{bmatrix}$ of matrix $\boldsymbol{D}$ represents the removal depth sequence of point i. The j-th column vector $\boldsymbol{d}_j = \begin{bmatrix} d_{1j}, & d_{2j}, & d_{3j}, & \cdots, & d_{mj} \end{bmatrix}^{\mathrm{T}}$ of matrix $\boldsymbol{D}$ represents the removal depth sequence of all the points generated by pulse j.

There are many process parameters in micro-EDM, such as parameters that determine the discharge energy per single pulse, including the open-circuit voltage, peak current and pulse width, and other parameters, such as the electrode material, workpiece material, dielectric fluid, machining polarity, rotation speed, geometric shape of the electrode, feed speed, path, pulse frequency, layer thickness and control threshold of the gap voltage. The various combinations of these parameters will result in different material removal volumes per discharge and lead to different spatial positions per discharge where the removal is located, which means that different process parameters determine different matrixes $\boldsymbol{D}$.

For a certain set of process parameters, the formation of the removal depth sequence $\boldsymbol{d}_j$ generated by pulse j will be initially influenced in the time domain by the accumulative result of the erosions generated by the previous j-1 pulses. Meanwhile, the surface morphologies of the tool electrode and the workpiece change due to the erosion of the j-th pulse. The spatial relative position of the tool electrode and workpiece is also changed by the relative movement between them during this time, which will also influence the composition of the next column vector, which is $\boldsymbol{d}_{j+1}$. Because of this accumulative effect in time and space, the accumulative removal amount may be different at different positions of the end surface of the tool electrode, which means that each row of matrix $\boldsymbol{D}$ has differences. If proper parameters are adopted so that the sum of $\boldsymbol{d}_i$ that can be represented by $\sum_{j=1}^{n} d_{ij}$ is approximately equal regardless of how much i is, then uniform wear of the electrode can be achieved. The material of the corresponding machining area of the workpiece can also be uniformly removed according to a "profile-copying" micro-EDM characteristic, such as the groove with a rectangular cross-section. However, if the parameters adopted make a relatively large difference in $\sum_{j=1}^{n} d_{ij}$ when i is different, then nonuniform wear of the electrode occurs. Likewise, this kind of nonuniform shape change will also be "copied" to the workpiece, resulting in nonuniform material removal of the workpiece, such as a groove with an arc cross-section.

Consequently, this accumulative difference in time and space can be adjusted by adjusting the process parameters. As a result, the wear amount at different positions on the bottom of the electrode and the final machining profile of the workpiece can be effectively controlled.

In the straight-line single path of micro-EDM milling, taking the common cylindrical tool electrode as an example, the different combinations of the feed speed and rotation speed of the electrode can reveal the difference in the spatial point pair relationship between the tool electrode and workpiece. The probability of pulse erosion at the different positions on the bottom surface of the electrode will thus be different. The motion trajectory of different positions on the bottom surface of the cylindrical electrode when the feed direction is in the x-direction is expressed by Equation (2).

$$\begin{cases} x = vt - r\sin\left(\frac{\pi n}{30}t\right) \\ y = r\cos\left(\frac{\pi n}{30}t\right) \end{cases} \quad (2)$$

where x and y are the x-coordinate and y-coordinate of the point, respectively, v is feed speed (μm/s), t is the time (s), r is the distance between the centre point of the bottom surface of the electrode and the point (μm) and n is the rotation speed (rpm).

According to the relationship between the linear feed speed and rotation speed, the motion trajectory of the point that is on the circle with a radius of $r = \frac{30v}{\pi n}$ is cycloid, and the motion trajectory of the other points (except the centre point of the bottom surface of the electrode) is trochoid.

The motion trajectories are shown in Figure 4. Figure 4a shows the motion trajectories of the points in different radii on the bottom surface of the electrode when the rotation speed of the electrode is 0, and Figure 4b–e show those when the rotation speed of the electrode is 30, 60, 120 and 300 rpm, respectively. It can be seen from Figure 4a that when the electrode is not rotating, the trajectories are parallel. The chord length of the bottom surface of the electrode along the machining direction varies with its distance from the machining centre axis. Obviously, the smaller the distance between the chord and the machining centre axis is, the longer the chord length is, and the more material removal of the workpiece is. When the electrode rotates, the motion trajectories of the points in different radii are quite different, so the probability of discharge at each point may also be different. When the feed speed remains unchanged, these trajectories become denser with increasing rotation speed, and the difference in the probability of discharge at each point will also change.

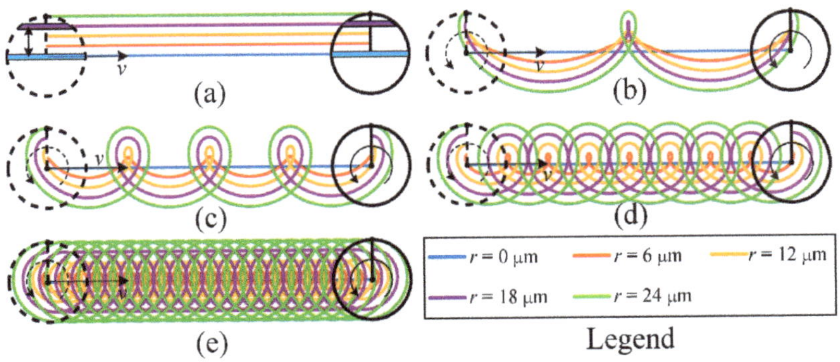

Figure 4. Differences in spatial motion trajectories of different points on the bottom surface of a cylindrical electrode when the rotation speed of the electrode is (**a**) 0; (**b**) 30; (**c**) 60; (**d**) 120; and (**e**) 300 rpm.

Through the above analysis, it can be concluded that the combination of various process parameters in micro-EDM milling will influence the composition of matrix $\boldsymbol{D}$, which generates the difference in material removal under different parameters. On the other hand, the matrix $\boldsymbol{D}$ may be different in time and space dimensions for a certain set of process parameters. This is the accumulative difference mechanism in time and space for material removal in micro-EDM milling.

3. Modelling and Simulation of Material Removal in Micro-EDM Milling

Micro-EDM is a complicated physical process that is affected by various influencing factors. It is difficult and laborious to simulate the entire micro-EDM milling machining process from the perspective of physical processes [39]. In some previous studies, researchers have proposed geometric simulation models to simulate various EDM processes [40–43].

The profiles of the electrode and workpiece generated in micro-EDM are accumulative results that are the accumulation of each discharge crater eroded by a single discharge pulse [39]. Based on this technique, a geometric simulation method will be employed in this paper [42], in which the single discharge crater eroded by a single discharge pulse is considered the elementary unit to calculate the amount of material removal. A schematic diagram of a single discharge crater is shown in Figure 5. The actual crater on the workpiece shown in Figure 5a can be simplified with a spherical crown model shown in Figure 5b, where h is the depth of the crater and d is the diameter of the crater.

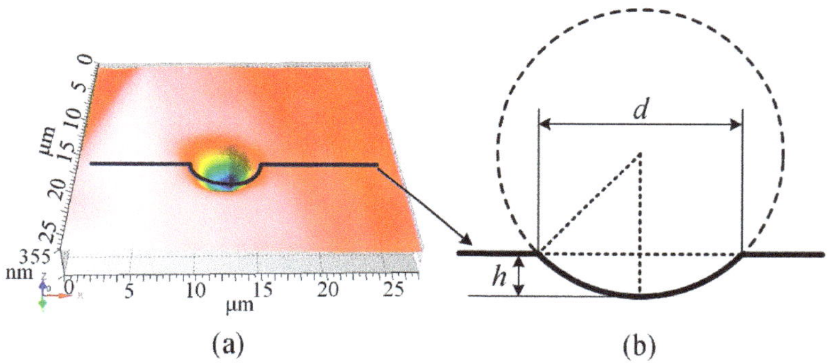

Figure 5. Crater eroded by single discharge: (**a**) actual crater on the workpiece, and (**b**) simplification of crater shape with a spherical crown model.

Therefore, the following assumptions are devised before modelling to simplify the simulation process: (1) the discharge occurs at the points between the tool electrode and the workpiece with the shortest distance that are no more than the critical discharge gap; (2) only one micro-energy electrical pulse is generated at each time step in the simulation; (3) the energy of each effective discharge spark is constant, and the material removal volume per spark is constant; and (4) the influences of the removed debris and the flushing of the dielectric fluid are ignored.

To describe the material removal through calculation, the two geometric models of the tool electrode and the workpiece and the time domain need to be discretised. Hence, the Z-map algorithm [42] is employed to define the geometric models, which represent the bottom surface of the tool electrode and the upper surface of the workpiece, and a set of square grids with the same size are used to mesh them, as shown in Figure 6. Therefore, the workpiece coordinate system is fixed, and the relative motion during machining can be expressed as the movement of the tool electrode coordinate system in the workpiece coordinate system. Therefore, the X-Y coordinates of the Z-map node of the tool electrode in the workpiece coordinate system will change with the movement and rotation of the tool electrode. The shape of a single discharge crater on the tool electrode and the workpiece can also be represented by a Z-map, as shown in Figure 7. In the time domain, the simulation time step Δt is determined by the pulse frequency f according to assumption (2), as shown in Equation (3).

$$\Delta t = \frac{1}{f} \quad (3)$$

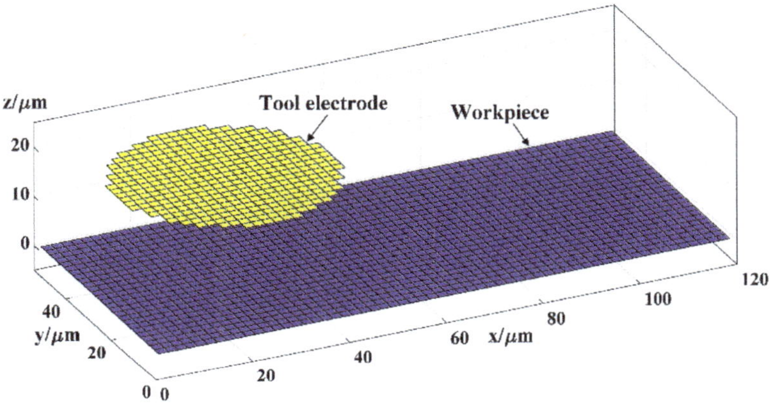

Figure 6. Meshing of the tool electrode and the workpiece.

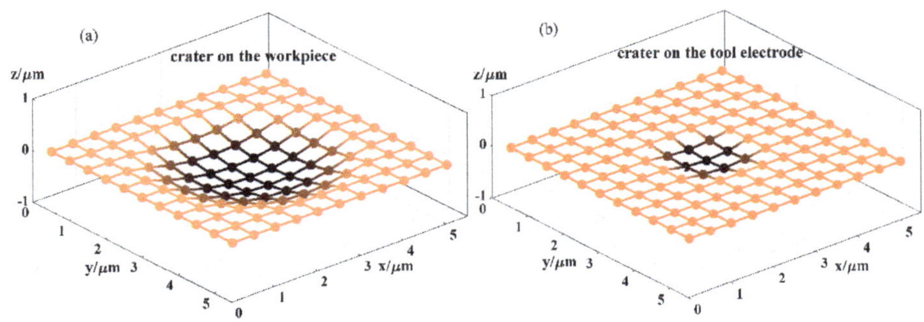

Figure 7. Schematic diagram of a single discharge crater represented by grids: (**a**) crater on the workpiece and (**b**) crater on the tool electrode.

Thus, the entire calculation and simulation processes of micro-EDM milling are described by the following steps: (1) calculating and updating the position coordinates of the grid node of the tool electrode in the workpiece coordinate system according to the state of motion of the electrode; (2) determining whether discharge occurs and finding the location where the discharge occurs on both the tool electrode and workpiece; and (3) generating the single discharge craters at the corresponding locations for the tool electrode and the workpiece, respectively. The three steps are repeated until the set goals are reached. The flow chart of the entire simulation process is shown in Figure 8.

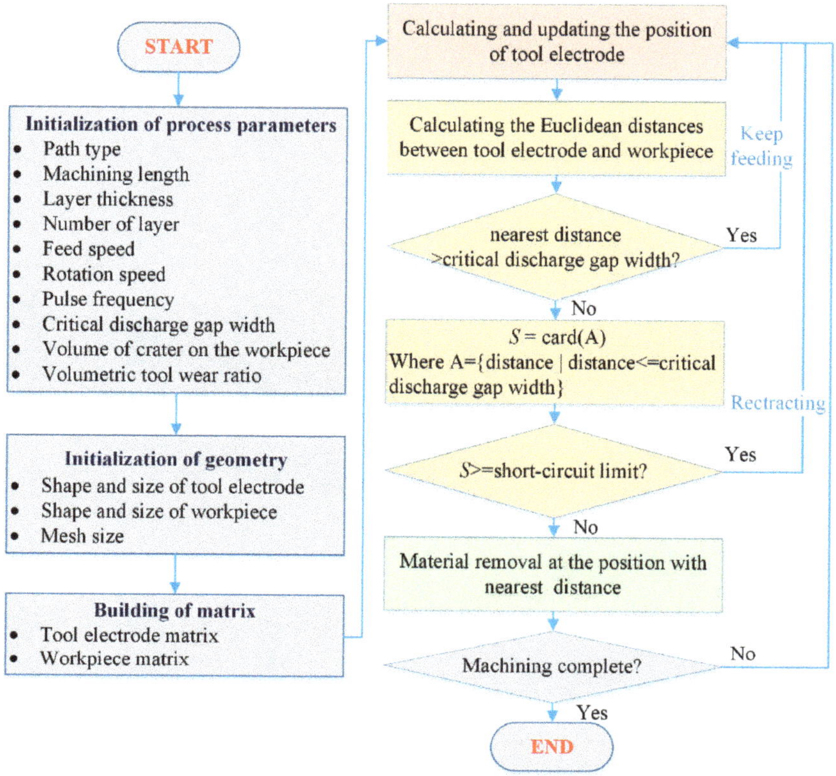

Figure 8. Schematic flowchart of the micro-EDM milling process simulation model.

As shown in the simulation flow chart, there may be three situations for step (2) according to assumption (1):

(a) The distance between the nearest point pair is larger than the critical discharge gap, which means that material removal is not generated by the current pulse during this time step. This is the open-circuit state in micro-EDM.

(b) There are S point pairs whose distances are less than or equal to the critical discharge gap, but S is larger than or equal to the number of short-circuit limits. Material removal is not generated by the current pulse in this time step either, and the tool electrode will retract along the original path. This is the short-circuit state in micro-EDM.

(c) Conversely, if S is less than the number of short-circuit limits, the point pair whose distance is the shortest will be selected as the locations where the discharge occurs. If there are multiple point pairs that have the same shortest distance, a pair will be randomly selected from them. Material removal is generated by the current pulse during this time step. This is the effective-spark state in micro-EDM. The Euclidean distances of the point pairs between the tool electrode and the workpiece can be calculated by Equation (4).

$$G(M \times N) = \sqrt{\begin{pmatrix} \|T_1\|_2^2 & \|T_1\|_2^2 & \cdots & \|T_1\|_2^2 \\ \|T_2\|_2^2 & \|T_2\|_2^2 & \cdots & \|T_2\|_2^2 \\ \vdots & \vdots & \ddots & \vdots \\ \|T_M\|_2^2 & \|T_M\|_2^2 & \cdots & \|T_M\|_2^2 \end{pmatrix} + \begin{pmatrix} \|W_1\|_2^2 & \|W_2\|_2^2 & \cdots & \|W_N\|_2^2 \\ \|W_1\|_2^2 & \|W_2\|_2^2 & \cdots & \|W_N\|_2^2 \\ \vdots & \vdots & \ddots & \vdots \\ \|W_1\|_2^2 & \|W_2\|_2^2 & \cdots & \|W_N\|_2^2 \end{pmatrix} - 2 \times TW^T} \quad (4)$$

where T_i represents the i-th row of matrix and $T(M \times 3)$ are the three-dimensional coordinates of all grid nodes on the tool electrode. W_i represents the i-th row of matrix $W(N \times 3)$, and $W(N \times 3)$ are the three-dimensional coordinates of all grid nodes on the workpiece. $G(M \times N)$ is the distance matrix of point pairs between the tool electrode and the workpiece.

Two types of electrodes, cylindrical electrodes and square electrodes, which are common in micro-EDM, are selected in the simulation experiment. There are two machining modes—the rotating and nonrotating mode—for the cylindrical electrode and only the nonrotating mode for the square electrode. In addition, there are also two types of paths—a unidirectional and reciprocating path—in straight-line path milling, as shown in Figure 9. The simulation parameters are presented in Table 1. The critical discharge gap, the shapes of the single discharge craters on the tool electrode and workpiece and the volumetric tool wear ratio depend on the machining parameters, which are determined through preliminary experiments. The electrical parameters and materials of the preliminary experiments are the same as those in Section 4.

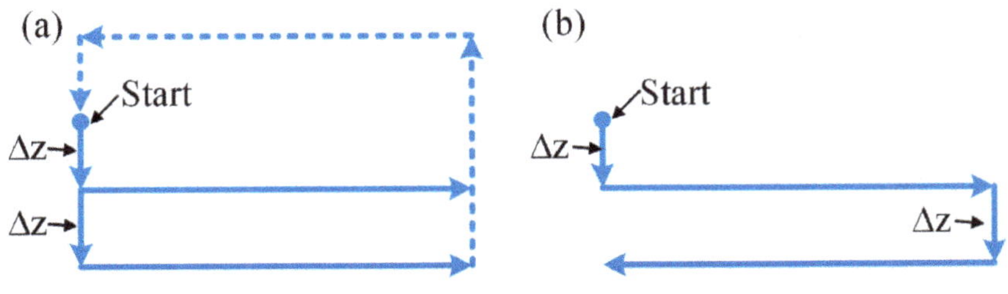

Figure 9. Schematic diagram of two types of paths in straight-line path milling: (**a**) unidirectional path and (**b**) reciprocating path.

Table 1. Simulation parameters.

Parameters	Serial Number					
	a	**b**	**c**	**d**	**e**	**f**
Tool type	cylindrical		square	cylindrical		square
Rotation speed (rpm)	300	0		300	0	
Path type	unidirectional path			reciprocating path		
Tool diameter or edge length (μm)	46					
Feed speed (μm/s)	30					
Mesh size (μm)	0.5					
Machining length (μm)	500					
Layer thickness (μm)	1					
Number of layers	150					
Pulse frequency (MHz)	0.67					
Critical discharge gap width (μm)	2					
Volume of crater on the workpiece (μm^3)	2.4066					
Volumetric tool wear ratio	0.082					

Figures 10 and 11 show the simulation machining results of the tool electrodes and microgrooves, respectively. It can be seen from Figures 10a,d and 11a,d that the sidewalls of the microgrooves are almost perpendicular to the bottom surface, and the bottom surface is relatively flat. In addition, there is no obvious nonuniform wear on the end of the cylindrical tool electrode. Figures 10b,e and 11b,e are the simulation machining results for the cylindrical tool electrode in nonrotating mode. The results show that the cross-sections of the microgrooves are no longer an approximate rectangle but a typical arc shape. Moreover, there are two clearly different profiles for the cylindrical tool electrode in two directions. In the direction parallel to the machining direction, the profile of the central cross-section is approximately rectangular, while it is also a typical arc shape similar to the cross-section of the microgroove in the direction that is perpendicular to the machining direction. Figures 10c,f and 11c,f show the simulation machining results of the square tool electrode in nonrotating mode. As shown in Figures 10c,f and 11c,f, although the electrodes do not rotate, the sidewalls of the microgrooves are perpendicular to the bottom surface, and the cross-sections are approximately rectangular. There is little difference between the central cross-sections of the square tool electrode in the two directions that are perpendicular and parallel to the machining direction, but there is no noteworthy arc phenomenon.

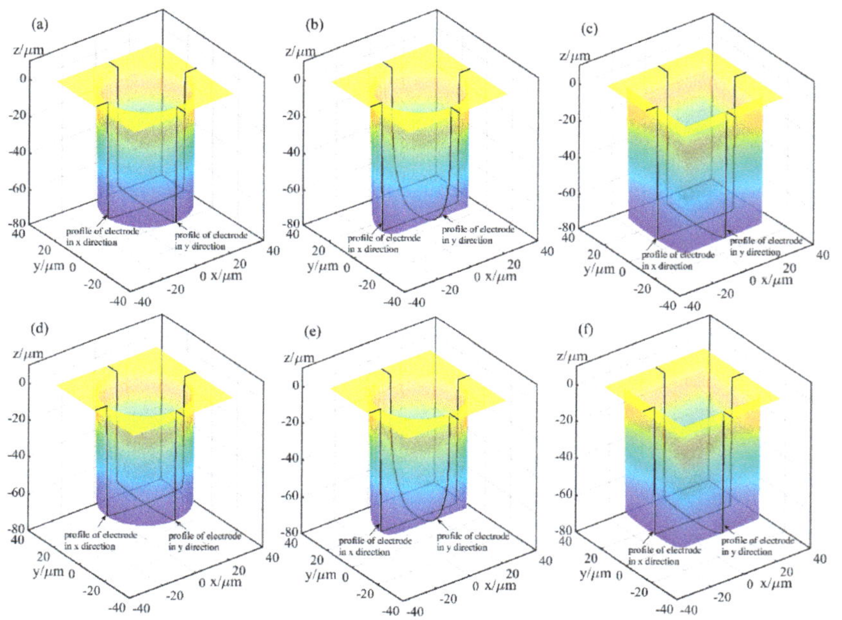

Figure 10. Simulation machining results of the tool electrodes: (**a**) cylindrical electrode-rotating mode-unidirectional path; (**b**) cylindrical electrode-nonrotating mode-unidirectional path; (**c**) square electrode-nonrotating mode-unidirectional path; (**d**) cylindrical electrode-rotating mode-reciprocating path; (**e**) cylindrical electrode-nonrotating mode-reciprocating path; (**f**) square electrode-nonrotating mode-reciprocating path.

The simulation results indicate that different rotation speeds and different electrode shapes generate different microgrooves, which demonstrates that the accumulative difference in the time and space of material removal is a critical reason for the nonuniform wear of the electrode and the generation of the corresponding cross-sectional profile of the microgroove. Furthermore, simulation experiments can be carried out based on the established simulation model to predict the machining profile so that the process parameters can be pre-determined, and the regulation and control of this accumulative difference can be achieved.

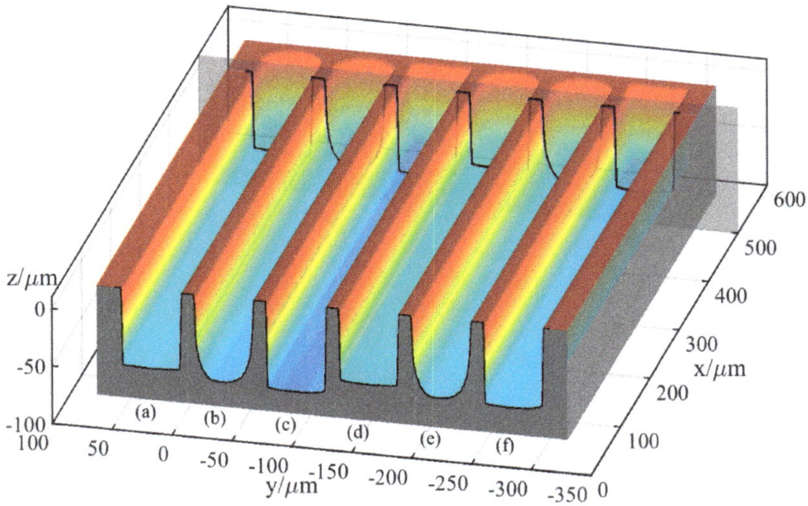

Figure 11. Simulation machining results of the microgrooves: (**a**) cylindrical electrode-rotating mode-unidirectional path; (**b**) cylindrical electrode-nonrotating mode-unidirectional path; (**c**) square electrode-nonrotating mode-unidirectional path; (**d**) cylindrical electrode-rotating mode-reciprocating path; (**e**) cylindrical electrode-nonrotating mode-reciprocating path; (**f**) square electrode-nonrotating mode-reciprocating path.

4. Experimental Design and Studies

4.1. Experimental Apparatus and Materials

The experiments were performed on a micro-EDM machine tool named μEM-200CDS2 (self-developed, Mianyang, China), as shown in Figure 12. Its maximum travelling length is 200 (*X*-axis) × 100 (*Y*-axis) × 200 mm (*Z*-axis). The resolution is 0.1 μm, and the repeatability is 1 μm for all axes. The tool electrode is clamped on a spindle, and the workpiece is fixed on a workpiece holder, which can also level the workpiece. The tool electrode and the workpiece are both copper, and the material properties are listed in Table 2. EDM oil is selected as the dielectric fluid.

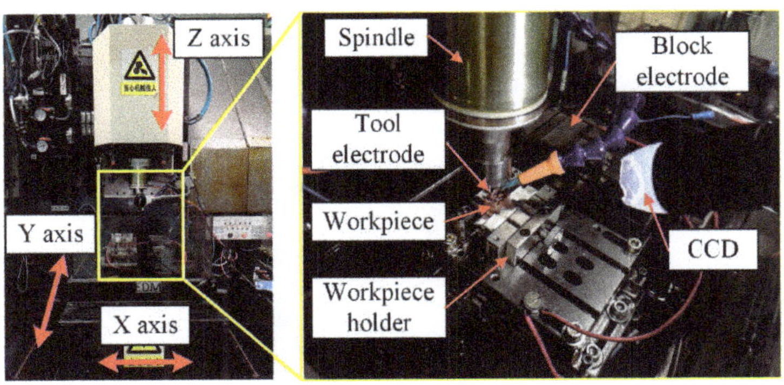

Figure 12. Photo of the experimental apparatus.

Table 2. Material properties.

Physical Properties (20 °C)	Copper
Melting point (°C)	1083
Boiling point (°C)	2600
Specific heat capacity (J/(kg·K))	400
Thermal conductivity (W/(m·K))	385

4.2. Experimental Procedure and Conditions

First, a set of verification experiments (named Exp. 1) of the mechanism and simulation model was undertaken. A total of six tool electrodes were fabricated in situ using the block electric discharge grinding (block-EDG) method, as shown in Figure 13. Among them, there are four cylindrical electrodes with diameters of 46 μm and two square electrodes with edge lengths of 46 μm. These tool electrodes were employed to perform straight-line path micro-EDM milling on the workpiece in sequence. The schematic diagram of the process is shown in Figure 14. The detailed process parameters are listed in Table 3, which correspond with the simulation parameters in Section 3.

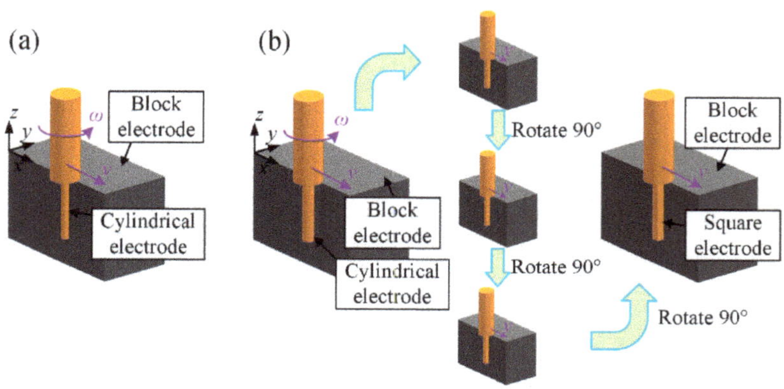

Figure 13. Schematic diagram of the in situ fabrication of the tool electrode: (a) cylindrical electrode and (b) square electrode.

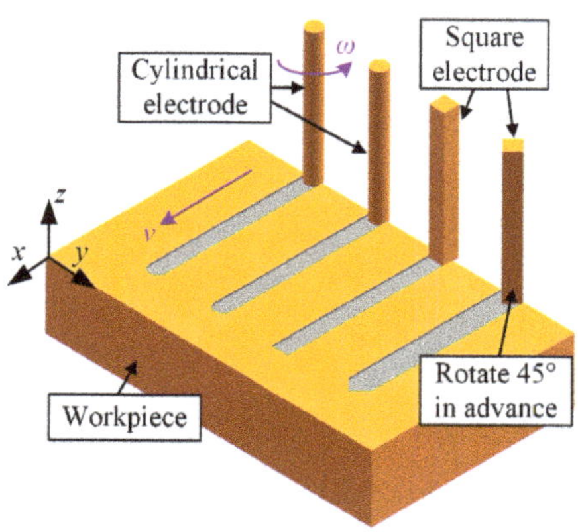

Figure 14. Schematic diagram of the machining process.

Table 3. Machining parameters for Exp. 1.

Parameters	Serial Number					
	a	b	c	d	e	f
Tool type	cylindrical		square	cylindrical		square
Rotation speed (rpm)	300	0		300	0	
Path type	unidirectional path			reciprocating path		
Tool diameter or edge length (μm)	46					
Feed speed (μm/s)	30					
Machining length (μm)	500					
Layer thickness (μm)	1					
Number of layers	150					
Pulse frequency (MHz)	0.67					
Open-circuit voltage (V)	65					
Pulse width (ns)	300					
Current limiting resistance (Ω)	50					

Two cylindrical electrodes with a diameter of 0.8 mm and two hollow electrodes with a diameter of 0.8 mm were fabricated in situ, and straight-line path micro-EDM milling was also performed (named Exp. 2). The processing parameters are listed in Table 4.

Table 4. Machining parameters for Exp. 2.

Parameters	Serial Number			
	a	b	c	d
Tool type	hollow	cylindrical	hollow	cylindrical
Number of layers	800	670	800	670
Rotation speed (rpm)	300		0	
Path type	unidirectional path			
Tool diameter (μm)	800			
Feed speed (μm/s)	500			
Machining length (μm)	1500			
Layer thickness (μm)	1			
Pulse frequency (MHz)	0.04			
Open-circuit voltage (V)	100			
Pulse width (ns)	5000			
Current limiting resistance (Ω)	10			

Finally, two application experiments of the material removal mechanism in time and space (named Exp. 3) were undertaken. A cylindrical electrode with a diameter of 46 μm and a square electrode with edge lengths of 46 μm were fabricated in situ again. For the cylindrical electrode, the number of machining layers is decreased. The square electrode was rotated 45 degrees before milling, as shown in Figure 14. The other parameters are the same as those in Exp. 1. The processing parameters are listed in Table 5.

4.3. *Measurements and Observation*

The panoramas of the machined microgrooves were observed using a scanning electron microscope (SEM, Zeiss, Jena, Germany). The tool electrodes and the entrance profiles of the microgrooves were observed using a digital microscope (VHX-6000, Keyence, Osaka, Japan). The data of the profile curves can be acquired through image processing, which can be compared with the simulation results.

Table 5. Machining parameters for Exp. 3.

Parameters	Serial Number	
	a	b
Tool type	square	cylindrical
Number of layers	150	55
Rotation speed (rpm)		0
Path type		unidirectional path
Tool diameter or edge length (μm)		46
Feed speed (μm/s)		30
Machining length (μm)		500
Layer thickness (μm)		1
Pulse frequency (MHz)		0.67
Open-circuit voltage (V)		65
Pulse width (ns)		300
Current limiting resistance (Ω)		50

5. Results and Discussion

Figure 15 shows the SEM images of the six microgrooves in Exp. 1. From the SEM images of the overall morphology, it can be seen that any cross-section of each microgroove in the direction that is perpendicular to the machining direction is highly consistent, which is also very consistent with the simulation results. Figures 16 and 17 show the optical images of the six electrodes and six entrance profiles of the microgrooves in Exp. 1, respectively. In Figure 16, 0 degrees represents that the observation direction of the electrode is parallel to the machining direction, so the observed cross-section of the electrode is perpendicular to the machining direction, and 90 degrees represents that the observation direction is perpendicular to the machining direction, so the observed cross-section of the electrode is parallel to the machining direction, which is also applicable to the following figures in this paper.

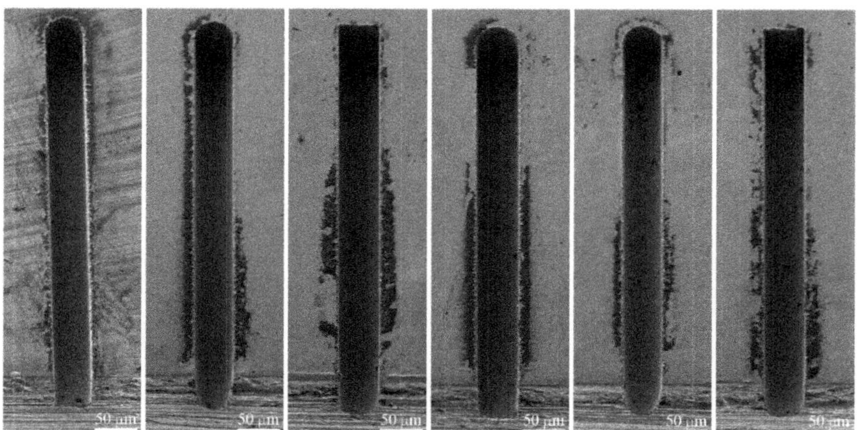

(a) (b) (c) (d) (e) (f)

Figure 15. SEM images of the six microgrooves in Exp. 1: (**a**) cylindrical electrode-rotating mode-unidirectional path; (**b**) cylindrical electrode-nonrotating mode-unidirectional path; (**c**) square electrode-nonrotating mode-unidirectional path; (**d**) cylindrical electrode-rotating mode-reciprocating path; (**e**) cylindrical electrode-nonrotating mode-reciprocating path; (**f**) square electrode-nonrotating mode-reciprocating path.

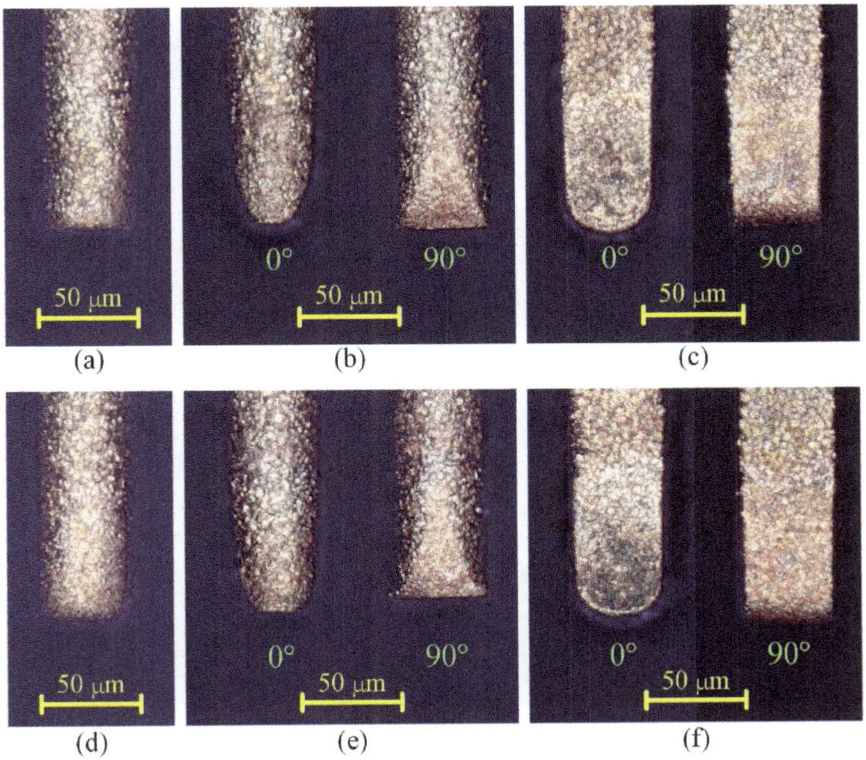

Figure 16. Optical images of the six electrodes in Exp. 1: (**a**) cylindrical electrode-rotating mode-unidirectional path; (**b**) cylindrical electrode-nonrotating mode-unidirectional path; (**c**) square electrode-nonrotating mode-unidirectional path; (**d**) cylindrical electrode-rotating mode-reciprocating path; (**e**) cylindrical electrode-nonrotating mode-reciprocating path; (**f**) square electrode-nonrotating mode-reciprocating path.

It can be seen in Figures 16a,d and 17a,d that when the cylindrical tool electrode rotates during machining, the end face of the cylindrical tool electrode is flat, and the sidewall of the machined microgroove is perpendicular to the bottom surface, which is also relatively flat. Figures 16b,e and 17b,e show that when the cylindrical tool electrode does not rotate during machining, the cross-section of the machined microgroove is a typical arc shape. Furthermore, there are two clearly different profiles of the cylindrical tool electrode in two directions. The profile of the central cross-section is approximately rectangular in the direction that is parallel to the machining direction, while it is also a typical arc shape similar to the cross-section of the microgroove in the direction that is perpendicular to the machining direction. Figures 16c,f and 17c,f show that when the square tool electrode does not rotate during machining, the sidewall of the machined microgroove is perpendicular to the bottom surface, but the bottom surface is not as flat as those in Figure 17a,d, and a slight arc shape appears. In addition, the profile of the central cross-section of the electrode is approximately rectangular in the direction that is parallel to the machining direction, while it is also a slight arc shape that is similar to the cross-section of the microgroove in the direction that is perpendicular to the machining direction.

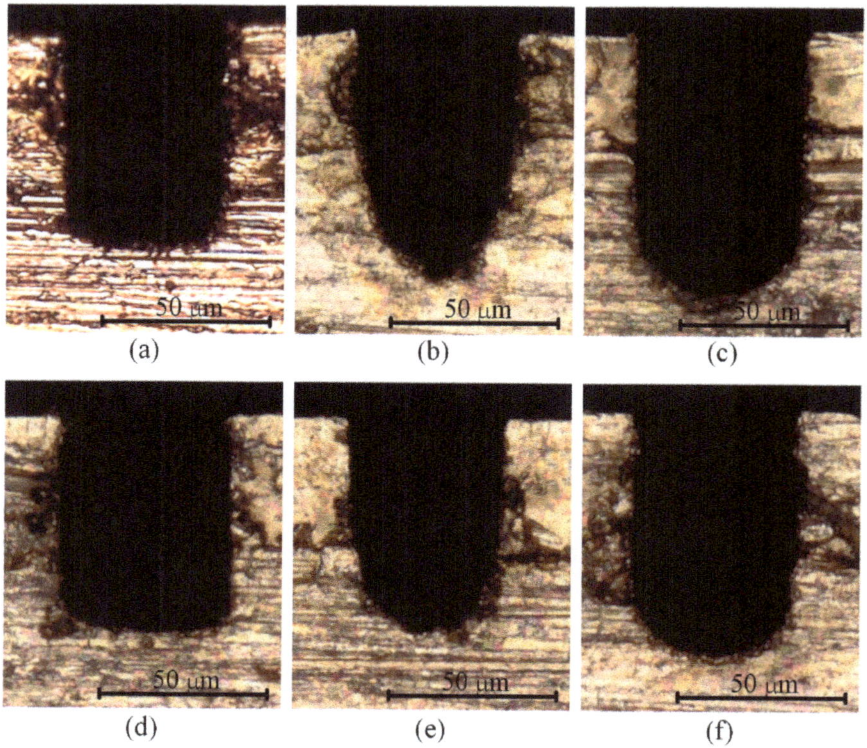

Figure 17. Optical images of the six entrance profiles of the microgrooves in Exp. 1: (**a**) cylindrical electrode-rotating mode-unidirectional path; (**b**) cylindrical electrode-nonrotating mode-unidirectional path; (**c**) square electrode-nonrotating mode-unidirectional path; (**d**) cylindrical electrode-rotating mode-reciprocating path; (**e**) cylindrical electrode-nonrotating mode-reciprocating path; (**f**) square electrode-nonrotating mode-reciprocating path.

The experimental results and the simulation results are compared together, as shown in Figure 18, which shows that the experimental results agree well with the simulation results overall. The depths of the machined microgrooves in the reciprocating path are slightly less than those in the unidirectional path, which also indicates that there is an accumulative difference in material removal due to the difference in the spatial relative motion relationship. In the machining results for the square electrode, the sidewall of the microgroove is vertical, which is consistent with the simulation. However, there is a certain difference between the bottom surfaces in the experiment and the simulation. Nonetheless, this is only a difference between the bottom surfaces, while the sidewall is still vertical, which is quite different from the machining result when the cylindrical electrode does not rotate during machining. Conversely, when the cylindrical tool electrode does not rotate during machining, as in Exp. 1 b and Exp. 1 e, as shown in Figure 18, the experimental profiles and the simulation profiles are approximately consistent. It should be noted that the debris removal condition will be not ideal if the process parameters are improper in the actual machining process [44], while the simulation is based on the assumption that the debris removal condition has no effect on the machining process.

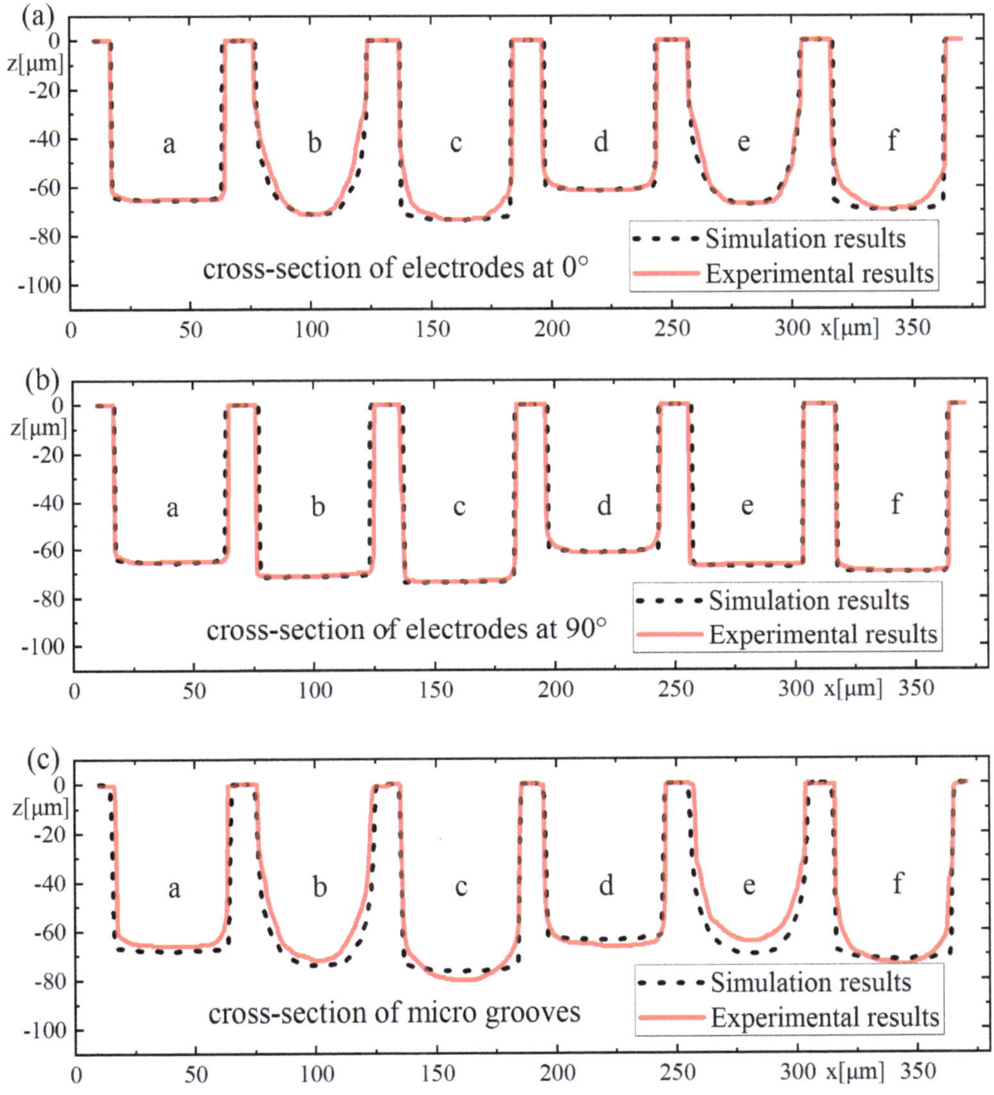

Figure 18. Comparison of the experimental results for Exp. 1 and the simulation results: (**a**) cross-section of electrodes at 0°, (**b**) cross-section of electrodes at 90°, and (**c**) cross-section of microgrooves.

In the experimental process, the machining process is very stable, and there are almost no abnormal discharges and short circuits when the cylindrical electrode rotates at high speed or does not rotate. This shows that the debris removal condition under these two conditions has no significant impact on the machining process, which satisfies the assumption in the simulation. It is easy to understand that a higher rotation speed can quickly take away the discharge debris. For the cylindrical electrode in nonrotating mode, the tool electrode and microgroove become arc-shaped quickly due to the accumulative difference in time and space, as shown in Figures 16b,e and 17b,e, which makes the working fluid take away the debris more unimpeded. However, for the square electrode in nonrotating mode, the tool electrode shape should remain unchanged from the perspective of the accumulative difference in time and space, as shown in Figure 10c,f. Actually, there is almost no channel for debris to leave the machining gap because the square electrode does not rotate, which is bound to cause a lot of abnormal discharge and short circuit states. This is the reason why the bottom surface in machining results is arc-shaped, which is different from the simulation results. Furthermore, it is why a higher rotation speed and the nonrotating mode are chosen for the verification of the simulation model.

By subtracting the z-coordinate value z_e of the experimental profile from the z-coordinate value z_s of the simulated profile in Figure 18c, the maximum deviation value, the average deviation value of each group and the percentage of average deviation relative to the maximum depth of the corresponding microgroove can be obtained, as shown in Table 6.

Table 6. Deviations of the microgroove profile between simulation and Exp. 1.

Deviations	Serial Number					
	a	b	c	d	e	f
Maximum deviation (μm)	4.10	12.17	4.33	3.08	14.93	6.79
Mean deviation (μm)	2.29	5.69	2.85	2.34	7.15	2.14
Mean relative deviation	3.47%	7.90%	3.56%	3.51%	11.09%	2.89%

Figures 19 and 20 show the optical images of the four electrodes and the top view and back view of the four grooves in Exp. 2, respectively. It can be seen from Figures 19a,b and 20a,b that whether using a cylindrical electrode or a hollow electrode, when the electrode rotates during machining, the sidewall of the machined groove is vertical to the bottom surface, and the bottom surface is very flat. The end face of the tool electrode is also flat. Figures 19c and 20c show that when the hollow tool electrode does not rotate during machining, the cross-section of the machined groove is a unique arc shape, which is asymmetric. The profile of the central cross-section is also a unique arc shape that is similar to the cross-section of the groove in the direction that is perpendicular to the machining direction, which is asymmetric as well. It is worth noting that when observing the end surface of the hollow electrode, it can be seen that the hole is eccentric in the direction that is perpendicular to the machining direction. Hence, the machining depth is deeper on the side farther from the hole, and the machining depth is shallower on the closer side to the hole. As a result, the cross-section of the groove and central cross-section of the hollow electrode in the direction that is perpendicular to the machining direction are all asymmetric. Figures 19d and 20d show that when the cylindrical tool electrode does not rotate during machining, the cross-section of the machined groove is a typical arc shape, which is symmetric. The profile of the central cross-section is also a typical arc shape that is similar to the cross-section of the groove in the direction that is perpendicular to the machining direction, which is symmetric as well. The difference in the groove profiles between Figure 20c,d is due to the end face of the hollow electrode lacking a part with respect to the cylindrical electrode. However, the machined groove profiles are very similar to each other as well when both of the electrodes rotate at a certain speed during micro-EDM milling, as shown in Figure 20a,b, which further demonstrates that different electrode shapes can generate accumulative differences in material removal, and this difference can be controlled by adjusting process parameters such as the rotation speed.

Figure 19. Optical images of the four electrodes in Exp. 2: (**a**) hollow electrode-rotating mode; (**b**) cylindrical electrode-rotating mode; (**c**) hollow electrode-nonrotating mode; (**d**) cylindrical electrode-nonrotating mode.

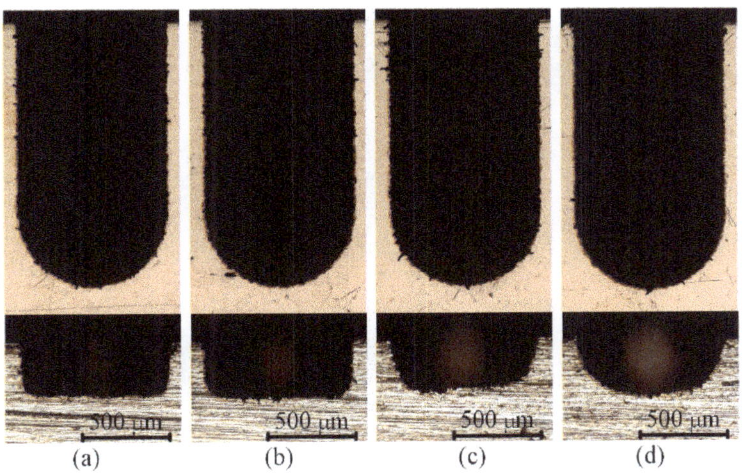

Figure 20. Optical images of the top view and back view of the four grooves in Exp. 2: (**a**) hollow electrode-rotating mode; (**b**) cylindrical electrode-rotating mode; (**c**) hollow electrode-nonrotating mode; (**d**) cylindrical electrode-nonrotating mode.

Figure 21 shows the SEM images of the two microgrooves in Exp. 3. It can also be seen from the SEM images of the overall morphology that any cross-section of each microgroove in the direction that is perpendicular to the machining direction is highly consistent. The length-width ratio of the microgroove is up to 10:1, which also indicates that a microgroove with a special-shaped cross-section and a large length-width ratio can be machined by a simple common electrode through utilising this mechanism so as to avoid the machining error caused by the wear of the forming tool electrode. Figures 22 and 23 show the optical images of the two electrodes and two entrance profiles of the microgrooves in Exp. 3, respectively.

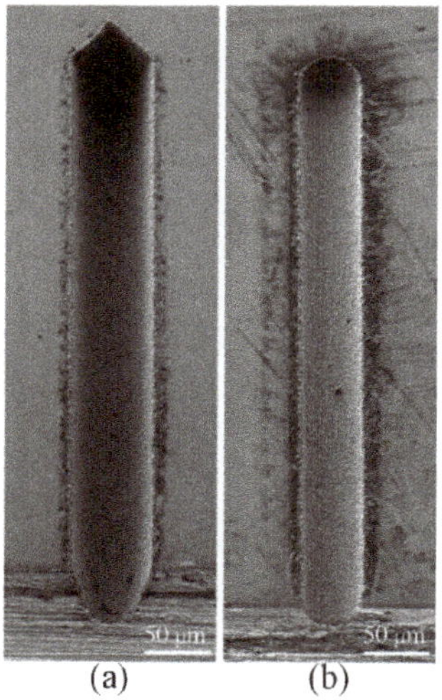

Figure 21. SEM images of the two microgrooves in Exp. 3: (**a**) square electrode-rotate 45° in advance-nonrotating mode; (**b**) cylindrical electrode-nonrotating mode.

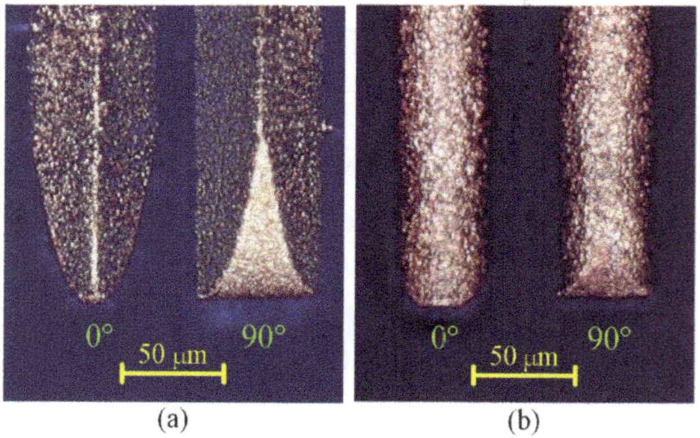

Figure 22. Optical images of the two electrodes in Exp. 3: (**a**) square electrode-rotate 45° in advance-nonrotating mode; (**b**) cylindrical electrode-nonrotating mode.

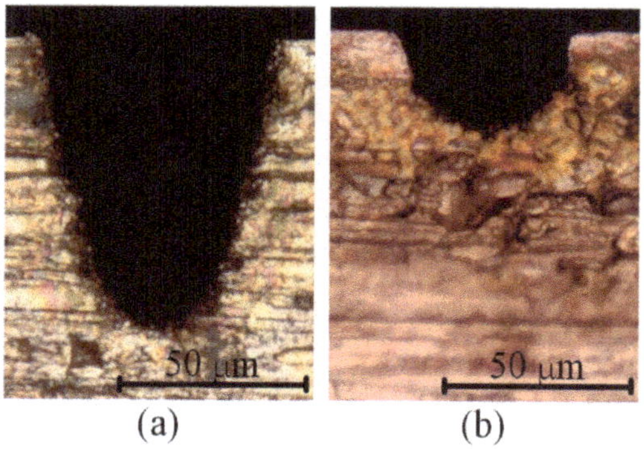

Figure 23. Optical images of the two entrance profiles of the microgrooves in Exp. 3: (**a**) square electrode-rotate 45° in advance-nonrotating mode; (**b**) cylindrical electrode-nonrotating mode.

It can be seen in Figures 22a and 23a that when the square electrode was rotated 45 degrees in advance and then micro-EDM milling was performed in nonrotating mode on the workpiece, the cross-section of the microgroove was similar to a V-shape. The profile of the central cross-section is also a V-shape that is similar to the cross-section of the microgroove in the direction that is perpendicular to the machining direction, and it is approximately rectangular in another direction. It can be seen in Figures 22b and 23b that when the cylindrical tool electrode does not rotate during machining, the cross-section of the machined microgroove is an approximately semi-circular shape because of the decrease in the number of machining layers.

The experimental results are also compared to the simulation results, as shown in Figure 24, which shows that the experimental results agree well with the simulation results. By subtracting the z-coordinate value z_e of the experimental profile from the z-coordinate value z_s of the simulated profile in Figure 24c, the maximum deviation value, the average deviation value of each group and the percentage of average deviation relative to the maximum depth of the corresponding microgroove can be obtained, as shown in Table 7.

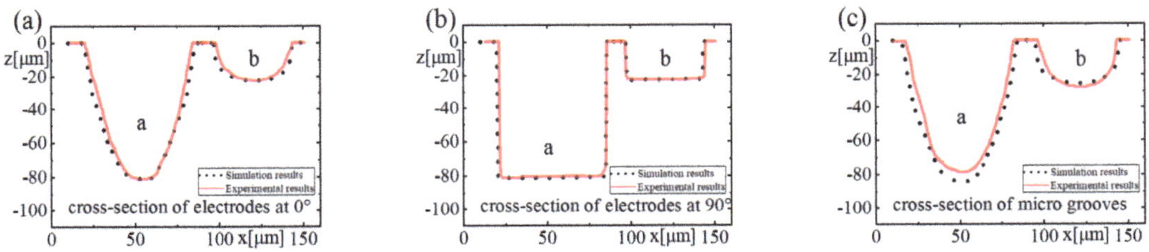

Figure 24. Comparison of the experimental results for Exp. 3 and the simulation results: (**a**) cross-section of electrodes at 0°, (**b**) cross-section of electrodes at 90°, and (**c**) cross-section of microgrooves.

Table 7. Deviations of the microgroove profile between simulation and Exp. 3.

Deviations	Serial Number	
	a	b
Maximum deviation (μm)	16.38	8.22
Mean deviation (μm)	8.02	2.31
Mean relative deviation	10.22%	8.40%

6. Conclusions

In this work, the nonuniform wear of electrodes and the phenomenon of arc shape appearing on the corresponding workpieces in micro-EDM milling were explained by the accumulative difference mechanism in time and space for material removal. Then, a simulation model was established to simulate straight-line single path micro-EDM milling, and experiments on the straight-line single path of micro-EDM milling were also undertaken. This work provides theoretical guidance for solving the problems of the machining accuracy of detail features, for instance, to machine a microgroove with an ideal rectangular cross-section. The following conclusions can be drawn from the results:

(1) The various combinations of process parameters in micro-EDM milling may lead to accumulative differences in the material removal at the tool electrode and the workpiece. In addition, for a certain set of process parameters, the material removal at the tool electrode and the workpiece may also have accumulative differences in both time and space dimensions, which may cause nonuniform wear of the tool electrode and thereby have an impact on the corresponding profile of the structure on the workpiece.

(2) The established simulation model provides a good verification of the existence of this accumulative difference caused by different process parameters. Furthermore, the

maximum mean relative deviation between the microgroove profiles of simulation results and those of experiments is 11.09%, and the profile shapes of simulations and experiments have a good consistency. Therefore, the established simulation model is capable of predicting the machining profile so that the proper process parameters can be pre-determined, and the modification and control of this accumulative difference can be achieved.

(3) The experimental results not only demonstrate the validities of the simulation model and the mechanism revealed but also verify that microgrooves with different cross-sections can be generated through the utilisation of this accumulative difference in time and space.

There are still some problems needing further research. For example, the debris removal condition may be terrible under some process parameters, which may also lead to the nonuniform wear of the tool electrode. Moreover, this will be coupled with the accumulative difference in time and space, which is complicated. Therefore, the influence of process parameters on chip removal will be modelled in the future, which can be introduced into the simulation model in this paper as a factor to provide a more complete model for the explanation of the nonuniform wear of the tool electrode in micro-EDM milling.

In fact, any machining method can be regarded as an accumulative process in time and space, whether it is subtractive manufacturing or additive manufacturing. The accumulative difference mechanism in time and space for material removal in micro-EDM milling proposed in this paper is a generic idea, which can be applied to some similar machining methods, such as pulsed laser machining, water jet machining.

Author Contributions: Q.J.: Conceptualization, Methodology, Investigation, Validation, and Writing—original draft; Y.Z.: Supervision, Project administration, Funding acquisition; L.K.: Writing–Review and Editing, Supervision, Funding acquisition; M.X.: Supervision, Funding acquisition; F.J.: Supervision, Resources. All authors have read and agreed to the published version of the manuscript.

Funding: This work was supported by the National Key R&D Program of China (2017YFA0701200), the Natural Science Foundation of China (Nos. 51475439, 52075100), the Science Challenging Program of CAEP (JDZZ2016006-0104), and the Shanghai Science and Technology Committee Innovation Grant (19ZR1404600).

Data Availability Statement: The data presented in this study are available in this published article.

Conflicts of Interest: The authors declare that they have no known competing financial interests or personal relationships that could have appeared to influence the work reported in this paper.

Nomenclature

micro-EDM	micro-electrical discharge machining
micro-ECM	micro-electrochemical machining
WEDG	wire electrical discharge grinding
UWM	uniform wear method
LCM	linear compensation method
CLU	method that combined the LCM and UWM
FLC	fixed-length compensation
EB	energy beam
TIF	tool influence function
block-EDG	block electric discharge grinding
z_T	the surface morphologies of the tool electrode
z_W	the surface morphologies of the workpiece
matrix $\boldsymbol{D}$	the removal depths of m points in time and space after n pulses
vector $\boldsymbol{d}_i$	the removal depth sequence of point i
vector $\boldsymbol{d}_j$	the removal depth sequence of all the points generated by pulse j

x	x-coordinate of the point
y	x-coordinate of the point
v	feed speed (μm/s)
t	time (s)
r	the distance between the centre point of the bottom surface of the electrode and the point (μm)
n	rotation speed (rpm).
Δt	simulation time step (s)
f	pulse frequency (Hz)
matrix $\boldsymbol{T}$	the three-dimensional coordinates of all grid nodes on the tool electrode
matrix $\boldsymbol{W}$	the three-dimensional coordinates of all grid nodes on the workpiece
matrix $\boldsymbol{G}$	the distance matrix of point pairs between the tool electrode and the workpiece
z_s	the z-coordinate value of the simulated profile
z_e	the z-coordinate value of the experimental profile

References

1. Hourmand, M.; Sarhan, A.A.D.; Sayuti, M. Micro-electrode fabrication processes for micro-EDM drilling and milling: A state-of-the-art review. *Int. J. Adv. Manuf. Technol.* **2017**, *91*, 1023–1056. [CrossRef]
2. Jahan, M.P.; Wong, Y.S.; Rahman, M. A study on the quality micro-hole machining of tungsten carbide by micro-EDM process using transistor and RC-type pulse generator. *J. Mater. Process. Technol.* **2009**, *209*, 1706–1716. [CrossRef]
3. Prakash, C.; Singh, S.; Pruncu, C.; Mishra, V.; Królczyk, G.; Pimenov, D.; Pramanik, A. Surface Modification of Ti-6Al-4V Alloy by Electrical Discharge Coating Process Using Partially Sintered Ti-Nb Electrode. *Materials* **2019**, *12*, 1006. [CrossRef] [PubMed]
4. Patel Gowdru Chandrashekarappa, M.; Kumar, S.; Jagadish, J.; Pimenov, D.Y.; Giasin, K. Experimental Analysis and Optimization of EDM Parameters on HcHcr Steel in Context with Different Electrodes and Dielectric Fluids Using Hybrid Taguchi-Based PCA-Utility and CRITIC-Utility Approaches. *Metals* **2021**, *11*, 419. [CrossRef]
5. Saleh, M.; Anwar, S.; El-Tamimi, A.; Khan Mohammed, M.; Ahmad, S. Milling Microchannels in Monel 400 Alloy by Wire EDM: An Experimental Analysis. *Micromachines* **2020**, *11*, 469. [CrossRef]
6. Korlos, A.; Tzetzis, D.; Mansour, G.; Sagris, D.; David, C. The delamination effect of drilling and electro-discharge machining on the tensile strength of woven composites as studied by X-ray computed tomography. *Int. J. Mach. Mach. Mater.* **2016**, *18*, 426. [CrossRef]
7. Rahman, M.; Asad, A.B.M.A.; Masaki, T.; Saleh, T.; Wong, Y.S.; Senthil Kumar, A. A multiprocess machine tool for compound micromachining. *Int. J. Mach. Tools Manuf.* **2010**, *50*, 344–356. [CrossRef]
8. Saxena, K.K.; Qian, J.; Reynaerts, D. A review on process capabilities of electrochemical micromachining and its hybrid variants. *Int. J. Mach. Tools Manuf.* **2018**, *127*, 28–56. [CrossRef]
9. Câmara, M.A.; Rubio, J.C.C.; Abrão, A.M.; Davim, J.P. State of the Art on Micromilling of Materials, a Review. *J. Mater. Sci. Technol.* **2012**, *28*, 673–685. [CrossRef]
10. Masuzawa, T.; Tönshoff, H.K. Three-Dimensional Micromachining by Machine Tools. *CIRP Ann.* **1997**, *46*, 621–628. [CrossRef]
11. Nguyen, N.; Huang, X.; Chuan, T.K. MEMS-Micropumps: A Review. *J. Fluids Eng.* **2002**, *124*, 384–392. [CrossRef]
12. Malek, C.K.; Saile, V. Applications of LIGA technology to precision manufacturing of high-aspect-ratio micro-components and -systems: A review. *Microelectron. J.* **2004**, *35*, 131–143. [CrossRef]
13. Kim, S.; Kim, B.H.; Chung, D.K.; Shin, H.S.; Chu, C.N. Hybrid micromachining using a nanosecond pulsed laser and micro EDM. *J. Micromech. Microeng.* **2010**, *20*, 15037. [CrossRef]
14. Hasan, M.; Zhao, J.; Jiang, Z. A review of modern advancements in micro drilling techniques. *J. Manuf. Process.* **2017**, *29*, 343–375. [CrossRef]
15. Ali, M.Y.; Maleque, M.A.; Banu, A.; Sabur, A.; Debnath, S. Micro Electro Discharge Machining of Non-Conductive Ceramic. *Mater. Sci. Forum* **2018**, *911*, 20–27. [CrossRef]
16. Jiang, B.; Lan, S.; Ni, J.; Zhang, Z. Experimental investigation of spark generation in electrochemical discharge machining of non-conducting materials. *J. Mater. Process. Technol.* **2014**, *214*, 892–898. [CrossRef]
17. Chavoshi, S.Z.; Luo, X. Hybrid micro-machining processes: A review. *Precis. Eng.* **2015**, *41*, 1–23. [CrossRef]
18. Masuzawa, T.; Fujino, M.; Kobayashi, K.; Suzuki, T.; Kinoshita, N. Wire Electro-Discharge Grinding for Micro-Machining. *CIRP Ann.* **1985**, *34*, 431–434. [CrossRef]
19. Uhlmann, E.; Roehner, M. Investigations on reduction of tool electrode wear in micro-EDM using novel electrode materials. *CIRP J. Manuf. Sci. Technol.* **2008**, *1*, 92–96. [CrossRef]
20. Yu, Z.Y.; Masuzawa, T.; Fujino, M. Micro-EDM for Three-Dimensional Cavities—Development of Uniform Wear Method. *CIRP Ann. Manuf. Technol.* **1998**, *47*, 169–172. [CrossRef]
21. Kar, S.; Patowari, P.K. Electrode wear phenomenon and its compensation in micro electrical discharge milling: A review. *Mater. Manuf. Process.* **2018**, *33*, 1491–1517. [CrossRef]
22. Yu, H.; Luan, J.; Li, J.; Zhang, Y.; Yu, Z.; Guo, D. A new electrode wear compensation method for improving performance in 3D micro EDM milling. *J. Micromech. Microeng.* **2010**, *20*, 55011. [CrossRef]

23. Li, J.; Xiao, L.; Wang, H.; Yu, H.; Yu, Z. Tool wear compensation in 3D micro EDM based on the scanned area. *Precis. Eng.* **2013**, *37*, 753–757. [CrossRef]
24. Pei, J.Y.; Deng, R.; Hu, D.J. Bottom Surface Profile of Single Slot and Fix-length Compensation Method in Micro-EDM Process. *J. Shanghai Jiaotong Univ.* **2009**, *43*, 42–46. [CrossRef]
25. Zhang, L.; Jia, Z.; Liu, W.; Wei, L. A study of electrode compensation model improvement in micro-electrical discharge machining milling based on large monolayer thickness. *Proc. Inst. Mech. Eng. Part B J. Eng. Manuf.* **2012**, *226*, 789–802. [CrossRef]
26. Zhang, L.; Du, J.; Zhuang, X.; Wang, Z.; Pei, J. Geometric prediction of conic tool in micro-EDM milling with fix-length compensation using simulation. *Int. J. Mach. Tools Manuf.* **2015**, *89*, 86–94. [CrossRef]
27. Pei, J.; Zhuang, X.; Zhang, L.; Zhu, Y.; Liu, Y. An improved fix-length compensation method for electrical discharge milling using tubular tools. *Int. J. Mach. Tools Manuf.* **2018**, *124*, 22–32. [CrossRef]
28. Yan, M.; Huang, K.; Lo, C. A study on electrode wear sensing and compensation in Micro-EDM using machine vision system. *Int. J. Adv. Manuf. Technol.* **2009**, *42*, 1065–1073. [CrossRef]
29. Pham, D.T.; Ivanov, A.; Bigot, S.; Popov, K.; Dimov, S. A study of micro-electro discharge machining electrode wear. *Proc. Inst. Mech. Eng. Part C J. Mech. Eng. Sci.* **2007**, *221*, 605–612. [CrossRef]
30. Karthikeyan, G.; Garg, A.K.; Ramkumar, J.; Dhamodaran, S. A microscopic investigation of machining behavior in μED-milling process. *J. Manuf. Process.* **2012**, *14*, 297–306. [CrossRef]
31. Nguyen, M.D.; Wong, Y.S.; Rahman, M. Profile error compensation in high precision 3D micro-EDM milling. *Precis. Eng.* **2013**, *37*, 399–407. [CrossRef]
32. Joye, C.D.; Calame, J.P.; Nguyen, K.T.; Garven, M. Microfabrication of fine electron beam tunnels using UV-LIGA and embedded polymer monofilaments for vacuum electron devices. *J. Micromech. Microeng.* **2012**, *22*, 15010. [CrossRef]
33. Patel, D.; Jain, V.K.; Ramkumar, J. Micro texturing on metallic surfaces: State of the art. *Proc. Inst. Mech. Eng. Part B J. Eng. Manuf.* **2016**, *232*, 941–964. [CrossRef]
34. Hung, J.; Yang, T.; Li, K. Studies on the fabrication of metallic bipolar plates—Using micro electrical discharge machining milling. *J. Power Sources* **2011**, *196*, 2070–2074. [CrossRef]
35. Axinte, D.; Billingham, J.; Guillerna, A.B. New models for energy beam machining enable accurate generation of free forms. *Sci. Adv.* **2017**, *3*. [CrossRef] [PubMed]
36. Axinte, D.; Billingham, J. Time-dependent manufacturing processes lead to a new class of inverse problems. *Proc. Natl. Acad. Sci. USA* **2019**, *116*, 5341–5343. [CrossRef]
37. Wan, S.; Zhang, X.; Xu, M.; Wang, W.; Jiang, X. Region-adaptive path planning for precision optical polishing with industrial robots. *Opt. Express* **2018**, *26*, 23782–23795. [CrossRef]
38. Wan, S.; Zhang, X.; Wang, W.; Xu, M.; Jiang, X. Edge control in precision robotic polishing based on space-variant deconvolution. *Precis. Eng.* **2019**, *55*, 110–118. [CrossRef]
39. Hinduja, S.; Kunieda, M. Modelling of ECM and EDM processes. *CIRP Ann.* **2013**, *62*, 775–797. [CrossRef]
40. Jeong, Y.H.; Min, B. Geometry prediction of EDM-drilled holes and tool electrode shapes of micro-EDM process using simulation. *Int. J. Mach. Tools Manuf.* **2007**, *47*, 1817–1826. [CrossRef]
41. Morimoto, K.; Kunieda, M. Sinking EDM simulation by determining discharge locations based on discharge delay time. *CIRP Ann.* **2009**, *58*, 221–224. [CrossRef]
42. Heo, S.; Jeong, Y.H.; Min, B.; Lee, S.J. Virtual EDM simulator: Three-dimensional geometric simulation of micro-EDM milling processes. *Int. J. Mach. Tools Manuf.* **2009**, *49*, 1029–1034. [CrossRef]
43. Kunieda, M.; Kaneko, Y.; Natsu, W. Reverse simulation of sinking EDM applicable to large curvatures. *Precis. Eng.* **2012**, *36*, 238–243. [CrossRef]
44. Ekmekci, B.; Sayar, A. Debris and consequences in micro electric discharge machining of micro-holes. *Int. J. Mach. Tools Manuf.* **2013**, *65*, 58–67. [CrossRef]

micromachines

MDPI

Article

Parametric Optimization and Influence of Near-Dry WEDM Variables on Nitinol Shape Memory Alloy

Rakesh Chaudhari [1], Aniket Kevalramani [1], Jay Vora [1,*], Sakshum Khanna [2], Vivek K. Patel [1], Danil Yurievich Pimenov [3] and Khaled Giasin [4,*]

1 Department of Mechanical Engineering, School of Technology, Pandit Deendayal Energy University, Raysan, Gandhinagar 382007, India; rakesh.chaudhari@sot.pdpu.ac.in (R.C.); aniket.kmc18@sot.pdpu.ac.in (A.K.); vivekp@sot.pdpu.ac.in (V.K.P.)
2 Journal of Visualized Experiments, Delhi 110016, India; sakshum.khanna@gmail.com
3 Department of Automated Mechanical Engineering, South Ural State University, Lenin Prosp. 76, 454080 Chelyabinsk, Russia; danil_u@rambler.ru
4 School of Mechanical and Design Engineering, University of Portsmouth, Portsmouth PO1 3DJ, UK
* Correspondence: jay.vora@sot.pdpu.ac.in (J.V.); khaled.giasin@port.ac.uk (K.G.)

Abstract: Nitinol-shape memory alloys (SMAs) are widely preferred for applications of automobile, biomedical, aerospace, robotics, and other industrial area. Therefore, precise machining of Nitinol SMA plays a vital role in achieving better surface roughness, higher productivity and geometrical accuracy for the manufacturing of devices. Wire electric discharge machining (WEDM) has proven to be an appropriate technique for machining nitinol shape memory alloy (SMA). The present study investigated the influence of near-dry WEDM technique to reduce the environmental impact from wet WEDM. A parametric optimization was carried out with the consideration of design variables of current, pulse-on-time (T_{on}), and pulse-off-time (T_{off}) and their effect were studied on output characteristics of material removal rate (MRR), and surface roughness (SR) for near-dry WEDM of nitinol SMA. ANOVA was carried out for MRR, and SR using statistical analysis to investigate the impact of design variables on response measures. ANOVA results depicted the significance of the developed quadratic model for both MRR and SR. Current, and T_{on} were found to be major contributors on the response value of MRR, and SR, respectively. A teaching–learning-based optimization (TLBO) algorithm was employed to find the optimal combination of process parameters. Single-response optimization has yielded a maximum MRR of 1.114 mm^3/s at T_{on} of 95 μs, T_{off} of 9 μs, current of 6 A. Least SR was obtained at T_{on} of 35 μs, T_{off} of 27 μs, current of 2 A with a predicted value of 2.81 μm. Near-dry WEDM process yielded an 8.94% reduction in MRR in comparison with wet-WEDM, while the performance of SR has been substantially improved by 41.56%. As per the obtained results from SEM micrographs, low viscosity, reduced thermal energy at IEG, and improved flushing of eroded material for air-mist mixture during NDWEDM has provided better surface morphology over the wet-WEDM process in terms of reduction in surface defects and better surface quality of nitinol SMA. Thus, for obtaining the better surface quality with reduced surface defects, near-dry WEDM process is largely suitable.

Keywords: shape memory alloys; nitinol; optimization; near-dry wire electric discharge machining (WEDM); teaching–learning based optimization (TLBO) algorithm

Citation: Chaudhari, R.; Kevalramani, A.; Vora, J.; Khanna, S.; Patel, V.K.; Pimenov, D.Y.; Giasin, K. Parametric Optimization and Influence of Near-Dry WEDM Variables on Nitinol Shape Memory Alloy. *Micromachines* **2022**, *13*, 1026. https://doi.org/10.3390/mi13071026

Academic Editor: Francesco Modica

Received: 23 May 2022
Accepted: 27 June 2022
Published: 28 June 2022

1. Introduction

The Shape memory alloys (SMA) are shape memory materials that can withstand immense deformations and yet return to their original shape by applied heat, stress or magnetic field [1,2]. This type of effect of regaining original shape is known as the shape memory effect (SME). SMAs exhibit superior thermomechanical properties [3–5]. Superelasticity of SMAs represents the property of regaining the original shape of material with the removal of applied external force. SMAs are widely preferred for engineering fields such as automobile, biomedical, aerospace, robotics [6,7]. Important applications for the industrial sector are fasteners and couplings generally for the military sector, cellular antennas, etc. [8]. The most commonly used SMAs are NiTi, CuZnAl, CoAl, NiMnGa, CuSn, FeMnSi, ZrCu, and CuZnNi. However, the instability and poor thermo-mechanic performance of these copper and iron-based SMAs have restricted their applications in certain areas [1]. Among others, nickel-titanium alloy also considered Nitinol was already employed in various engineering and industrial fields due to its enhanced characteristics such as high corrosion and wear resistance, biocompatibility, Superelasticity, SME, high strength, etc. [9–11]. Nitinol is also employed in multiple applications of biomedical fields, automotive sector, sensors, MEMS devices, actuators, structural elements, oil industries, robotics, aerospace components, etc. [12,13]. Therefore, precise machining of Nitinol SMA plays a vital role in achieving better surface roughness, higher productivity and geometrical accuracy for the manufacturing of devices [14,15]. The machinability of Nitinol with conventional methods was observed to be a non-effective technique due to the formation burrs at the machined surface, poor surface roughness, and high tool wear [16,17]. One of the prime reasons behind this was the higher strength and hardening of the nitinol SMA [18,19]. Another reason which possesses the difficulties includes high chemically active material which in turn results in tool failure, and low thermal conductivity [20,21].

To overcome difficulties faced during machining of SMAs using traditional techniques, non-conventional machining techniques are considered good alternatives since the workpiece and tool are not in contact with each other. Among other methods, Wire electric discharge machining (WEDM) has proven to be an appropriate technique for machining nitinol SMAs [22,23]. The WEDM technique works based on spark generation and erosion between electrode and workpiece [24]. In WEDM process, the material is melted and vaporized by repeated electrical discharges in presence of a suitable dielectric medium [25,26]. It uses wire as an electrode and the dielectric fluid which firstly acts as an insulator and later gets ionized by increasing the amount of voltage [27]. This further increases the electrical discharges (sparks) which in turn helps in increasing the material removal rate (MRR). Numerical control of the wire electrode has made WEDM process to be vastly suitable for creating the complex shape profiles of the workpiece [28,29]. Kulkarni et al. [30] employed WEDM process to study surface integrity aspects for Nitinol SMA. They utilized RSM models to generating the relationship between design variables and responses. Higher erosion with good surface morphology was obtained at optimal parameter settings. Thus, the monitoring and controlling of the machining process should be carried out properly for better machining efficiency, and to prevent wire breakage and surface quality. WEDM can provide a good surface finish, and good machining efficiency for machining complex shapes [31,32]. However, WEDM operation utilizes dielectric fluid which is a key factor in environmental issues. To overcome this issue, near-dry machining process can be an efficient and effective way by means of providing negligible health hazards [33]. Near dry machining replaces the EDM oil with a mixture of compressed air and water [34]. In addition to the environmental issues, machining characteristics such as productivity and surface quality should not be compromised. Near dry WEDM (NDWEDM) process was found to be capable of providing enhanced machining results for machining of hard materials [35,36]. NDWEDM process makes use mixture of minimal dielectric fluid and gas/air. Minimal use of deionized water along with larger proportion of compressed air/gas was found to be effective to enhance the NDWEDM performance under eco-friendly atmosphere [37]. Researchers have reported several studies to examine the implications of NDWEDM technique on machining. A suitable amalgamation of design variables for WEDM can be achieved by finding an optimal solution to opposing responses [38,39]. To find an optimal solution for contradictory responses, new optimization techniques were invented wherein amendment of algorithm-specific parameters is not required [40]. Teaching–learning-based optimization (TLBO) algorithm is one such technique which does not require fine-tuning of variables [41,42] and is found to be easy to execute [43]. Researchers have successfully used this techniques in various fields along with problems related to manufacturing sectors [44,45].

Kumar et al. [33] studied the effect of NDWEDM on nickel-based superalloy-Monel. They utilized a blend of compressed air and deionized water at a suitable proportion for obtaining the near-dry condition. Effect of pulse-on-time (T_{on}), voltage, pulse-off-time (T_{off}), and wire feed has been studied for responses of material removal rate (MRR), and surface roughness (SR). Lower values of T_{on} are key influencing factors for desired better surface finish. Comparison of NDWEDM with wet WEDM results yielded substantial improvement in SR for NDWEDM process. Liu et al. [46] concluded that the machined trim cut samples of Nitinol using WEDM machining have lower SR and minimal white layer as compared to main cut Nitinol samples. Dhakar et al. [47] studied near dry EDM and wet-EDM with inputs of T_{off}, current, T_{on} to evaluate the MRR of high-speed steel (HSS). A correlation between design variables and output parameters for NDEDM and wet EDM has been developed. The current was established as the largest dominating variable for enhancing the MRR during near-dry EDM and wet EDM processes. Wet EDM process was found to produce more gas emission concentration in comparison with NDEDM. For achieving the required desirable machining performance, Kao et al. [48] have performed NDWEDM in which a liquid and gas mixture was used as a dielectric fluid and also has the advantage to modify properties of the dielectric medium and liquid concentration. WEDM and EDM drilling were examined under all three variants of the EDM process such as dry, wet, and near-dry EDM. Their obtained results have depicted higher MRR for near dry WEDM process. Yu et al. [49] compared the dry-EDM machining performance of cemented carbide with wet EDM technique. Their obtained results have shown an improvement in machining efficiency and drop-in tool wear rate by implementing dry EDM process. For obtaining the good machining efficiency at minimal discharge energy and simultaneously better surface quality with low environmental problems, Boopathi et al. [50] used near dry EDM for conducting experiments on HSS-M2 with a mixture of liquid with air and liquid with oxygen as a dielectric medium. The effect of design variables has been studied on MRR and SR by employing Taguchi method. Their obtained results have shown that the use of a moderate proportion of air-mist pressure increases MRR with subsequently reduces SR. Gholipoor et al. [51] have compared output characteristics of MRR, TWR, and SR obtained by near-dry EDM with wet EDM and dry EDM for machining of SPK steel. Scanning electron microscopy (SEM) was used to analyse the surface integrity of this process and compare it to wet and dry EDM processes. SEM micrographs demonstrated that the surface morphology of obtained surface by NDEDM was better in comparison with the surfaces obtained at dry and wet EDM process as the surface has largely reduced micro-cracks and craters with the use of NDEDM technique. Boopathi and Sivakumar [52] optimized the performance of near-dry WEDM process of HSS by using a multi-objective evolutionary algorithm. They have utilized air-mist dielectric condition to study the influence of design variables such as T_{on}, gap voltage, current, T_{off}, and current on MRR and SR. ANOVA results has shown that current was having highest impact on deciding the values of MRR and SR. Moderate air-mist pressure was found to have substantial effect for increase in MRR and simultaneous improvement in surface quality. Regression equations were developed to find correlation between design variables and responses. They employed Pareto fronts to solve the contradictory situation among responses of MRR, and SR.

Till now, most of the research has been performed on studying the effect of near-dry WEDM variables and their impact on machining characteristics for steels and other alloys. However, to the best of the authors' knowledge, experimental investigations, and multi-objective optimization of near-dry WEDM process for nitinol SMA not yet conveyed. The current study investigated the performance of near-dry WEDM process with consideration of WEDM parameters of T_{on}, T_{off}, and current for Nitinol SMA. Box–Behnken design was utilized to conduct the experiments and mathematical correlations were developed between output characteristics (MRR and SR) and design variables. ANOVA was carried out for MRR, and SR using statistical analysis to investigate the impact of design variables on response measures. ANOVA results depicted the significance of the developed quadratic model for both MRR and SR. Current, and T_{on} were found to be major contributors on the response value of MRR, and SR, respectively. TLBO algorithm has been executed for single-objective and multi-objective optimization of MRR, and SR. Near-dry WEDM process yielded an 8.94% reduction in MRR in comparison with wet-WEDM, while the performance of SR has been substantially improved by 41.56%. Lastly, scanning electron microscopy was utilized to study the surface morphology of obtained surfaces from near-dry WEDM and wet WEDM. Thus, for obtaining the better surface quality with reduced surface defects, near-dry WEDM process is largely suitable. Authors believes that current study will be useful for machining of nitinol SMA for acquiring good surface quality.

2. Materials and Methods

Concord WEDM DK 7732 machine was employed in the present work to conduct the experiments by using near-dry WEDM process which is an advanced variant of WEDM technique. Nitinol rod with 10 mm diameter were considered as work material in the present study. The selected work material of nitinol contains 55.8% of nickel and reminder is titanium. Molybdenum wire having a diameter of 0.18 mm was selected as a tool material. With respect to the use of dielectric medium, WEDM process consists of three main process namely, wet-WEDM, dry WEDM, and near dry WEDM. In case wet WEDM process, only dielectric has been used, while dry WEDM makes use of only compresses gas as dielectric medium. Near dry WEDM process consist of both these medium, i.e., minimum quantity of dielectric fluid and compressed gas. In current study, a mixture of dielectric fluid (minimum quantity of liquid) and compressed air was used as an air-mist dielectric medium. Figure 1 shows a schematic diagram of near-dry WEDM experimental setup.

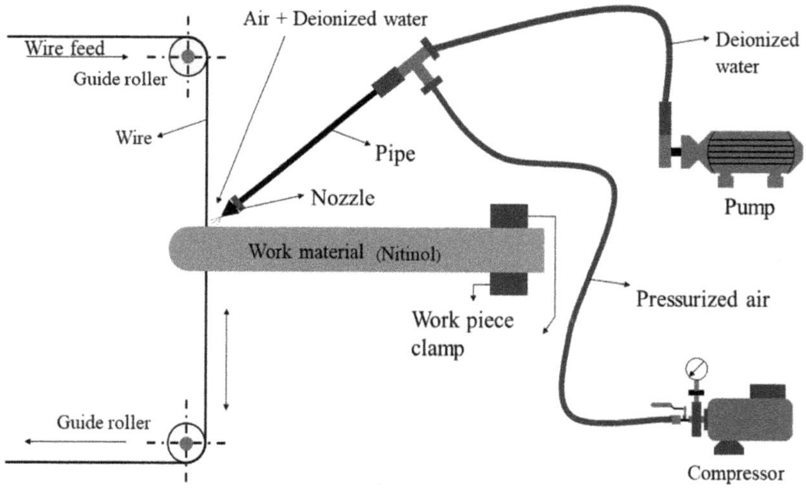

Figure 1. Schematic representation of near-dry WEDM setup.

Based on the machine limits, preliminary trials, and recent literature conducted on machining of near-dry WEDM and nitinol SMA, and preliminary experimentations, T_{on}, T_{off} and Current were selected as design variables for studying their effects on MRR, and SR. The three levels of design variables for T_{on} includes 35, 65, and 95 μs; T_{off} includes 9, 18, and 27 μs; Current includes 2, 4, and 6 A. The experimental matrix was formed by using the Box–Behnken design (BBD) technique. By following the BBD matrix, 15 trails were completed with the variation in design variables at three levels. BBD design of RSM was used to obtain an optimum response by using a series of designed experiments. Another purpose of implementing BBD design was to develop mathematical correlations between input and output parameters [53]. RSM was employed for the reduction in the experimental trials which avoids additional time and cost required for material [54,55]. To study the statistical analysis of design variables for responses of MRR, and SR, Minitab 17 was employed.

By following the BBD design matrix, experimental trials were performed thrice by taking average value of repetitions. As per Equation 1, the material erosion rate was evaluated in mm^3/s.

$$\mathrm{MRR} = \frac{\Delta \mathrm{W} * 1000}{\rho * \mathrm{t}} \quad (1)$$

where ΔW represents eroded material in gram, t depicts the time in second, and ρ represents the density of the nitinol SMA (6.5 g/cm^3).

SR was determined by employing Mitutoyo make Surftest SJ-410 with the consideration of 0.8 mm as cut-off length. Measurement of SR was performed thrice at various locations by taking average value of repetitions. SEM was employed to investigate the effect of near-dry WEDM and wet WEDM processes on surface morphology. TLBO algorithm developed by Vivek and Vimal [56] was employed to find the optimal combination of process parameters. TLBO operates on the principle of teaching and learning activities of students in a group. During the execution of the algorithm, the teacher tries to achieve the performance of class students adjacent to the student securing the highest grade by means of shifting the means of topper student grade. During the teacher phase, teacher guides the students of class. Learner phase consists of an interaction of students among themselves. Working principle of TLBO technique was depicted in Figure 2.

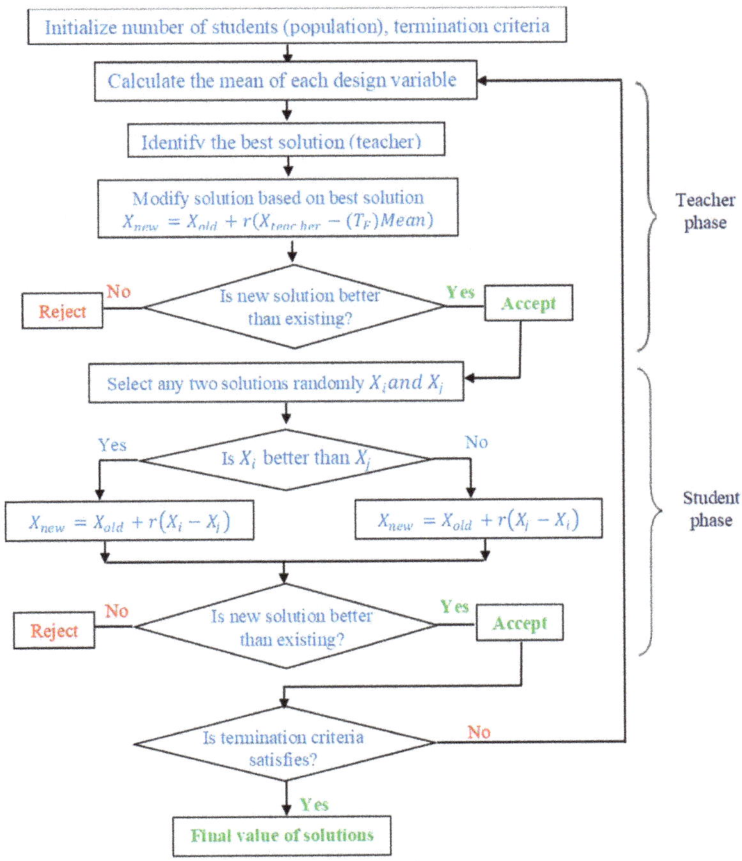

Figure 2. TLBO algorithm [57].

3. Results and Discussions

Table 1 depicts the experimental matrix as per the selected Box–Behnken design, design variables with their levels and evaluated readings of MRR, and SR. A mathematical relationship between design variables and responses has been developed by using RSM approach and by employing Minitab v17 (Bangalore, India). ANOVA was then carried out by using Minitab v17 for statistical analysis and to investigate the impact of design variables on response measures. Further, main effect plots were used to understand the influence of design variables on deciding the values of MRR, and SR. These main effect plots highlight the suitable levels of design variable for a specifically required output.

Table 1. The experimental matrix as per BBD and response measures of MRR, and SR.

Run Order	T_{on} (μs)	T_{off} (μs)	Current (A)	MRR (mm^3/s)	SR (μm)
1	65	27	6	0.91295	4.77
2	65	27	2	0.73005	3.08
3	35	18	2	0.68045	3.41
4	35	27	4	0.62155	3.32
5	65	9	2	0.90365	3.59
6	65	9	6	0.88660	5.14
7	95	27	4	0.96255	4.52
8	65	18	4	0.77655	3.92
9	95	18	2	0.99355	3.86
10	35	18	6	0.77655	4.72
11	65	18	4	0.78895	3.81
12	35	9	4	0.72230	4.36
13	65	18	4	0.77035	3.86
14	95	18	6	1.04005	5.19
15	95	9	4	1.03695	4.41

3.1. Generation of Non-Linear Regression Equations for Responses

A mathematical correlation has been developed between design variables and response measured with the help of RSM approach. Regression equations were generated by using Minitab v17 software. Equations (2) and (3) depicts the generated regression equations for MRR, and SR, respectively, by employing the stepwise approach which eliminates the non-significant terms from the model as they do not have any meaningful impact on response values.

$$\begin{aligned} MRR = 1.0813 \quad & +0.00079 \cdot T_{on} - 0.02512 \cdot T_{off} - 0.1338 \cdot Current \\ & +0.000040 \cdot T_{on} \cdot T_{on} + 0.000265 \cdot T_{off} \cdot T_{off} + 0.01456 \\ & \cdot Current \cdot Current - 0.000207 \cdot T_{on} \cdot Current + 0.002777 \\ & \cdot T_{off} \cdot Current \end{aligned} \tag{2}$$

$$\begin{aligned} SR = 5.379 - 0.04107 \quad & \cdot T_{on} - 0.0944 \cdot T_{off} - 0.046 \cdot Current + 0.000238 \cdot T_{on} \\ & \cdot T_{on} + 0.0517 \cdot Current \cdot Current + 0.001065 \cdot T_{on} \cdot T_{off} \end{aligned} \tag{3}$$

3.2. ANOVA for MRR and SR

ANOVA was carried out for MRR, and SR by using Minitab v17 for statistical analysis and to investigate the impact of design variables on response measures. A confidence level of 5% was employed to investigate the effect of design variables [58]. To have a significance of an input variable on the output variable, it is desired to have the P-value be less than 0.05 [59,60]. Table 2 shows the ANOVA for the response measure of MRR. A stepwise approach with α value equivalent to 0.15 was developed which eliminates the non-significant terms having from the model as they do not have any meaningful impact on response values. ANOVA results of Table 2 describe the statistical significance of the quadratic model of MRR as the regression model term, linear model, square interaction, and 2-way interactions are all significant. In addition to this, the non-significance of lack of fit with a P-value of 0.257 signified the robustness and adequacy of the developed model for MRR [57]. According to P-values, statistically significant factors include all the linear terms such as T_{on}, T_{off}, current; all square terms $T_{on} \times T_{on}$, $T_{off} \times T_{off}$, current × Current; interaction term $T_{off} \times$ Current. A major contributor to deciding the response value of MRR was found to be T_{on} followed by T_{off}, and current. R^2 value adjacent to one is considered as acceptability of regressions to predict the response value. The obtained values of R^2 with 0.99447 and Adj. R^2 with 0.9876 depicts the adequacy and fitness of the model. The standard deviation of 0.0147 has been observed for MRR response. It reveals that theoretical maximum deviation for MRR will be only 0.0147 from the mean value of MRR.

Table 2. ANOVA for MRR.

Source	DF	SS	MS	F	P	Significance
Model	8	0.242278	0.030285	139.89	0.000	#
Linear	3	0.214690	0.071563	330.56	0.000	#
T_{on}	1	0.189805	0.189805	876.73	0.000	#
T_{off}	1	0.012993	0.012993	60.01	0.000	#
Current	1	0.011893	0.011893	54.93	0.000	#
Square	3	0.016978	0.005659	26.14	0.001	#
$T_{on} \times T_{on}$	1	0.004727	0.004727	21.83	0.003	#
$T_{off} \times T_{off}$	1	0.001698	0.001698	7.84	0.031	#
Current × Current	1	0.012530	0.012530	57.88	0.000	#
2-Way Interaction	2	0.010610	0.005305	24.50	0.001	#
T_{on} × Current	1	0.000615	0.000615	2.84	0.143	*
T_{off} × Current	1	0.009995	0.009995	46.17	0.000	#
Error	6	0.001299	0.000216			#
Lack of fit	4	0.001120	0.000280	3.12	0.257	*
Pure error	2	0.000179	0.000090			
Total	14	0.243577				

R^2 = 99.47%; R^2 (Adj.) = 98.76%; # = Significant term; * = Non-Significant term.

Statistical analysis from ANOVA for SR was shown in Table 3. Non-significant terms from the model have been eliminated by following the stepwise approach with an α value equivalent to 0.15 as these eliminated terms do not have any meaningful impact on response values. ANOVA results of Table 3 describe the statistical significance of the quadratic model of SR as the regression model term, linear model, square interaction, and 2-way interactions are all significant. In addition to this, the non-significance of lack of fit with a P-value of 0.193 signified the robustness and adequacy of the developed model for MRR [61]. According to P-values, statistically significant factors include all the linear terms T_{on}, T_{off}, current; square terms $T_{on} \times T_{on}$, current × current; interaction term $T_{on} \times T_{off}$. A major contributor to deciding the response value of MRR was found to be currently followed by T_{on} and then T_{off}. R^2 value close to unity is considered as acceptability of regressions to predict the response value. The obtained values of R^2 (0.9855) and Adj. R^2 (0.9746) closed to unity has depicted the adequacy and fitness of the model. The standard deviation of 0.1048 has been observed for SR response. It reveals that theoretical maximum deviation for MRR will be only 0.1048 from the mean value of SR. These obtained results from ANOVA for both the responses of MRR and SR have suggested the suitability of the developed model for the prediction of upcoming response measures. However, it is mandatory to validate the results obtained from ANOVA by generating the residual plots.

Table 3. ANOVA for SR.

Source	DF	SS	MS	F	P	Significance
Model	6	5.95782	0.99297	90.40	0.000	#
Linear	3	5.31992	1.77331	161.45	0.000	#
T_{on}	1	0.58861	0.58861	53.59	0.000	#
T_{off}	1	0.40951	0.40951	37.28	0.000	#
Current	1	4.32180	4.32180	393.48	0.000	#
Square	2	0.30727	0.15364	13.99	0.002	#
$T_{on} \times T_{on}$	1	0.17047	0.17047	15.52	0.004	#
Current × Current	1	0.15874	0.15874	14.45	0.005	#
2-Way Interaction	1	0.33063	0.33063	30.10	0.001	#
$T_{on} \times T_{off}$	1	0.33063	0.33063	30.10	0.001	#
Error	8	0.08787	0.01098			#
Lack of fit	6	0.08180	0.01363	4.49	0.193	*
Pure error	2	0.00607	0.00303			
Total	14	6.04569				

R^2 = 98.55%; R^2 (Adj.) = 97.46%; # = Significant term; * = Non-Significant term.

3.3. Residual Plots for MRR and SR

Figure 3a,b depict the residual plots for response variables. Successful verification of residual plots produces the successful outcomes from ANOVA results [62,63]. Residual plots consist of four plots such as normal probability, versus fits, histogram, and versus order plot. From Figure 3a,b, the normal probability shows the plot between the percentages versus the residual. Normality plot verifies that entire the residuals are on a straight line. This shows that the model assumptions are correct, and the errors are normally distributed [64]. Randomized residuals were observed in the versus plot which suggests the suitability of the test [65]. Figure 3a,b validate the results of versus fits plot for both the responses. The histogram has shown a parabolic curve which depicted verification of ANOVA results [66]. In the last plot of residual versus observation orders, the absence of any pattern fulfils the key requirement of significant ANOVA [67]. Figure 3a,b do not depict any kind of formation of pattern for all responses which suggest good ANOVA results. Therefore, ANOVA test results can now be treated as effective and fit for developed regression models as residual plots has fulfilled the assumptions.

3.4. Effect of WEDM Variables on Responses

Main effect plots were derived by using Minitab v17 to investigate the impact of WEDM parameters on MRR, and SR. Desired output performance (maximum/minimum) of the responses in the selected levels can be efficiently represented by these main effect plots. By considering the requirement of higher productivity, the objective for MRR response was considered as maximization. Lower SR is anticipated for a better surface quality. So, minimization criteria were assigned to the SR response. The *X*-axis depicts the individual variable while *Y*-axis represents the output responses of MRR and SR.

Figure 4a shows the influence of the T_{on} on MRR and SR. Both the selected responses MRR, and SR were observed to be increased with the rise in T_{on} from 35 μs to 95 μs. As per ANOVA, T_{on} was having most dominating factor in affecting the MRR response. MRR was increased from 0.7002 mm^3/s to 1.0082 mm^3/s with a subsequent rise in T_{on}. The reason for the increase in MRR is the efficient flushing at the interelectrode gap (IEG) owing to the substantial flushing pressure during NEWEDM process [68]. This efficient flushing further enhanced spark formations which in turn increased the MRR [69]. Recurring spark formation leads to the melting and vaporization of work material thereby, erosion rate during machining [70]. Increased T_{on} subsequently enriches thermal energy which in turn enhances the sparking frequency [71,72]. This is the additional factor for obtaining higher MRR. An extensive conducted by Boopathi [34] has concluded similar observations for increased MRR with rise in T_{on} value. SR was increased with rise in T_{on} value. Production of higher frequency sparks and thermal energy due to the escalation in the value of T_{on}

has generated larger and deeper craters on the work surface [73,74]. Kumar et al. [33] has found a similar trend of increased SR with rise in T_{on}. The formations of larger craters diminish the surface quality and thus, SR also increased during the NDWEDM process [75]. This showed the different levels of T_{on} for acquiring the higher MRR and lower SR. For obtaining higher MRR and lower SR, desired levels of T_{on} were established as 95 μs, and 35 μs, respectively.

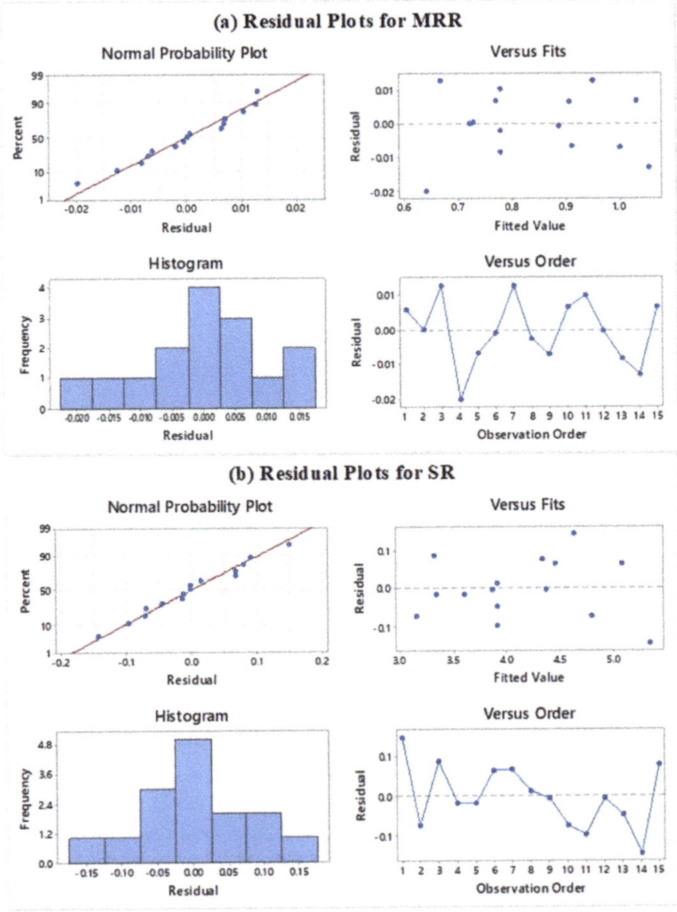

Figure 3. Residual plots for (**a**) MRR, and (**b**) SR.

The impact of the design variable T_{off} on MRR and SR can be observed in Figure 4b. The declining trend can be seen for both responses of MRR, and SR with an increasing value of T_{off}. Increasing pulse duration T_{off} from 9 μs to 27 μs has reduced MRR from 0.8873 mm^3/s to 0.8067 mm^3/s and improved SR from 4.37 μm to 3.92 μm. The reason for such declined value of MRR was due to the reduction in sparking frequency. T_{off} depicts the interval between the occurrences of two successive sparks [76]. Thus, an increase in T_{off} will have a negative effect on sparking between IEG. Reduction in sparking subsequently reduces the melting and vaporization of work material and thus, the erosion rate diminishes by leading to lower MRR [77]. Results obtained in present work are in line with the conclusion drawn by Manjaiah et al. [8] for drop in MRR. On the other hand, a rise in T_{off} has a positive effect on the SR of work material. Declined sparking in IEG also drops the temperature owing to a rise in T_{off}. This will further reduce the thermal and discharge energy and will create smaller craters [78]. Due to this reason, the quality of the work surface has been improved and a smooth surface was obtained by observing a drop in SR value [79]. Fuse et al. [80] has shown a similar trend of drop in SR value with increase in T_{off}. This showed the different levels of T_{off} for acquiring the higher MRR and lower SR. The desired levels of T_{off} were established as 9 μs, and 27 μs for MRR, and SR, respectively.

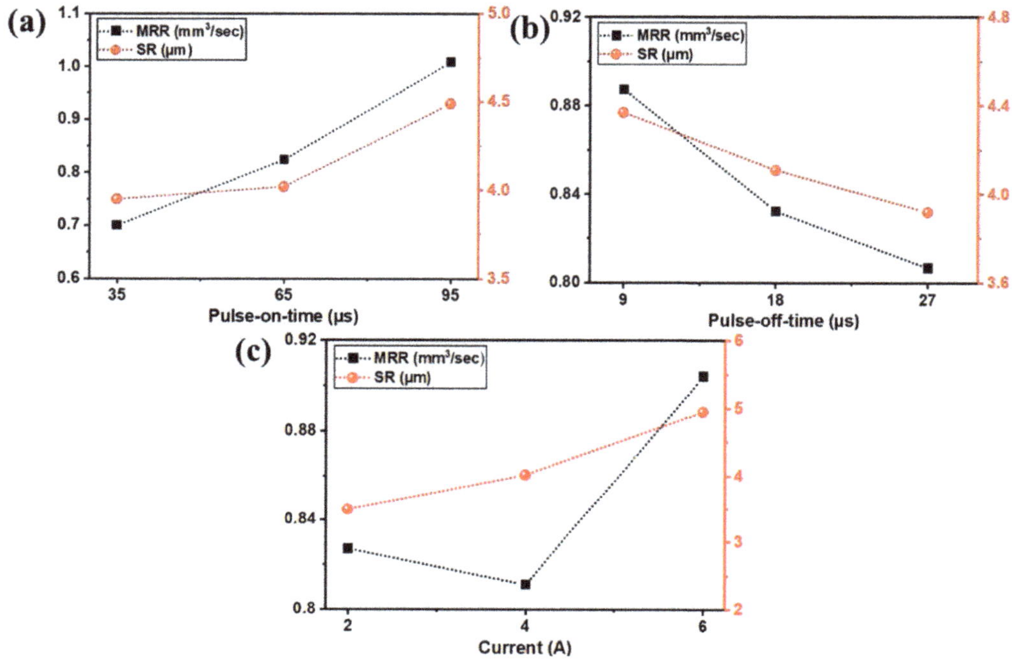

Figure 4. Impact of WEDM variables on MRR, and SR for (**a**) Pulse-on-time, (**b**) Pulse-off-time, and (**c**) Current.

Figure 4c represented the influence of the current on MRR and SR. Both the selected responses MRR, and SR were observed to be increased with the rise in current from 2 A to 6 A. As per ANOVA, the current was the most influential factor in the SR response. Increasing current from 2 A to 6 A has improved MRR from 0.8113 mm^3/s to 0.9041 mm^3/s and decreased the quality of the surface by increasing SR from 3.49 μm to 4.95 μm. The reason behind the improvement in MRR values is discharge energy. Enhancement in current further improved the discharge energy. It is further converted into thermal energy which enhances the sparking frequency during NDWEDM [81]. The formation of recurring spark leads to the melting and vaporization of work material thereby, erosion rate during machining [82]. Thus, MRR was improved with a rise in current. Similar conclusion was drawn in the study carried out by Dhakar et al. [47]. For SR response, the current was found to be the highest contributing factor. A negative effect of a rise in the current on SR can be seen in Figure 4c. As escalation in current gives rise to thermal energy, bigger and deeper craters get formed on work material [83]. Thus, a drop in SR with rising in current was depicted due to the formation of tiny craters. This main effect plot has shown the different levels of current for acquiring the higher MRR and lower SR. For obtaining higher MRR and lower SR, desired levels of current were established as 6 A, and 2 A, respectively.

3.5. Optimization Using TLBO Technique

TLBO algorithm has been executed for single-objective and multi-objective optimization of MRR, and SR. TLBO is one such technique which does not require fine-tuning of variables, and found to be easy to execute. Results of main effect plots have depicted extreme opposite levels of design variables to attain anticipated levels of responses. In the present study, the objective for MRR response was considered as maximization by considering the requirement of higher productivity. On the other hand, minimization criteria were assigned to SR response as lower SR is always desirable to acquire a better quality of the machined components. TLBO algorithm is fast and easy to implement. During the execution of the algorithm, MRR, and SR were considered positive entities. Levels of design variables employed during execution of TLBO include T_{on}: 35 μs $\leq T_{on} \geq$ 95 μs; T_{off}: 9 μs $\leq T_{off} \geq$ 27 μs; Current: 2 A $\leq$ current $\geq$ 6 A.

Results of single-response optimization have been represented in Table 4. Single-response optimization has yielded a maximum MRR of 1.114 mm^3/s at T_{on} of 95 μs, T_{off} of 9 μs, current of 6 A. Least SR was obtained at T_{on} of 35 μs, T_{off} of 27 μs, current of 2 A with the predicted value of 2.81 μm. Validation trials of these optimized results were carried out by performing the experiments at the obtained design variables. Predicted and actual determined values from trials were represented in Table 4. It can be observed that all the experimentally obtained response measures were in line with the predicted results showing a minimum error within the acceptable range. This has shown acceptability of proposed regression models with TLBO for the near-dry WEDM process. However, single-response optimal results have shown extreme opposite levels of design variables for attaining maximum MRR, and minimum SR. The suitable amalgamation of design variables for WEDM can be achieved by finding an optimal solution to opposing responses. To find the optimal solution for contradictory responses, a set of non-dominated optimum solutions provided by Pareto fronts has proven to be very effective. From Pareto fronts, user can select any optimal value as per their requirement with near-dry WEDM.

Table 4. TLBO results for individual response objectives.

Criteria	Design Variables			Predicted Results		Experimental Results		% Deviation	
	T_{on}	T_{off}	Current	MRR	SR	MRR	SR	MRR	SR
Maximization of MRR	95	9	2	1.114	3.80	1.119	3.69	4.55	2.98
Minimization of SR	35	27	2	0.599	2.81	0.608	2.85	1.54	1.75

Multi-response TLBO (MOTLBO) algorithm has been utilized to produce the simultaneous optimal levels of MRR, and SR. Fifty Pareto optimal points were generated and each Pareto point depicts the distinctive optimal result. Table 5 represents the generated solutions from the MOTLBO algorithm along with the values of design variables. Pareto curve has also been generated to understand the behavior of variation of MRR, and SR response measures. Figure 5 denotes the generated Pareto graph from unique and independent values of MRR, and SR. The nature of the Pareto curve depicts the conflicting nature between MRR and SR. Pursuant to this Pareto points will be useful selecting the corresponding levels of NDWEDM variables. By picking five random points from Table 5, validation trials of these optimized results were carried out. For all the performed experiments, an acceptable error of less than 5% was noticed among predicted and experiment trials. Thus, the obtained results have established an acceptability of the developed regression models with TLBO technique for near-dry WEDM process.

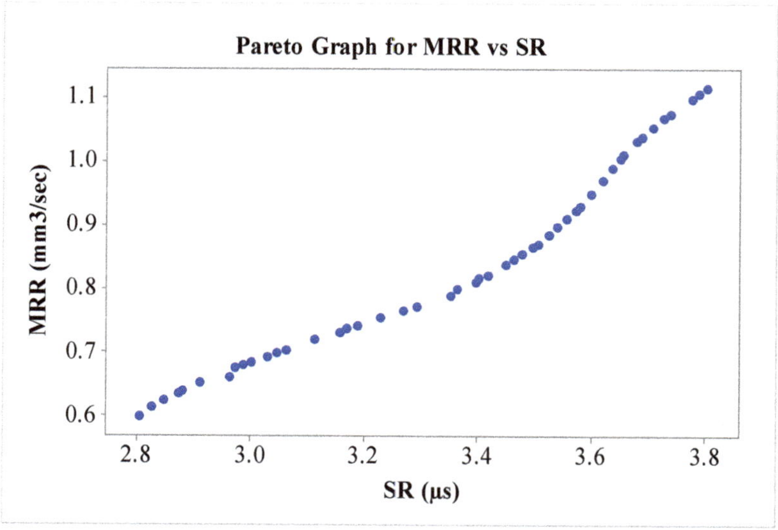

Figure 5. Pareto graph for MRR vs. SR.

Table 5. Non-dominated unique solutions obtained from TLBO.

Sr. No.	T_{on} (µs)	T_{off} (µs)	Current (A)	MRR (mm^3/s)	SR (µm)
1	35	27	2	0.599	2.80
2	95	9	2	1.114	3.80
3	93	9	2	1.098	3.78
4	90	9	2	1.075	3.74
5	76	11	2	0.948	3.60
6	78	9	2	0.990	3.64
7	87	9	2	1.053	3.71
8	42	27	2	0.623	2.85
9	39	27	2	0.612	2.83
10	64	26	2	0.730	3.16
11	84	9	2	1.031	3.68
12	71	22	2	0.798	3.36
13	81	9	2	1.010	3.66
14	74	13	2	0.909	3.56
15	75	12	2	0.929	3.58
16	77	10	2	0.969	3.62
17	49	27	2	0.651	2.91
18	58	27	2	0.693	3.03
19	46	27	2	0.639	2.88
20	56	27	2	0.683	3.00
21	71	16	2	0.855	3.48
22	67	18	2	0.810	3.40
23	85	9	2	1.038	3.69

Table 5. *Cont.*

Sr. No.	T_{on} (µs)	T_{off} (µs)	Current (A)	MRR (mm^3/s)	SR (µm)
24	74	15	2	0.885	3.52
25	45	27	2	0.635	2.87
26	50	26	2	0.661	2.97
27	69	27	2	0.753	3.23
28	73	18	2	0.846	3.46
29	67	27	2	0.742	3.19
30	75	17	2	0.869	3.51
31	72	13	2	0.897	3.54
32	74	12	2	0.922	3.57
33	80	9	2	1.003	3.65
34	94	9	2	1.106	3.79
35	71	27	2	0.765	3.27
36	54	27	2	0.674	2.97
37	63	27	2	0.719	3.11
38	67	17	2	0.821	3.42
39	60	27	2	0.703	3.06
40	66	27	2	0.736	3.17
41	59	27	2	0.698	3.05
42	74	26	2	0.789	3.35
43	70	17	2	0.838	3.45
44	55	27	2	0.679	2.99
45	71	15	2	0.866	3.50
46	71	20	2	0.815	3.41
47	72	27	2	0.771	3.29
48	89	9	2	1.068	3.73

3.6. Comparison Study Near-Dry WEDM with Wet WEDM Process

To investigate the performance of the NDWEDM process with wet WEDM, a case study has been considered with the objective function as represented in the equation. For assigning the identical significance to both the responses MRR, and SR, a multi-response optimization methodology was utilized with an equal weight of 0.5 to output responses by using the TLBO algorithm.

$$\text{Obj} = w_1 \cdot (\text{MRR}) + w_2 \cdot (\text{SR}) \quad (4)$$

The objective function has yielded optimized values of MRR, and SR as 0.815 mm^3/s, and 3.41 µm, respectively, at design variables of T_{on} of 71 µs, T_{off} of 20 µs, current of 2 A. validation trial was again conducted for the verification of these results and it has shown the actual MRR, and SR of 0.829 mm^3/s, and 3.29 µm, respectively. Now, for the comparison of these obtained results from the NDWEDM process, another experiment was carried out at the same set of parameters by using the wet-WEDM process. During the wet-WEDM process supply of air has been removed and only deionized water was used as a medium. Experimental results obtained from the wet-WEDM process have produced MRR, and SR of 0.761 mm^3/s, and 5.63 µm, respectively. A small reduction in MRR with a decrease of 8.94%

has been observed for the NDWEDM process in comparison with wet-WEDM. Higher MRR for the wet-WEDM process was due to the fact that dielectric fluid is having higher thermal conductivity as compared to the air-mist mixture [51]. Lower thermal conductive materials are having less impact on melting and vaporization during the machining process [33]. This in turn reduces the rate of erosion and thus, MRR. Another reason for higher MRR in the case of wet-WEDM is that it has improved sparking frequency as compared to near-dry WEDM. The reason for this is that NDWEDM provides dielectric fluid in the form of small droplets [48]. However, the performance of SR has been substantially improved by 41.56% with the use of the NDWEDM process. This is due to the fact that the lower viscosity of the NDWEDM process reduces the current density [33]. This in turn results in the formation of tiny shallow craters and produces better surface quality [84]. Another reason for lower SR during NDWEDM was due to improved flushing of debris particles from IEG [85,86].

3.7. Surface Morphology of Near-Dry WEDM and Wet WEDM Process

The surface morphology of the machined surface plays an important role to understand the significance of design variables and the machining process. Machined surfaces obtained at design variables of T_{on} of 71 µs, T_{off} of 20 µs and current of 2 A were selected to study the surface morphology of both the processes of NDWEDM and wet-WEDM. Figure 6a,b depict the SEM images for the machined surface obtained by using the wet-WEDM and NDWEDM processes, respectively. Figure 6a shows the large presence of surface defects such as globules and deposition of solidified material, micro-voids, and micro-cracks. This was due to the high thermal energy generated during the wet-WEDM process [34]. This high thermal energy generated enhances the intensity of the spark and thus, it produces a higher temperature at IEG. This in turn evaporates more material and generates high surface deviations in the form of micro-voids, deposition of solidified material, and micro-cracks [87,88]. On the other hand, the machined surface produced by using the NDWEDM process, as per Figure 6b, depicts lower surface deviations. This is due to the fact that the lower viscosity of the NDWEDM process reduces the current density [89]. This in turn results in the formation of tiny shallow craters and produces better surface quality by reducing the surface defects such as micro-voids, deposition of solidified material, and micro-cracks [84]. Another reason behind this was the improved flushing of debris particles from IEG [51]. Therefore, low viscosity, reduced thermal energy at IEG, and improved flushing of eroded material for air-mist mixture during NDWEDM have provided better surface morphology over the wet-WEDM process in terms of reduction in surface defects and better surface quality of nitinol SMA.

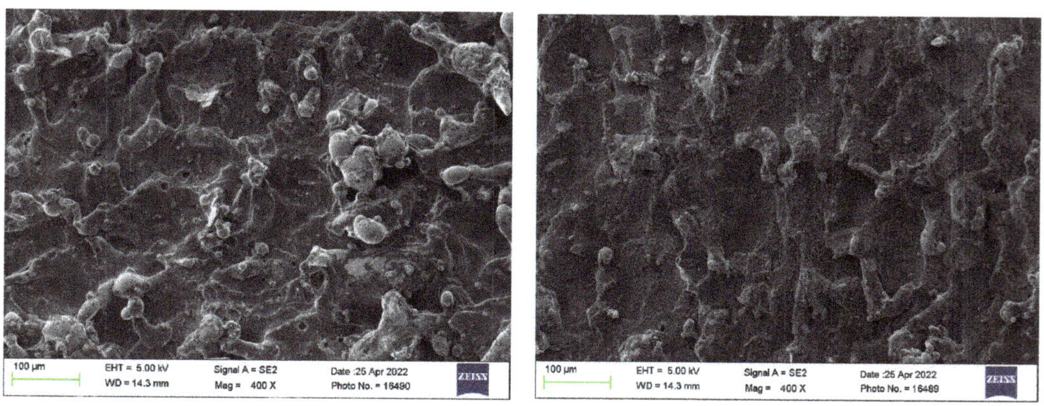

(a) (b)

Figure 6. SEM micrograph at T_{on} of 71 µs, T_{off} of 20 µs, current of 2 A for (**a**) wet-WEDM, and (**b**) near-dry WEDM.

4. Conclusions

In current study, near-dry machining process was used to overcome the environmental issues by means of providing negligible health hazards. Parametric optimization was carried out by employing the TLBO algorithm. The influence of near-dry WEDM technique was studied to relieve environmental issues related to wet WEDM with the consideration of T_{on}, T_{off}, and current as design variables. Following significant conclusions can be drawn from the present study.

- The mathematical non-linear regression equations obtained from experimental results were found to be effective for prediction of responses.
- ANOVA results depicted the statistical significance of the quadratic model for both responses MRR, and SR as the regression model term, linear model, square interaction, and 2-way interactions are all significant. A major contributor to deciding the response value of MRR was found to be T_{on} followed by T_{off}, and current, while for SR, the current was having a major contributing element followed by T_{on} and then T_{off}.
- R^2 values closed to unity signified the adequacy and fitness of the MRR, and SR model. Non-significance of lack of fit for both MRR and SR has again signified the robustness and adequacy of the developed model. All four residual plots for MRR and SR have verified the good statistical analysis for ANOVA and the outcome of developed regression equations.
- TLBO algorithm has been executed for single-objective and multi-objective optimization of MRR, and SR. Single-response optimization has yielded a maximum MRR of 1.114 mm^3/s at T_{on} of 95 µs, T_{off} of 9 µs, current of 6 A. Least SR was obtained at T_{on} of 35 µs, T_{off} of 27 µs, current of 2 A with the predicted value of 2.81 µm.
- Pareto fronts presented a trade-off between two conflicting objectives, and manufacturers can select any point on the front.
- The objective function for near-dry WEDM has yielded optimized values of MRR, and SR as 0.815 mm^3/s, and 3.41 µm, respectively, at design variables of T_{on} of 71 µs, T_{off} of 20 µs, current of 2 A. Experimental results obtained from the wet-WEDM process have produced MRR, and SR of 0.761 mm^3/s, and 5.63 µm, respectively.
- Near-dry WEDM process yielded a small reduction in MRR with an 8.94% decrease in comparison with wet-WEDM. However, the performance of SR has been substantially improved by 41.56%.
- SEM micrographs were used to study the surface morphology of obtained surfaces from near-dry WEDM and wet WEDM. Low viscosity, reduced thermal energy at IEG, and improved flushing of eroded material for air-mist mixture during NDWEDM has provided better surface morphology over the wet-WEDM process in terms of reduction in surface defects and better surface quality of nitinol SMA.
- Thus, for obtaining the better surface quality with reduced surface defects, near-dry WEDM process is largely suitable. Authors believes that current study will be useful for machining of nitinol SMA for acquiring good surface quality.

Author Contributions: Conceptualization, R.C., A.K. and J.V.; methodology, R.C., J.V.; software, V.K.P. and S.K.; validation, J.V., S.K., K.G. and D.Y.P.; formal analysis, R.C., S.K. and J.V.; investigation, R.C., J.V., K.G. and D.Y.P.; resources, V.K.P. and S.K; data curation, R.C.; writing—original draft preparation, R.C., A.K. and J.V.; writing—review and editing, V.K.P., S.K., K.G. and D.Y.P.; visualization, R.C.; supervision, J.V. and K.G. All authors have read and agreed to the published version of the manuscript.

Funding: This research received no external funding.

Conflicts of Interest: The authors declare no conflict of interest.

Nomenclature

ANOVA	Analysis of variance
BBD	Box–Behnken design
DF	Degree of freedom
DOE	Design of Experiments
EDM	Electrical Discharge Machining
IEG	Inter-electrode gap
MOTLBO	Multi-objective teaching–learning based optimization
MRR	Material removal rate (mm^3/s)
NDEDM	Near dry electrical discharge machining
NDWEDM	Near dry wire electrical discharge machining
RSM	Response surface methodology
SEM	Scanning electron microscope
SMA	Shape memory alloy
SMAs	Shape memory alloys
SME	Shape memory effect
SR	Surface roughness (μm)
TEM	Transmission electron microscope
TLBO	Teaching–Learning based optimization
T_{on}	Pulse on time (μs)
T_{off}	Pulse off time (μs)
t	Time in seconds
WEDM	Wire electric discharge machine
ρ	Density in g/cm^3

References

1. Jani, J.M.; Leary, M.; Subic, A.; Gibson, M.A. A review of shape memory alloy research, applications and opportunities. *Mater. Des.* **2014**, *56*, 1078–1113. [CrossRef]
2. Rajput, G.S.; Vora, J.; Prajapati, P.; Chaudhari, R. Areas of recent developments for shape memory alloy: A review. *Mater. Today Proc.* **2022**, *15*, 2152–2169. [CrossRef]
3. Chaudhari, R.; Vora, J.J.; Prabu, S.M.; Palani, I.; Patel, V.K.; Parikh, D. Pareto optimization of WEDM process parameters for machining a NiTi shape memory alloy using a combined approach of RSM and heat transfer search algorithm. *Adv. Manuf.* **2019**, *9*, 64–80. [CrossRef]
4. Khanna, S.; Marathey, P.; Patel, R.; Paneliya, S.; Chaudhari, R.; Vora, J.; Ray, A.; Banerjee, R.; Mukhopadhyay, I. Unravelling camphor mediated synthesis of TiO_2 nanorods over shape memory alloy for efficient energy harvesting. *Appl. Surf. Sci.* **2021**, *541*, 148489. [CrossRef]
5. Liang, L.; Xu, M.; Chen, Y.; Zhang, T.; Tong, W.; Liu, H.; Wang, H.; Li, H. Effect of welding thermal treatment on the microstructure and mechanical properties of nickel-based superalloy fabricated by selective laser melting. *Mater. Sci. Eng. A* **2021**, *819*, 141507. [CrossRef]
6. Wang, Y.; Venezuela, J.; Dargusch, M. Biodegradable shape memory alloys: Progress and prospects. *Biomaterials* **2021**, *279*, 121215. [CrossRef]
7. Chaudhari, R.; Vora, J.J.; Parikh, D. A review on applications of nitinol shape memory alloy. *Recent Adv. Mech. Infrastruct. Proc. ICRAM* **2020**, 123–132. [CrossRef]
8. Manjaiah, M.; Narendranath, S.; Basavarajappa, S. Review on non-conventional machining of shape memory alloys. *Trans. Nonferrous Met. Soc. China* **2014**, *24*, 12–21. [CrossRef]
9. Khanna, S.; Marathey, P.; Paneliya, S.; Vinchhi, P.; Chaudhari, R.; Vora, J. Fabrication of graphene/Titania nanograss composite on shape memory alloy as photoanodes for photoelectrochemical studies: Role of the graphene. *Int. J. Hydrogen Energy* **2022**. [CrossRef]
10. Velmurugan, C.; Senthilkumar, V.; Dinesh, S.; Arulkirubakaran, D. Machining of NiTi-shape memory alloys—A review. *Mach. Sci. Technol.* **2018**, *22*, 355–401. [CrossRef]
11. Khanna, S.; Marathey, P.; Paneliya, S.; Chaudhari, R.; Vora, J. Fabrication of rutile–TiO_2 nanowire on shape memory alloy: A potential material for energy storage application. *Mater. Today Proc.* **2021**, *50*, 11–16. [CrossRef]
12. Bisaria, H.; Shandilya, P. Experimental studies on electrical discharge wire cutting of Ni-rich NiTi shape memory alloy. *Mater. Manuf. Process.* **2018**, *33*, 977–985. [CrossRef]
13. Guo, Y.; Klink, A.; Fu, C.; Snyder, J. Machinability and surface integrity of Nitinol shape memory alloy. *CIRP Ann.* **2013**, *62*, 83–86. [CrossRef]

14. Zhong, Y.; Xie, J.; Chen, Y.; Yin, L.; He, P.; Lu, W. Microstructure and mechanical properties of micro laser welding NiTiNb/Ti6Al4V dissimilar alloys lap joints with nickel interlayer. *Mater. Lett.* **2022**, *306*, 130896. [CrossRef]
15. Vora, J.; Khanna, S.; Chaudhari, R.; Patel, V.K.; Paneliya, S.; Pimenov, D.Y.; Giasin, K.; Prakash, C. Machining parameter optimization and experimental investigations of nano-graphene mixed electrical discharge machining of nitinol shape memory alloy. *J. Mater. Res. Technol.* **2022**, *19*, 653–668. [CrossRef]
16. Zadafiya, K.; Kumari, S.; Chattarjee, S.; Abhishek, K. Recent trends in non-traditional machining of shape memory alloys (SMAs): A review. *CIRP J. Manuf. Sci. Technol.* **2021**, *32*, 217–227. [CrossRef]
17. Vora, J.; Jain, A.; Sheth, M.; Gajjar, K.; Abhishek, K.; Chaudhari, R. A Review on Machining Aspects of Shape Memory Alloys. In *Recent Advances in Mechanical Infrastructure*; Springer: Berlin/Heidelberg, Germany, 2022; pp. 449–458.
18. Hassan, M.; Mehrpouya, M.; Dawood, S. Review of the machining difficulties of nickel-titanium based shape memory alloys. In *Applied Mechanics and Materials*; Trans Tech Publications Ltd.: Freienbach, Switzerland, 2014; pp. 533–537.
19. Khanna, S.; Patel, R.; Marathey, P.; Chaudari, R.; Vora, J.; Banerjee, R.; Ray, A.; Mukhopadhyay, I. Growth of titanium dioxide nanorod over shape memory material using chemical vapor deposition for energy conversion application. *Mater. Today Proc.* **2020**, *28*, 475–479. [CrossRef]
20. Sharma, N.; Raj, T.; Jangra, K.K. Parameter optimization and experimental study on wire electrical discharge machining of porous Ni40Ti60 alloy. *Proc. Inst. Mech. Eng. Part B J. Eng. Manuf.* **2017**, *231*, 956–970. [CrossRef]
21. Zhao, Y.; Liu, K.; Hou, H.; Chen, L.-Q. Role of interfacial energy anisotropy in dendrite orientation in Al-Zn alloys: A phase field study. *Mater. Des.* **2022**, *216*, 110555. [CrossRef]
22. Chaudhari, R.; Vora, J.J.; Mani Prabu, S.; Palani, I.; Patel, V.K.; Parikh, D.; de Lacalle, L.N.L. Multi-response optimization of WEDM process parameters for machining of superelastic nitinol shape-memory alloy using a heat-transfer search algorithm. *Materials* **2019**, *12*, 1277. [CrossRef]
23. Chaudhari, R.; Vora, J.J.; Patel, V.; López de Lacalle, L.; Parikh, D. Surface analysis of wire-electrical-discharge-machining-processed shape-memory alloys. *Materials* **2020**, *13*, 530. [CrossRef] [PubMed]
24. Rathi, P.; Ghiya, R.; Shah, H.; Srivastava, P.; Patel, S.; Chaudhari, R.; Vora, J. Multi-response Optimization of $N_{55.8}$Ti Shape Memory Alloy Using Taguchi–Grey Relational Analysis Approach. In *Recent Advances in Mechanical Infrastructure*; Springer: Berlin/Heidelberg, Germany, 2020; pp. 13–23.
25. Sheth, M.; Gajjar, K.; Jain, A.; Shah, V.; Patel, H.; Chaudhari, R.; Vora, J. Multi-objective optimization of inconel 718 using Combined approach of taguchi—Grey relational analysis. In *Advances in Mechanical Engineering*; Springer: Berlin/Heidelberg, Germany, 2021; pp. 229–235.
26. Mandal, A.; Dixit, A.R.; Chattopadhyaya, S.; Paramanik, A.; Hloch, S.; Królczyk, G. Improvement of surface integrity of Nimonic C 263 super alloy produced by WEDM through various post-processing techniques. *Int. J. Adv. Manuf. Technol.* **2017**, *93*, 433–443. [CrossRef]
27. Sen, B.; Hussain, S.A.I.; Gupta, A.D.; Gupta, M.K.; Pimenov, D.Y.; Mikołajczyk, T. Application of Type-2 Fuzzy AHP-ARAS for Selecting Optimal WEDM Parameters. *Metals* **2021**, *11*, 42. [CrossRef]
28. Vakharia, V.; Vora, J.; Khanna, S.; Chaudhari, R.; Shah, M.; Pimenov, D.Y.; Giasin, K.; Prajapati, P.; Wojciechowski, S. Experimental investigations and prediction of WEDMed surface of Nitinol SMA using SinGAN and DenseNet deep learning model. *J. Mater. Res. Technol.* **2022**, *18*, 325–337. [CrossRef]
29. Gupta, N.K.; Somani, N.; Prakash, C.; Singh, R.; Walia, A.S.; Singh, S.; Pruncu, C.I. Revealing the WEDM process parameters for the machining of pure and heat-treated titanium (Ti-6Al-4V) alloy. *Materials* **2021**, *14*, 2292. [CrossRef]
30. Kulkarni, V.N.; Gaitonde, V.N.; Mallaiah, M.; Karnik, R.S.; Davim, J.P. Tool Wear Rate and Surface Integrity Studies in Wire Electric Discharge Machining of NiTiNOL Shape Memory Alloy Using Diffusion Annealed Coated Electrode Materials. *Machines* **2022**, *10*, 138. [CrossRef]
31. Ulas, M.; Aydur, O.; Gurgenc, T.; Ozel, C. Surface roughness prediction of machined aluminum alloy with wire electrical discharge machining by different machine learning algorithms. *J. Mater. Res. Technol.* **2020**, *9*, 12512–12524. [CrossRef]
32. Basak, A.; Pramanik, A.; Prakash, C.; Shankar, S.; Debnath, S. Understanding the Micro-Mechanical Behaviour of Recast Layer Formed during WEDM of Titanium Alloy. *Metals* **2022**, *12*, 188. [CrossRef]
33. Kumar, N.A.; Babu, A.S.; Sathishkumar, N. Influence of near-dry ambiance on WEDM of Monel superalloy. *Mater. Manuf. Processes* **2021**, *36*, 827–835. [CrossRef]
34. Boopathi, S. An extensive review on sustainable developments of dry and near-dry electrical discharge machining processes. *J. Manuf. Sci. Eng.* **2022**, *144*. [CrossRef]
35. Dhakar, K.; Dvivedi, A. Parametric evaluation on near-dry electric discharge machining. *Mater. Manuf. Processes* **2016**, *31*, 413–421. [CrossRef]
36. Kannan, E.; Youssef, T.; Sampath, B.; Sivapragasam, A. Influences of cryogenically treated work material on near-dry wire-cut electrical discharge machining process. *Surf. Topogr. Metrol. Prop.* **2022**, *10*, 015027. [CrossRef]
37. Singh, N.K.; Pandey, P.M.; Singh, K.; Sharma, M.K. Steps towards green manufacturing through EDM process: A review. *Cogent Eng.* **2016**, *3*, 1272662. [CrossRef]
38. Chaudhari, R.; Vora, J.; Lacalle, L.; Khanna, S.; Patel, V.K.; Ayesta, I. Parametric Optimization and Effect of Nano-Graphene Mixed Dielectric Fluid on Performance of Wire Electrical Discharge Machining Process of $Ni_{55.8}$Ti Shape Memory Alloy. *Materials* **2021**, *14*, 2533. [CrossRef]

39. Lenin, N.; Sivakumar, M.; Selvakumar, G.; Rajamani, D.; Sivalingam, V.; Gupta, M.K.; Mikolajczyk, T.; Pimenov, D.Y. Optimization of Process Control Parameters for WEDM of Al-LM25/Fly Ash/B4C Hybrid Composites Using Evolutionary Algorithms: A Comparative Study. *Metals* **2021**, *11*, 1105. [CrossRef]
40. Fuse, K.; Chaudhari, R.; Vora, J.; Patel, V.K.; de Lacalle, L.N.L. Multi-Response Optimization of Abrasive Waterjet Machining of Ti6Al4V Using Integrated Approach of Utilized Heat Transfer Search Algorithm and RSM. *Materials* **2021**, *14*, 7746. [CrossRef]
41. Chaudhari, R.; Khanna, S.; Vora, J.; Patel, V.K.; Paneliya, S.; Pimenov, D.Y.; Giasin, K.; Wojciechowski, S. Experimental investigations and optimization of MWCNTs-mixed WEDM process parameters of nitinol shape memory alloy. *J. Mater. Res. Technol.* **2021**, *15*, 2152–2169. [CrossRef]
42. Suresh, S.; Elango, N.; Venkatesan, K.; Lim, W.H.; Palanikumar, K.; Rajesh, S. Sustainable friction stir spot welding of 6061-T6 aluminium alloy using improved non-dominated sorting teaching learning algorithm. *J. Mater. Res. Technol.* **2020**, *9*, 11650–11674. [CrossRef]
43. Sharma, P.; Chakradhar, D.; Narendranath, S. Measurement of WEDM performance characteristics of aero-engine alloy using RSM-based TLBO algorithm. *Measurement* **2021**, *179*, 109483. [CrossRef]
44. Ma, Y.; Zhang, X.; Song, J.; Chen, L. A modified teaching–learning-based optimization algorithm for solving optimization problem. *Knowl. Based Syst.* **2021**, *212*, 106599. [CrossRef]
45. Li, Z.; Huang, J.; Wang, J.; Ding, M. Development and application of hybrid teaching-learning genetic algorithm in fuel reloading optimization. *Prog. Nucl. Energy* **2021**, *139*, 103856. [CrossRef]
46. Liu, J.; Guo, Y.; Butler, T.; Weaver, M. Crystallography, compositions, and properties of white layer by wire electrical discharge machining of nitinol shape memory alloy. *Mater. Des.* **2016**, *109*, 1–9. [CrossRef]
47. Dhakar, K.; Chaudhary, K.; Dvivedi, A.; Bembalge, O. An environment-friendly and sustainable machining method: Near-dry EDM. *Mater. Manuf. Process.* **2019**, *34*, 1307–1315. [CrossRef]
48. Kao, C.; Tao, J.; Shih, A.J. Near dry electrical discharge machining. *Int. J. Mach. Tools Manuf.* **2007**, *47*, 2273–2281. [CrossRef]
49. Yu, Z.; Jun, T.; Masanori, K. Dry electrical discharge machining of cemented carbide. *J. Mater. Process. Technol.* **2004**, *149*, 353–357. [CrossRef]
50. Boopathi, S.; Sivakumar, K. Experimental comparative study of near-dry wire-cut electrical discharge machining (WEDM). *Eur. J. Sci. Res.* **2012**, *75*, 472–481h.
51. Gholipoor, A.; Baseri, H.; Shabgard, M.R. Investigation of near dry EDM compared with wet and dry EDM processes. *J. Mech. Sci. Technol.* **2015**, *29*, 2213–2218. [CrossRef]
52. Boopathi, S.; Sivakumar, K. Experimental investigation and parameter optimization of near-dry wire-cut electrical discharge machining using multi-objective evolutionary algorithm. *Int. J. Adv. Manuf. Technol.* **2013**, *67*, 2639–2655. [CrossRef]
53. Vora, J.; Patel, V.K.; Srinivasan, S.; Chaudhari, R.; Pimenov, D.Y.; Giasin, K.; Sharma, S. Optimization of Activated Tungsten Inert Gas Welding Process Parameters Using Heat Transfer Search Algorithm: With Experimental Validation Using Case Studies. *Metals* **2021**, *11*, 981. [CrossRef]
54. Chaudhari, R.; Vora, J.J.; Pramanik, A.; Parikh, D. Optimization of Parameters of Spark Erosion Based Processes. In *Spark Erosion Machining*; CRC Press: Boca Raton, FL, USA, 2020; pp. 190–216.
55. Chaudhari, R.; Vora, J.; Parikh, D.; Wankhede, V.; Khanna, S. Multi-response Optimization of WEDM Parameters Using an Integrated Approach of RSM–GRA Analysis for Pure Titanium. *J. Inst. Eng. Ser. D* **2020**, *101*, 117–126. [CrossRef]
56. Patel, V.K.; Savsani, V.J. A multi-objective improved teaching–learning based optimization algorithm (MO-ITLBO). *Inf. Sci.* **2016**, *357*, 182–200. [CrossRef]
57. Chaudhari, R.; Prajapati, P.; Khanna, S.; Vora, J.; Patel, V.K.; Pimenov, D.Y.; Giasin, K. Multi-Response Optimization of Al_2O_3 Nanopowder-Mixed Wire Electrical Discharge Machining Process Parameters of Nitinol Shape Memory Alloy. *Materials* **2022**, *15*, 2018. [CrossRef] [PubMed]
58. Wankhede, V.; Jagetiya, D.; Joshi, A.; Chaudhari, R. Experimental investigation of FDM process parameters using Taguchi analysis. *Mater. Today Proc.* **2020**, *27*, 2117–2120. [CrossRef]
59. Chaurasia, A.; Wankhede, V.; Chaudhari, R. Experimental investigation of high-speed turning of INCONEL 718 using PVD-coated carbide tool under wet condition. In *Innovations in Infrastructure*; Springer: Berlin/Heidelberg, Germany, 2019; pp. 367–374.
60. Vora, J.; Prajapati, N.; Patel, S.; Sheth, S.; Patel, A.; Khanna, S.; Ayesta, I.; Lacalle, L.; Chaudhari, R. Multi-response Optimization and Effect of Alumina Mixed with Dielectric Fluid on WEDM Process of Ti6Al4V. In *Recent Advances in Mechanical Infrastructure*; Springer: Berlin/Heidelberg, Germany, 2022; pp. 277–287.
61. Al-Amin, M.; Abdul-Rani, A.M.; Ahmed, R.; Shahid, M.U.; Zohura, F.T.; Abd Rani, M.D.B. Multi-objective optimization of process variables for MWCNT-added electro-discharge machining of 316L steel. *Int. J. Adv. Manuf. Technol.* **2021**, *115*, 179–198. [CrossRef]
62. Kanlayasiri, K.; Boonmung, S. Effects of wire-EDM machining variables on surface roughness of newly developed DC 53 die steel: Design of experiments and regression model. *J. Mater. Process. Technol.* **2007**, *192*, 459–464. [CrossRef]
63. Vora, J.; Parikh, N.; Chaudhari, R.; Patel, V.K.; Paramar, H.; Pimenov, D.Y.; Giasin, K. Optimization of Bead Morphology for GMAW-Based Wire-Arc Additive Manufacturing of 2.25 Cr-1.0 Mo Steel Using Metal-Cored Wires. *Appl. Sci.* **2022**, *12*, 5060. [CrossRef]
64. Bobbili, R.; Madhu, V.; Gogia, A. Effect of wire-EDM machining parameters on surface roughness and material removal rate of high strength armor steel. *Mater. Manuf. Process.* **2013**, *28*, 364–368. [CrossRef]

65. Vora, J.; Chaudhari, R.; Patel, C.; Pimenov, D.Y.; Patel, V.K.; Giasin, K.; Sharma, S. Experimental Investigations and Pareto Optimization of Fiber Laser Cutting Process of Ti6Al4V. *Metals* **2021**, *11*, 1461. [CrossRef]
66. Joshi, A.Y.; Banker, V.J.; Patel, K.K.; Patel, K.S.; Joshi, D.M.; Purohit, M.R. Experimental Investigation in Wire Cut EDM of Inconel 718 Superalloy. In *Recent Advances in Manufacturing Processes and Systems*; Springer: Berlin/Heidelberg, Germany, 2022; pp. 445–455.
67. Majumder, H.; Maity, K. Application of GRNN and multivariate hybrid approach to predict and optimize WEDM responses for Ni-Ti shape memory alloy. *Appl. Soft Comput.* **2018**, *70*, 665–679. [CrossRef]
68. Singh, N.K.; Singh, Y.; Sharma, A.; Singla, A.; Negi, P. An environmental-friendly electrical discharge machining using different sustainable techniques: A review. *Adv. Mater. Process. Technol.* **2021**, *7*, 537–566. [CrossRef]
69. Kulkarni, A.A. Comparative Analysis of Dry-EDM and Conventional EDM in machining of Hastelloy. *Turk. J. Comput. Math. Educ.* **2021**, *12*, 3538–3543.
70. Chaudhari, R.; Shah, H.; Ayesta, I.; Lacalle, L.; Vora, J. Experimental Investigations and Optimization of WEDM Parameters Using Taguchi Analysis of Pure Titanium. In *Recent Advances in Mechanical Infrastructure*; Springer: Berlin/Heidelberg, Germany, 2022; pp. 349–358.
71. Yadav, V.K.; Singh, R.; Kumar, P.; Dvivedi, A. Performance enhancement of rotary tool near-dry EDM process through tool modification. *J. Braz. Soc. Mech. Sci. Eng.* **2021**, *43*, 1–16. [CrossRef]
72. Chaudhari, R.; Patel, H.; Sheth, M.; Prajapati, N.; Fuse, K.; Abhishek, K.; Vora, J. Effect of Different Tool Electrodes (Wire) of WEDM Process of Inconel 718. In *Recent Advances in Mechanical Infrastructure*; Springer: Berlin/Heidelberg, Germany, 2022; pp. 317–327.
73. Chaudhari, R.; Sheth, M.; Patel, H.; Fuse, K.; Ayesta, I.; Lacalle, L.; Vora, J. Multi-response Optimization of Alumina Powder-Mixed WEDM Process Using Taguchi-TOPSIS Approach of Nitinol SMA. In *Recent Advances in Mechanical Infrastructure*; Springer: Berlin/Heidelberg, Germany, 2022; pp. 359–367.
74. Boopathi, S.; Myilsamy, S. Material removal rate and surface roughness study on Near-dry wire electrical discharge Machining process. *Mater. Today Proc.* **2021**, *45*, 8149–8156. [CrossRef]
75. Khundrakpam, N.S.; Brar, G.S.; Deepak, D. Grey-Taguchi optimization of near dry EDM process parameters on the surface roughness. *Mater. Today Proc.* **2018**, *5*, 4445–4451. [CrossRef]
76. Çakıroğlu, R.; Günay, M. Comprehensive analysis of material removal rate, tool wear and surface roughness in electrical discharge turning of L2 tool steel. *J. Mater. Res. Technol.* **2020**, *9*, 7305–7317. [CrossRef]
77. Bai, X.; Zhang, Q.-H.; Yang, T.-Y.; Zhang, J.-H. Research on material removal rate of powder mixed near dry electrical discharge machining. *Int. J. Adv. Manuf. Technol.* **2013**, *68*, 1757–1766. [CrossRef]
78. Gupta, A.; Dwivedi, V. Analysis of Output Parameters of EDM: A Review. *Adv. Mech. Mater. Technol.* **2022**, 825–839. [CrossRef]
79. Farooq, M.U.; Ali, M.A.; He, Y.; Khan, A.M.; Pruncu, C.I.; Kashif, M.; Ahmed, N.; Asif, N. Curved profiles machining of Ti6Al4V alloy through WEDM: Investigations on geometrical errors. *J. Mater. Res. Technol.* **2020**, *9*, 16186–16201. [CrossRef]
80. Fuse, K.; Dalsaniya, A.; Modi, D.; Vora, J.; Pimenov, D.Y.; Giasin, K.; Prajapati, P.; Chaudhari, R.; Wojciechowski, S. Integration of Fuzzy AHP and Fuzzy TOPSIS Methods for Wire Electric Discharge Machining of Titanium (Ti6Al4V) Alloy Using RSM. *Materials* **2021**, *14*, 7408. [CrossRef]
81. Li, Z.; Liu, Y.; Cao, B.; Li, W. Modeling of material removal morphology and prediction of surface roughness based on WEDM successive discharges. *Int. J. Adv. Manuf. Technol.* **2022**, *120*, 2015–2029. [CrossRef]
82. Han, F.; Jiang, J.; Yu, D. Influence of discharge current on machined surfaces by thermo-analysis in finish cut of WEDM. *Int. J. Mach. Tools Manuf.* **2007**, *47*, 1187–1196. [CrossRef]
83. Han, F.; Jiang, J.; Yu, D. Influence of machining parameters on surface roughness in finish cut of WEDM. *Int. J. Adv. Manuf. Technol.* **2007**, *34*, 538–546. [CrossRef]
84. Jia, Y.; Kim, B.; Hu, D.; Ni, J. Parametric study on near-dry wire electrodischarge machining of polycrystalline diamond-coated tungsten carbide material. *Proc. Inst. Mech. Eng. Part B J. Eng. Manuf.* **2010**, *224*, 185–193. [CrossRef]
85. Tao, J.; Shih, A.J.; Ni, J. Near-dry EDM milling of mirror-like surface finish. *Int. J. Electr. Mach.* **2008**, *13*, 29–33. [CrossRef]
86. Yadav, V.K.; Kumar, P.; Dvivedi, A. Investigations on rotary tool near-dry electric discharge machining. In *Applications of Process Engineering Principles in Materials Processing, Energy and Environmental Technologies*; Springer: Berlin/Heidelberg, Germany, 2017; pp. 327–334.
87. Soni, H.; Mr, R. Experimental investigation on effects of wire electro discharge machining of $Ti_{50}Ni_{45}Co_5$ shape memory alloys. *Silicon* **2018**, *10*, 2483–2490. [CrossRef]
88. Chaudhari, R.; Vora, J.J.; Patel, V.; Lacalle, L.; Parikh, D. Effect of WEDM Process Parameters on Surface Morphology of Nitinol Shape Memory Alloy. *Materials* **2020**, *13*, 4943. [CrossRef]
89. Dhakar, K.; Dvivedi, A.; Dhiman, A. Experimental investigation on effects of dielectric mediums in near-dry electric discharge machining. *J. Mech. Sci. Technol.* **2016**, *30*, 2179–2185. [CrossRef]

micromachines

MDPI

Article

Efficiency Optimization of the Electroerosive Process in μ-WEDM of Steel MS1 Sintered Using DMLS Technology

Ľuboslav Straka [1,*], Miroslav Gombár [2], Alena Vagaská [3] and Patrik Kuchta [1]

1 Department of Automobile and Manufacturing Technologies, The Technical University of Kosice, Sturova 31, 080 01 Presov, Slovakia
2 Department of Management, University of Presov, 080 01 Presov, Slovakia
3 Department of Natural Sciences and Humanities, The Technical University of Kosice, Bayerova 1, 080 01 Presov, Slovakia
* Correspondence: luboslav.straka@tuke.sk; Tel.: +421-55-602-6365

Abstract: Although the application of mathematical optimization methods for controlling machining processes has been the subject of much research, the situation is different for μ-WEDM. This fact has prompted us to fill the gap in this field in conjunction with investigating μ-WEDM's very low productivity and overall process efficiency, since the current trend is oriented towards achieving high quality of the machined area at a high manufacturing productivity. This paper discusses in detail the application of non-linear programming (NLP) methods using MATLAB to maximize the process performance of μ-WEDM maraging steel MS1 sintered using direct metal laser sintering (DMLS) technology. The novelty of the solution lies mainly in the selection of efficient approaches to determine the optimization maximum on the basis of a solution strategy based on multi-factor analysis. The main contribution of this paper is the obtained mathematical-statistical computational (MSC) model for predicting high productivity and quality of the machined area with respect to the the optimal efficiency of the electrical discharge process in the μ-WEDM of maraging steel MS1 material. During the experimental research and subsequent statistical processing of the measured data, a local maximum of 0.159 $mm^3 \cdot min^{-1}$ for the *MRR* parameter and a local minimum of 1.051 μm for the *Rz* parameter were identified simultaneously during μ-WEDM maraging steel MS1, which was in the range of the predicted optimal settings of the main technological parameters (MTP).

Keywords: efficiency; micromachining; optimization; performance; surface roughness; quality

Citation: Straka, Ľ.; Gombár, M.; Vagaská, A.; Kuchta, P. Efficiency Optimization of the Electroerosive Process in μ-WEDM of Steel MS1 Sintered Using DMLS Technology. *Micromachines* **2022**, *13*, 1446. https://doi.org/10.3390/mi13091446

Academic Editors: Irene Fassi and Francesco Modica

Received: 29 July 2022
Accepted: 26 August 2022
Published: 1 September 2022

1. Introduction

Electrical discharge machining is a progressive machining technology that is often used in technical practice for machining materials characterized by high hardness or significantly complicated machined surface shape [1]. The main advantage of this progressive machining technology over other technologies is the high quality of the machined surface, while the mechanical properties of the material to be machined impose almost no limits [2]. It is, therefore, possible to machine any material regardless of its mechanical properties. However, a limiting condition for the machinability of these materials is the minimum value of their electrical conductivity [3]. Although this machining technology has several advantages, it also has its shortcomings. The most significant drawback—which makes this technology much less popular in comparison to the vast majority of conventional, but also progressive, machining technologies—is the very low productivity and poor overall efficiency of the production process. The current trend aims to achieve high quality of the machined surface with high manufacturing productivity [4]. Thus, the primary reasons for the low productivity and efficiency of the production process are the very physical nature of the process of removing material particles from the machined surface and the current approaches in the established way of controlling the electrical discharge process [5]. Another reason is the absence of advanced research in this area. So far, these have been primarily oriented towards the area of improving the quality of the machined surface. Therefore, the use of this progressive machining technology has been isolated to special purpose applications where a high quality of the machined surface is required in terms of geometric accuracy, but especially in terms of the roughness parameters of the machined surface [6]. At the same time, not enough attention has been paid to the productivity of manufacturing; therefore, this technology is nowhere near being able to compete with commonly available machining technologies. Some experimental research also points to this fact [7]. It is therefore ideal to search for ways to increase the productivity and overall efficiency of the electroerosive process, while maintaining a high quality standard of the machined surface [8]. In this pursuit, one of the appropriate methods is probably to search for a unique combination of levels of important factors that can, in all circumstances, have a favourable impact on improving the productivity and overall efficiency of the electro-etching process while maintaining a high quality level of the machined surface [9–11]. The problem in finding a suitable solution in this area is the large number of factors that are involved in the machining process. At the same time, a significant number of researchers have studied the influence of the main technological and process parameters on the quality of the machined surface after μ-WEDM in such a way that only one input factor of the process was changed while the others were kept at a constant value [12–14]. However, as is known, there are interactions between the factors in the electroerosive process, so process conditions determined by such a method may not be optimal. At the same time, these input factors participate in the contraindication of quality and productivity in different percentages. In some specific cases, even the combination of input parameters significantly participates in the accumulation of adverse effects, which negatively affects the overall productivity and efficiency of the electroerosive process as a result [15,16]. It is often possible to eliminate these negative effects only by significant intervention in the main technological parameters (MTP) management method itself, as well as the process parameters. Hybrid or combined approaches in the way of managing the technological system, as well as MTP and process parameters, also appear to be a suitable methods of solving the aforementioned problems. They could significantly contribute to the elimination of the natural barrier of low productivity and overall efficiency of the electroerosive process. One of the special ways to increase the quality and productivity of electrical discharge machining was introduced by Zhu et al. in their conducted experimental research [17]. By adding TiC powder to the dielectric fluid, they were able to significantly increase the productivity of the electrical discharge machining process though the roughness parameter of the machined surface *Ra* was around 2 μm. However, the disadvantage of the given maximization of the *MRR* parameter in μ-WEDM is the need to maintain the optimal concentration of TiC in the dielectric fluid.

Another option is to take into account the interactions between factors and subsequently create a mathematical-statistical computational (MSC) model using the methodology of the design of experiments (DOE) in conjunction with the correct statistical analysis and evaluation of the experimentally obtained data. Based on the given MSC model and taking into account nonlinearities, the optimization procedure can be performed using suitable mathematical optimization methods and algorithms with the support of suitable software [18,19]. Some attempts using this method have been made by Pradhan et al. [20]. As part of their research, they carried out the optimization of the electrical discharge process in Ti-6Al-4V machining by using the response approach of selected input parameters on the quality of the machined surface. They found that peak discharge current and pulse duration were the most influential parameters in terms of material removal rate. In addition, Meena et al. [21] performed a mutual optimization of the material removal rate (*MRR*) and tool wear rate (*TWR*) as a function of the selected input factors, which were peak discharge current, voltage, and frequency, using the Taguchi method. They found that voltage has a significant effect on the output power of the electrical discharge process. In their experiments, Somashekhar et al. [22] described an artificial neural network method as suitable for optimizing the input parameters of μ-EDM in terms of *MRR*. At the same time, several researchers have attempted to investigate the effects of different input factors and their levels on the response variables such as *MRR*, *TWR*, and surface modification in μ-WEDM [23–27]. However, their research has been limited to μ-WEDM of selected materials in terms of machining performance parameters by modelling the material properties of the workpiece and the tool electrode. In doing so, only limited effort has been devoted to the optimization of these input parameters for the selected output parameters.

To meet this objective, it is necessary to monitor the performance of the machining process itself. This, as already mentioned, is closely related to the properties of the material being machined, among other things [28,29]. A special group of materials that cause problems associated with low performance are the highly structurally non-homogeneous materials. Maraging steel MS1 is one of these materials. Therefore, this material was included in the experimental research on the optimization of the quality of the machined area and the productivity of the machining process in μ-WEDM. It is a material that is produced by one of the very flexible advanced additive manufacturing technologies, namely direct metal laser sintering (DMLS) [30,31]. This is a technology that has many advantages, based on its ability to produce very complex parts in the workspace of a single machine. However, it has a number of disadvantages in addition to its number of advantages. The most important disadvantage is that due to the many limiting factors that affect the final quality of the machined surface, satisfactory results cannot always be achieved [32,33]. This is even more so when it comes to micromachining technology. This is because in the production of a given material by laser sintering of powdered metal, there are changes in the temperature and condition of the material, which often cause deformation of its surface [34]. This surface deformation causes geometric surface variations, which according to Sarafan et al. are at the level of 100 μm [35]. At the same time, the machined surface roughness parameter *Ra* of such surfaces is at the level of 12 μm. This is—from a precision manufacturing point of view—unacceptable. Therefore, additional finishing operations are required. Since high-strength materials are used in the production of this technology, it is not possible to post-machine them using conventional machining technologies. At the same time, in the vast majority of cases, the products of this technology are characterized by highly complex shapes, which precludes the additional application of many other technologies [36,37]. Therefore, in this regard, the application of μ-WEDM technology is a suitable way forward for post-machining. Here, however, the aforementioned problem arises, which is associated with the non-homogeneity of the material structure as a negative consequence of the given additive manufacturing technology [38]. This is due to the anisotropic mechanical properties between the direction of the layer increment and the plane of the build layer, which are mainly due to the principle of the technology, where the powder is sintered layer by layer. In addition, non-homogeneity of the individual layers also often occurs, which again causes significant problems in electrical discharge machining associated with a substantial reduction in the productivity of the electrical discharge machining [39]. The above reasons have therefore led us to carry out experimental research aimed to make significant progress in solving the above-mentioned problems. At the same time, the research carried out contributes to filling the gap in the field of productivity maximization, as well as improving the overall efficiency of the electrical discharge process. Thus, by predicting specific settings of the main input factors, it is possible to maximize productivity while maintaining a high quality of the machined surface. This also results in a substantial increase in the overall efficiency of the machining process, which can bring the technology closer to being a serious competitor. A multi-objective optimization technique based on the need for a specific approach and a genetic algorithm determine the optimal combination of μ-WEDM process input parameters for machining maraging steel MS1 material with respect to high productivity and quality of the machined area. Thus, the main contribution of the experimental research carried out is the obtained MSC model, predicting the MTP settings with respect to maximizing the efficiency of the electrical discharge process, which, of course, is also reflected in the reduction of machining times. On the basis of the MSC model, it is possible to determine the optimal values of the factors acting during the electrical discharge process on the quality of the machined surface and the productivity of the electrical discharge process, namely the maximum peak current, pulse-on time duration, pulse-off time duration, and maximum voltage of discharge. Their specific values, determined by means of nonlinear programming (NLP) methods, make it possible to minimize the roughness parameters of the machined surface and maximize the productivity of the electrical discharge process, thereby reducing machining time and increasing the overall efficiency of the electrical discharge process.

2. Materials and Methods

2.1. Mathematical Modeling and Optimization of the μ-WEDM Process Efficiency

The first step in the process of finding a suitable solution to a given problem in the form of mathematical modelling followed by optimization is the transformation of the physical structure into an MSC model. The MSC model describes the results of the process accordingly based on a detailed definition of the quantifiable input parameters [40]. The formulation of the objective function is a key step toward proper optimization, and its selection requires deep experience in mathematical modelling issues in the given research area. Only then can one proceed to the selection of an appropriate optimization algorithm and its implementation in a suitable software environment in order to obtain a suitable solution for the optimization of the specified problem [41]. Deterministic methods can be applied to find a suitable solution to the given problem through mathematical modelling and optimization [42]. Due to the fact that classical gradient-based methods are subject to rigorous mathematical logic, they are considered suitable for performing optimization of the vast majority of processes. These include gradient-based methods such as the steepest descent method (SDM), quasi-newton methods (QNM), the interior point method (IPM), and sequential quadratic programming (SQP). IPM and SQP have been successfully used to solve large-scale engineering problems. Meanwhile, the SQP algorithm is generally applied for transforming the original problem into a sequence of subproblems of quadratic programming. However, each quadratic programming subproblem contains Jacobian and Hessian matrices, and these must be computed for each Newton iteration of the SQP loop, which can lead to a significant increase in the computational burden. Therefore, IPM was developed as an alternative to the gradient SQP method. However, in certain specific cases, the classical gradient-based method may no longer be reliable because it is difficult to obtain the required gradient information for objective functions or special constraints. In these cases, stochastic and metaheuristic approaches provide some advantages because no inferred information is needed to implement evolution-based methods. Because these methods do not suffer from the difficulty of computing Jacobian and Hessian matrices, they are suitable for obtaining the optimum. Moreover, compared to classical gradient-based methods, stochastic and metaheuristic approaches introduce a random step size within the numerical iteration-based computation. This means that, in many cases, algorithms in this category do not require any initial estimation value due to random initialization [43]. At the same time, there are many types of evolutionary-based algorithms, commonly known as global optimization methods, that are suitable and convenient for finding the optimum in solving any problem. These evolutionary algorithms essentially use the "survival of the fittest" principle. Hence, the determination of a global minimum or maximum tends to be more likely when stochastic algorithms are applied than by applying classical deterministic methods. Furthermore, the popularity of implementing metaheuristic methods is increasing in conjunction with the continuous progress and development of computing technology. They have also been successfully applied in complex and multivariate optimization of processes characterized by a high degree of nonlinearity. A genetic algorithm was then designed by mimicking natural evolution using selection, which includes the operations of crossover, mutation, and selection. Although the feasibility of using meta-heuristic methods to solve engineering optimization problems has been demonstrated, there are some difficulties in validating the optimality of the solution and they are still not considered "standard" optimization algorithms. Recently, a method based on convexification has started to attract attention. For example, linear programming as a convex optimization procedure has been successfully used to determine optimal cutting parameters in machining processes [44–46]. Engineering optimization problems are usually nonconvex, so it is necessary to transform the original problem formulation into the form of a convex optimization procedure using a convexification technique [47] before applying the convex method. Thus, from the above, it is evident that for the purpose of optimizing the productivity of an electrical discharge process, it is appropriate to apply static nonlinear programming, with Design Expert, R, QC Expert, Minitab, Statistica, and MATLAB software finding suitable applications in the design of experiments, statistical analysis of data, and optimization.

Based on a preliminary analysis of the state of the field, it was evident that the optimization of the response of the machined surface based on suitability analysis was the ideal technique to find the optimum conditions for machining maraging steel MS1 material using μ-WEDM technology. The optimization criterion in this case was the maximization of the material removal rate (*MRR*) parameter and the minimization of the machined surface roughness parameter *Rz*, where the predicted response is *y* and the desired function is *MRR* and *Rz*. The goodness of fit value varies from 0 to 1. If the goodness of fit value is 0, it means that the predicted value is completely undesirable. If the suitability value is 1, the predicted value is ideal. The requirement for an appropriate response increases with the value of the *MRR* and *Rz* parameters. Formula (1) describes the one-sided transformation maximization function for *MRR*, and Formula (2) describes the minimization function for the selected surface roughness parameter *Rz*.

$$MRR = \left\{ \left(\frac{y - y_{min}}{y_{max} - y_{min}} \right)^{vol} \begin{cases} 0 \to y \leq y_{max} \\ y_{min} \leq y \leq y_{max} \\ 1 \to y \leq y_{max} \end{cases} \right. \tag{1}$$

$$Rz = \left\{ \left(\frac{y - y_{max}}{y_{min} - y_{max}} \right)^{vol} \begin{cases} 0 \to y \leq y_{min} \\ y_{min} \leq y \leq y_{max} \\ 1 \to y \leq y_{max} \end{cases} \right. \tag{2}$$

In these equations, *MRR* and *Rz* are the desired value of function *y* and the parameters $y_{min/max}$ are the lower/upper response limit values of *y*. *Vol* is the volume, which can be varied from 0.01 to 10 to adjust the shape of the desired function. The total desired function D ($0 \leq D \leq 1$) is then defined as the geometric mean of the individual desired functions. The multi-objective function is then the geometric mean of all the transformed responses of the single objective problem shown in Equation (3). The higher the value of *D*, the better the need for combined response levels.

$$D = (MRR \times Rz)^{1/n} \tag{3}$$

Multi-response optimization can be performed using the desirability function in conjunction with the machined surface response methodology [48–50]. The process input parameters were the maximum peak current, the pulse-on time duration, the pulse-off time duration, and the maximum voltage of discharge. The objective was to maximize the *MRR* and minimize the machined surface roughness parameter *Rz*. Volume values were assigned for both *MRR* and *Rz*, with equal importance assigned to each. Using the statistical software Design Expert, a set of optimal solutions were derived for the specified spatial design constraints, namely for the *MRR* and the machined surface roughness parameter *Rz*. The set of constraints with the highest desirability value was selected as the optimal constraint for the desired responses.

2.2. *Identification of MTP in Relation to MRR and Roughness Parameter Rz in the μ-WEDM Process*

The *MRR* in the μ-WEDM process is significantly affected by the MTP. This parameter can generally be considered to be an important quantitative output parameter of the μ-

WEDM process, which comprehensively characterizes the material removal rate from the workpiece as well as the performance of the electrical discharge process itself. The *MRR* parameter describes the amount of material removed in a specific operation in a specific time unit. It is expressed mathematically by relation (4):

$$MRR = \frac{Volume\ of\ Material\ Taken}{Time\ Taken} \quad (4)$$

In addition to the quantitative output power parameter *MRR*, the output quality parameter *Rz* is also important in the μ-WEDM optimization process. This characterizes the quality of the machined surface after μ-WEDM in terms of its roughness [51]. It is an objective parameter that serves to assess the surface roughness of the μ-WEDM surface accurately, as it determines the amount of profile roughness without any averaging quantities. In the case of μ-WEDM, this parameter exceeds the arithmetic mean value of the surface roughness Ra of the machined surface by a factor of approximately 5. The roughness of the machined surface after μ-WEDM is, as in the case of *MRR*, significantly influenced by the MTP setting. Of these, the maximum peak current *I* (A), the pulse-on time duration t_{on} (μs), and the associated pause for discharge channel recovery, i.e., the pulse-off time duration t_{off} (μs), and the electrical discharge voltage *U* (V) have a decisive influence on the roughness of the machined surface. An overview of the MTPs in μ-WEDM that significantly affect the quantitative performance parameter *MRR* and the qualitative parameter of machined surface roughness *Rz*, including their range of settings within the experiment, is given in Table 1.

Table 1. The range of MTP settings for μ-WEDM and their influence on the output parameters *MRR* and Rz.

MTP	μ-WEDM Operation	Setting Range	Influence of MTP on *Rz*	Influence of MTP on *MRR*
Maximum peak current *I* (A)	roughing semifinishing finishing	6.0–8.0 4.0–6.0 2.0–4.0	As the value of parameter *I* increases, the surface roughness deteriorates [13].	As the value of parameter *I* increases, the *MRR* increases [22].
Pulse-on time duration t_{on} (μs)	roughing semifinishing finishing	20.0–40.0 10.0–20.0 5.0–10.0	As the value of the parameter t_{on} increases, the surface roughness deteriorates [14].	As the value of the parameter t_{on} increases, the *MRR* increases [22].
Pulse-off time duration t_{off} (μs)	roughing semifinishing finishing	9.0–15.0 6.0–9.0 3.0–6.0	As the value of the parameter t_{off} increases, the surface roughness improves [23].	As the value of the parameter t_{off} increases, the *MRR* decreases [23].
Voltage of discharge *U* (V)	roughing semifinishing finishing	85–90 75–80 70–75	As the value of parameter *U* increases, the surface roughness improves [40].	As the value of the parameter *U* increases, the *MRR* decreases [51].

2.3. Conditions of the Experiment

A galvanized wire tool electrode, made of drawn CuZn37 brass wire with a diameter of 0.070 mm, a tensile strength of 1000 MPa, and an elongation >1%, was used in the experiment. The experimental specimens with dimensions of 50.0 mm × 15.0 mm × 15.0 mm were made of high-alloy steel with the designation of maraging steel MS1 in powder form. The sintering of the fine particles of the material was carried out by direct metal laser sintering (DMLS) technology using a 3D metal printer with a laser power of 400 W and a wavelength of 1060–1100 nm. The sintering process consisted of successive sintering of Maraging Steel MS1 powder material using laser technology, which was deposited in fine layers until the desired shape and dimension of the experimental specimen were achieved. The detailed procedure for the manufacture of experimental samples using DMLS technology is described in detail by Simkulet et al. in their work [30,33]. The metal powder was produced by atomization and its chemical composition corresponded to US 18% Ni maraging 300 grade steel with material number 1.2709 and the German steel standard X3NiCoMoTi18-9-5. Table 2 shows the chemical composition and Table 3 shows the mechanical and physical properties of the high-alloy steel designated maraging steel MS1.

Table 2. Chemical composition of steel MS1 [52].

Fe	Ni	Co	Mo	Ti	Al	Cr	C	Mn, Si	P, S
Zb.	17–19%	8.5–9.5%	4.5–5.2%	0.6–0.8%	0.05–0.15%	≤0.5	≤0.03	each ≤ 0.1%	each ≤ 0.01%

Table 3. Mechanical and physical properties of steel MS1 [52].

Mechanical properties of maraging steel MS1		
Parameter	As built	After age hardening
Tensile strength (MPa)	1200 ± 100	1950 ± 100
Yield strength *Rp* 0.2% (MPa)	1100 ± 100	1900 ± 100
Elongation at break (%)	8 ± 3	2 ± 1
Hardness (HRC)	33–37	50–54
Ductility (J)	45 ± 10	11 ± 4
Modulus of elasticity (GPa)	150 ± 25	
Physical properties of maraging steel MS1		
Density ($g{\cdot}cm^{-3}$)	8.0–8.1	
Electrical conductivity ($Siemens{\cdot}m{\cdot}mm^{-2}$)	2.25	
Thermal conductivity ($W{\cdot}m^{-1}{\cdot}{}^{\circ}C$)	15 ± 0.8	
Specific heat capacity ($J{\cdot}kg^{-1}{\cdot}{}^{\circ}C$)	450 ± 20	
Melting temperature (°C)	1370–1400	

Based on the mechanical and physical properties listed in Table 3, it is evident that maraging steel MS1 has high tensile strength and a relatively low thermal conductivity (15 $Wm^{-1}K^{-1}$) at 20 °C. At the same time, it has a favourable electrical conductivity (2.25 $Siemens{\cdot}m{\cdot}mm^{-2}$) and is therefore suitable for μ-WEDM machining. This steel is also characterized by very good mechanical properties and heat treatability after the atomization process. Heat treatment at 820 °C followed by age hardening at 490 °C results in a hardness of the final material of more than 50 HRC, which precludes the suitability of many post-machining technologies. Figure 1 shows an experimental sample after sintering prepared on μ-WEDM.

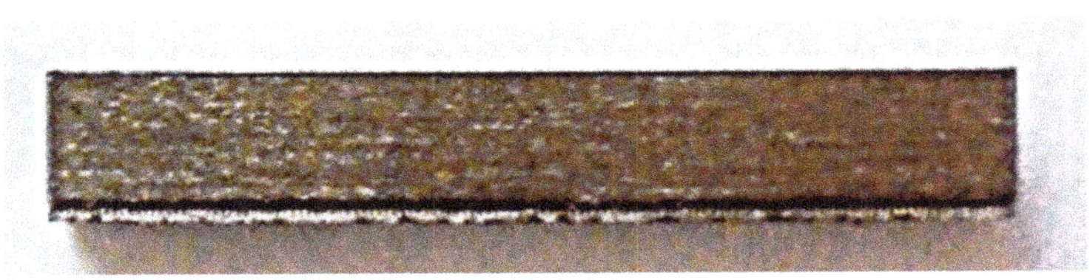

Figure 1. Experimental sample of maraging steel MS1 after DMLS.

Subsequently, the prepared experimental samples were subjected to electrical discharge machining by μ-WEDM technology (Figure 2a,b) using a CNC electroerosive machine CHMER G32F (Figure 2c). Eroding was carried out in a dielectric fluid based on deionized water with an electrical conductivity of less than 10 $\mu S{\cdot}cm^{-1}$.

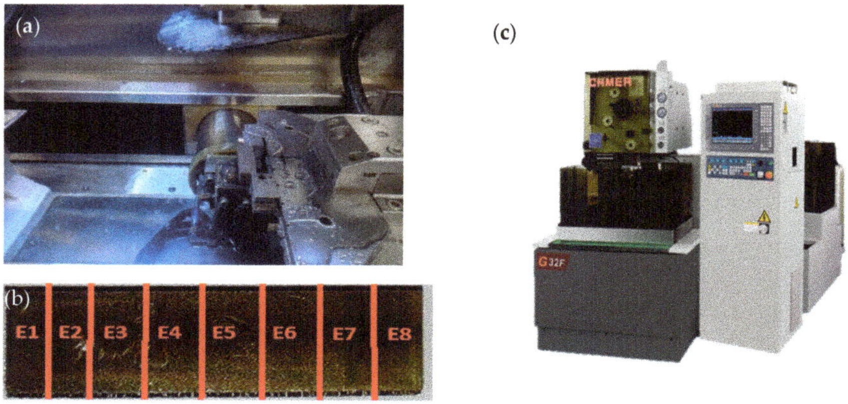

Figure 2. Made of experimental samples from maraging steel MS1 by μ-WEDM. (**a**) μ-WEDM process of the specimens; (**b**) marking of roughing sections E1 to E8, (**c**) the used electroerosive machine CHMER G32F.

The quantitative performance parameters of the *MRR* electroerosive process were determined for the individual experimental section cuts based on the time t required to execute them and the total mass loss of the sample m_{wi} during the making of the individual sections using accurate laboratory scales with an average of five repetitions. The measurement of the qualitative parameter *Rz* of the machined area in the individual sections of the experimental samples was carried out using a Mitutoyo SJ 210 contact roughness meter with a measuring range of −200 μm to +160 μm, again with an average of five repetitions.

3. Results and Discussion

3.1. DoE Statistical Analysis of Experimentally Measured MRR and Rz Data at μ-WEDM

As is well known, experimental measurement results are commonly characterized in practice by a highly asymmetrical distribution and an unconventional scatter. However, the violation of the basic requirements of the dataset being evaluated is no exception. Therefore, when evaluating the obtained results of experimental measurements of the quantitative performance parameter *MRR*, as well as the qualitative parameter of the roughness of the machined surface *Rz*, it was not necessary to implement a series of sequential steps [53]. The first step was an exploratory data analysis (EDA), which allowed to exclude certain anomalies of the obtained results of the experimental measurements. These were mainly specificities in the shape of the data distribution, the exclusion of outliers, and the detection of the local concentration of the measured data. In the next step, it was unavoidable to carry out the verification of the requirements for the set of measured data due to the considerable non-homogeneity of the material of the experimental samples [54–57]. Finally, through confirmatory data analysis (CDA), verification of the measured data was performed with the application of the estimation of the position, scattering, and shape parameters. The sampling analysis procedure was aimed at determining the objective mean of a representative sample from the experimental measurements of the output parameters *MRR* and *Rz* of the μ-WEDM. The results of the individual measurements of these parameters were firstly evaluated by standard statistical methods (Shapiro–Wilk test), which aimed to examine the normality of the dataset and then to identify outliers and extremes (Grubs and Dixon tests). This analysis was applied to the results of all recorded data for both *MRR* and *Rz*. In the case of recorded data where the analysis confirmed the presence of outliers or extremes and a normal distribution could not be established, even in cases where the normality of the data distribution was not demonstrated even after exclusion of confirmed outliers, exponential and Box–Cox transformations were performed. This ensured the accuracy of further statistical analyses of the experimentally obtained data.

3.2. Design and Validation of the MSC Model for the Prediction of MRR and Rz at μ-WEDM

An important step in the process of optimizing the performance of the μ-WEDM process of machining maraging steel MS1 sintered by DMLS technology is the design of the MSC model for the prediction of the output quantitative process performance parameters (*MRR*) and the output qualitative parameters (*Rz*) of the machined area [58,59]. It is important that the MSC model is designed with suitable prediction capability. This can only be achieved if data evaluation, regression analysis, and model interpretation have been performed in a statistically correct manner [60,61].

The basic statistical analysis of the used general model (17) for predicting the investigated response y depending on the change of the independently investigated variables x_i (maximum peak current, pulse-on time duration, pulse-off time duration, and voltage of dis-charge) was carried out using analysis of variance (ANOVA). ANOVA for the investigated parameter y represents a basic statistical analysis of the appropriateness of the used general model (17). Using ANOVA, it was analysed whether the variability caused by random errors is significantly smaller than the variability of the measured values explained by the model. The second statistical use of ANOVA results from its basic nature, where the null statistical hypothesis that none of the effects used in the model (I, t_{on}, t_{off}, U) have a significant impact on the investigated variable (*MRR*; *Rz*) is tested. The basic general ANOVA table is shown in Table 4.

Table 4. ANOVA table.

Source	DF	Sum of Squares	Mean Square	*F* Ratio	*Prob* > *F*
Model	$DF_{Model} = a - 1$	S_{Model}	$MS_{Model} = S_{Model}/DF_{Model}$	$F = MS_{Model}/MS_{Error}$	p_M
Error	$DF_{Error} = N - a$	S_{Error}	$MS_{Error} = S_{Error}/DF_{Error}$		
C.Total	$DF_{C.Total} = N - 1$	$S_{C.Total}$	$MS_{C.Total} = S_{C.Total}/DF_{C.Total}$		

The sum of squares of model S_{Model} represents the sum of squares of the differences between the sample means of individual groups from the overall mean and is calculated as:

$$S_{Model} = \sum_{i=1}^{a} n_i(\overline{y}_i - \overline{y})^2 \quad (5)$$

where a is the number of groups of factor A and n_i is the number of subjects in the i-th group.

It is also possible to calculate the residual sum of squares sum of squares S_{Error}, which represents the sum of squares of the difference between the observed values and the respective group averages, i.e., the amount of data fluctuation around the mean value in individual selections according to the relationship (6):

$$S_{Error} = \sum_{i=1}^{a}\sum_{j=1}^{N} (y_{ij} - \overline{y}_i)^2 \quad (6)$$

Finally, the total sum of squares $S_{C.Total}$ describes the overall variability of the measured quantity by comparing individual observations with the overall average and is calculated according to the relationship (7):

$$S_{C.Total} = \sum_{i=1}^{a}\sum_{j=1}^{N} (y_{ij} - \overline{y})^2 \quad (7)$$

For the sums of squares, $S_{C.Total} = S_{Model} + S_{Error}$ is valid and thus represents the decomposition of the total variability of the measured quantity between the variability of the model and the residual variability of the error. The p_M value (Table 4) represents the resulting calculated value of the level of significance reached. For the value of the Fisher test statistic:

$$F > F_{1-\alpha}(a-1, N-a) \tag{8}$$

where α represents the chosen level of significance ($\alpha = 0.05$).

If relation (8) is valid, then we reject the null statistical hypothesis of the agreement of the mean values of individual groups of subjects and we can say that the variability caused by random errors is significantly smaller than the variability of the measured values explained by the model. The used model (17) is adequate, based on the Fisher–Snedecor test criterion. The second conclusion of relation (8) is that at least one of the used input factors x_i is different from zero and thus has a significant influence on the change in the value of the investigated parameter y.

The actual calculation of specific values of the regression coefficients $b = (b_0, b_1 \ldots, b_{1234})$ of the general model is possible from a known relationship. It is based on the least squares method of deviations between the original and model-determined response values y in matrix form (9):

$$b = \left(X^T \cdot X\right)^{-1} \cdot X^T \cdot y = M^{-1} \cdot X^T \cdot y = V \cdot X^T \cdot y \tag{9}$$

where M s the so-called moment matrix and V is the variation matrix (inverse matrix of the moment matrix M). The fact that the calculated regression coefficient of the model has a non-zero value does not mean that it is statistically different from zero at the chosen significance level α. This can only be assessed after determining its inaccuracy, which is characterized by directional deviation. The variance of the regression coefficient b_j, $j = 0 \ldots p$ (if we only consider the first-order model without interactions), where p is the number of regressors, can be determined using the relationship (10):

$$s^2(b_j) = s^2(e) \cdot V_{jj} \tag{10}$$

where Vjj is the main diagonal of the variation matrix V, i.e., $diag(V)$, and $s^2(e)$ is the residual (unexplained) variance. This unexplained variance can be calculated from the residual sum of squares (RSC), i.e., from the sum of squared residuals (model error) e:

$$e = y - \tilde{y} \tag{11}$$

$$RSC = e^T \cdot e \tag{12}$$

$$s^2(e) = \frac{RSC}{n-(p+1)} = \frac{e^T \cdot e}{n-p-1} \tag{13}$$

where n is the number of measured values and $p + 1$ is the number of model regressors including the absolute term. The standard deviation of the regression coefficient is the square root of the variance:

$$s(b_j) = \sqrt{s^2(b_j)} \tag{14}$$

where for the test t-statistic of the statistical significance of the respective regression coefficient the ratio applies:

$$t(b_j) = \frac{b_j}{s(b_j)} \tag{15}$$

As long as the inequality holds

$$|t(b_j)| \geq t\left(1 - \frac{\alpha}{2}, n-p-1\right) \tag{16}$$

where $t(1-\frac{\alpha}{2}, n-p-1)$ is the quantile of the Student's *t*-distribution, then we reject the hypothesis of statistical insignificance of the regression coefficient.

When conducting an experiment, the estimate of the investigated dependent variable is generally described by a model of the form (17):

$$\hat{y} = b_0.x_0 + \sum_{j=1}^{N} b_j.x_j + \sum_{\substack{u,j=1 \\ u \neq j}}^{N} b_{uj}.x_u.x_j + \sum_{\substack{u,j=1 \\ u \neq j}}^{N} b_{uj}.x_u^2.x_j + \sum_{\substack{u,j=1 \\ u \neq j}}^{N} b_{uj}.x_u.x_j^2 + \sum_{j=1}^{N} b_{jj}.x_j^2 \quad (17)$$

where b_0, b_j, b_{uj}, b_{jj} are the respective regression coefficients and x_j are the respective independent variables, factors.

Table 5 shows the estimated parameters of the MSC model for the prediction of the output quantitative performance parameters *MRR* and the output qualitative parameters *Rz* of the machined area.

Table 5. Estimated MSC model parameters for prediction of *MRR* and *Rz*.

Term for *MRR*	**Estimate**	**Std Error**	***t*-Ratio**	***Prob* > |*t*|**
Intercept (x_0)	−0.010813	0.020121	−0.54	0.5929
x_1	0.0165482	0.000279	59.26	0.0001 *
x_2	0.0012834	7.19×10^{-5}	17.86	0.0001 *
x_3	−0.001947	0.00014	−13.94	0.0001 *
x_4	0.0008016	0.000237	3.38	0.0013 *
$(x_1-5)\cdots(x_1-5)$	0.0128791	0.003127	4.12	0.0001 *
$(x_1-5)\cdots(x_2-22.8767)$	−0.000387	0.000016	−24.17	0.0001 *
$(x_2-22.8767)\cdots(x_2-22.8767)$	−0.000275	8.03×10^{-5}	−3.42	0.0011 *
$(x_2-22.8767)\cdots(x_3-9)$	0.0000673	0.000008	8.40	0.0001 *
$(x_2-22.8767)\cdots(x_4-80)$	0.0000419	4.70×10^{-6}	8.93	0.0001 *
$(x_1-5)\cdots(x_1-5)\cdots(x_4-80)$	−0.000165	2.62×10^{-5}	−6.32	0.0001 *
$(x_1-5)\cdots(x_1-5)\cdots(x_1-5)\cdots(x_1-5)$	−0.000438	5.73×10^{-5}	−7.63	0.0001 *
Term for *Rz*	**Estimate**	**Std Error**	***t*-Ratio**	***Prob* > |*t*|**
Intercept (x_0)	4.312305	0.462060	9.33	<0.0001 *
x_1	1.263985	0.017863	70.76	<0.0001 *
x_2	0.108301	0.003054	35.46	<0.0001 *
x_3	−0.115140	0.009508	−12.11	<0.0001 *
x_4	−0.025370	0.005077	−5.00	<0.0001 *
$(x_1-5)\cdots(x_1-5)$	−0.459620	0.148820	−3.09	0.0030 *
$(x_1-5)\cdots(x_2-22.8767)$	−0.025260	0.001029	−24.55	<0.0001 *
$(x_1-5)\cdots(x_3-9)$	0.113915	0.035306	3.23	0.0020 *
$(x_2-22{,}8767)\cdots(x_3-9)$	−0.031130	0.011999	−2.59	0.0119 *
$(x_2-22{,}8767)\cdots(x_4-80)$	−0.009710	0.003622	−2.68	0.0094 *
$(x_1-5)\cdots(x_1-5)\cdots(x_1-5)\cdots(x_1-5)$	0.051696	0.016525	3.13	0.0027 *
$(x_1-5)\cdots(x_1-5)\cdots(x_1-5)\cdots(x_4-80)$	0.013128	0.004660	2.82	0.0065 *

x_1—maximum peak current *I* (A); x_2—pulse-on time duration t_{on} (μs); x_3—pulse off duration time t_{off} (μs); x_4—voltage *U* (V), *—statistically significant at the significance level α = 0.05.

The regression coefficients of the MSC models listed separately for the prediction of the *MRR* and *Rz* parameters in the estimate column are not scaled but refer to the original scale of measurement of each factor. As can be seen in Table 5, the largest effect on explaining response variability, i.e., on *MRR* and *Rz*, is penetration (x_0), also referred to as the absolute term of the model. In terms of the four considered input factors of the electrical discharge process, the peak discharge current *I* (factor x_1) has a major influence of 37.513% on *MRR* as well as on *Rz* with a contribution of 41.466%. Another significant member of the MSC model is pulse-on time duration t_{on} (factor x_2) with an influence weight of 11.306% on the variation in *MRR* value and with an influence weight of 20.814% on *Rz*. In a similar way to the peak discharge current, increasing the value of pulse-on time duration results in an increase in the *MRR* parameter but also in *Rz*. The interaction of

MTP maximum peak current and pulse-on time duration also has a significant effect on *MRR* and *Rz*. Their interaction has an impact weight of 15.3% on *MRR* and 14.347% on *Rz*. Pulse-off time duration t_{off} (factor x_3) also contributes significantly to the change in *MRR* and *Rz* parameters. The latter influences the output process performance parameter *MRR* with a weight of 9.51% and the quality output parameter *Rz* of the machined area after μ-WEDM with a weight of 12.76%. By increasing its value, we achieve a decrease in both the *MRR* parameter and the *Rz* parameter. The last significant factor, acting as the main factor in the electrical discharge process, is the applied voltage of discharge *U* (factor x_4). The effect of factor x_4 on the change of output power parameter *MRR* is 5.18% and its effect on the change of output quality parameter *Rz* is 7.42%. As in the case of pulse-on time duration t_{off}, the response decreases as the value of the applied voltage of discharge increases. Consequently, based on the estimation of the parameters presented in Table 5, it was possible to construct MSC models expressing the relationship between the input factors (x_1–x_4) and the response to a change in the output quality parameter *MRR* according to relation (18) and the response to a change in the output quality parameter *Rz* according to relation (19):

$$\begin{aligned} MRR = {} & 4.868 \cdot 10^{-2} \cdot I + 1.209 \cdot 10^{-2} \cdot t_{on} - 3.489 \cdot 10^{-3} \cdot t_{off} - 4.407 \cdot 10^{-3} \cdot U - 3.900 \cdot 10^{-4} \cdot I \cdot t_{on} + \\ & + 1.700 \cdot 10^{-3} \cdot I \cdot U + 6.730 \cdot 10^{-5} \cdot t_{on} \cdot t_{off} + 4.190 \cdot 10^{-5} \cdot t_{on} \cdot U - 1.700 \cdot 10^{-4} \cdot I^2 \cdot U - 4.632 \cdot 10^{-2} \cdot I^2 + \\ & + 8.800 \cdot 10^{-3} \cdot I^3 - 2.800 \cdot 10^{-4} \cdot t_{on}^2 - 4.400 \cdot 10^{-4} \cdot I^4 + 0.106 \end{aligned} \quad (18)$$

$$\begin{aligned} Rz = {} & 1.292 \cdot t_{on} - 99.204 \cdot I + 2.734 \cdot 10^{-2} \cdot t_{off} - 1.444 \cdot U - 2.526 \cdot 10^{-2} \cdot I \cdot t_{on} + 0.114 \cdot I \cdot t_{off} + \\ & + 0.985 \cdot I \cdot U - 3.113 \cdot 10^{-2} \cdot t_{on} \cdot t_{off} - 9.711 \cdot 10^{-3} \cdot t_{on} \cdot U - 0.197 \cdot I^2 \cdot U + 1.313 \cdot 10^{-2} \cdot I^3 \cdot U + \\ & + 23.048 \cdot I^2 - 2.084 \cdot I^3 + 5.169 \cdot 10^{-2} \cdot I^4 + 134.468 \end{aligned} \quad (19)$$

A graphical representation of the most significant effects in terms of Table 5 on the change in the value of the variable under study, i.e., the *MRR* process performance ($mm^3 \cdot min^{-1}$) as a function of the change in the maximum peak current *I* parameter, at different values of the pulse-on time duration t_{on} in the minimum and maximum values of the pulse-off time duration t_{off} and voltage from discharge *U* parameters is shown in Figure 3.

The above-presented graphical dependencies confirmed the increasing trend of the output quality parameter of the electrical discharge process *MRR* depending on the increasing values of the input parameters *I* and t_{on}. Its lowest value of 0.005 $mm^3 \cdot min^{-1}$ was obtained at the combination of MTP I = 2.5 A, t_{on} = 3 μs, t_{off} = 15 μs, and U = 90 V. Conversely, its highest value of 0.190 $mm^3 \cdot min^{-1}$ was achieved at a combination of MTP I = 8.0 A, t_{on} = 40 μs, t_{off} = 3 μs, and U = 70 V.

A graphical representation of the most significant effects in terms of Table 5 on the change in the value of the variable under study, i.e., the *Rz* machined surface roughness (μm) as a function of the change in the maximum peak current *I* parameter, at different values of the pulse-on time duration t_{on} at the minimum and maximum values of the pulse-off time duration t_{off} and voltage from discharge *U* parameters is shown in Figure 4.

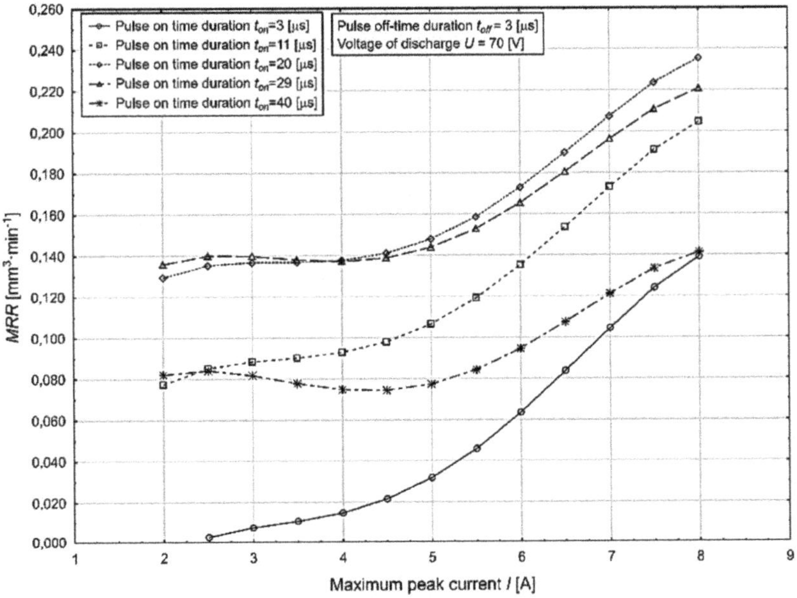

(a)

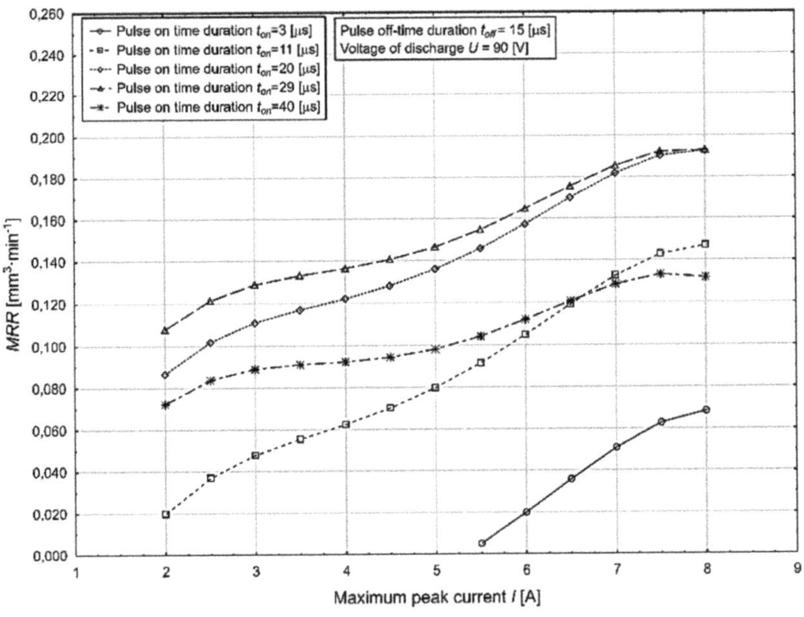

(b)

Figure 3. Dependence of the output power parameter of the *MRR* process on the change in the value of the maximum peak current *I* parameter at different values of t_{on}, t_{off}, and *U*. (**a**) Impact on *MRR* at minimum values of the parameter t_{off} and *U*. (**b**) Impact on *MRR* at maximum values of t_{off} and *U*.

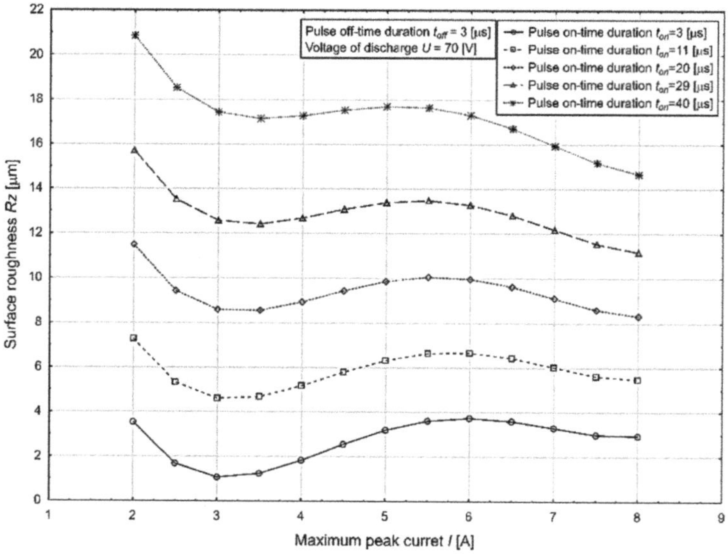

(a)

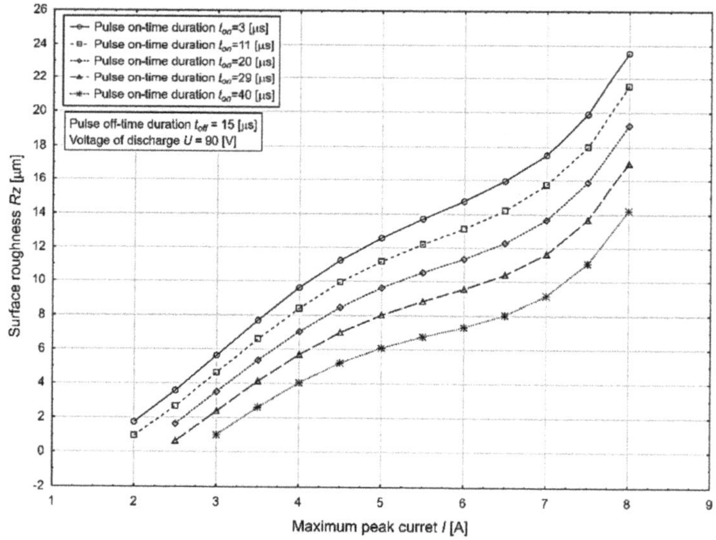

(b)

Figure 4. Dependence of the output qualitative parameter of the Rz process on the change in the value MTP of the maximum peak current I parameter at different values of t_{on}, t_{off}, and U. (**a**) Impact on Rz at minimum values of the parameter t_{off} and U. (**b**) Impact on Rz at maximum values of t_{off} and U.

The above presented graphical dependencies confirmed the increasing trend of the output quality parameter of the electrical discharge process Rz depending on the increasing value of the input parameters I and t_{on}. Its lowest value of 0.09 μm was obtained at the combination of MTP I = 2.5 A, t_{on} = 3 μs, t_{off} = 15 μs, and U = 90 V. Conversely, its highest value of 23.50 μm was achieved at a combination of MTP I = 8.0 A, t_{on} = 40 μs, t_{off} = 3 μs, and U = 70 V.

In terms of confirming the accuracy and suitability of the proposed MSC models (18) and (19), it was necessary to verify the residuals, the difference between the actual measured and predicted values, using the prediction model in terms of their distribution. The value of the Shapiro–Wilks test at the chosen significance level $\alpha = 5\%$ for the MSC model (18) represents a value of 2.311% ($p = 0.0519$) and for the MSC model (19) represents a value of 1.756% ($p = 0.0883$). Therefore, the null statistical hypothesis that there is no autocorrelation can be accepted. The achieved significance level of the Shapiro–Wilk test indicates a Gaussian distribution of the residuals on a basis of which it can be concluded that the predictive MSC models (18) and (19) were designed correctly in terms of statistical and numerical accuracy.

The model error for the investigated variables *MRR* and *Rz* were calculated as the difference between the experimentally obtained value and the value predicted by model (18) for the variable *MRR* and model (19) for the variable *Rz*. In general, we can express the model error by the relation (20):

$$Model\ error\ = \frac{y_e - y_m}{y_e} \cdot 100\ [\%] \tag{20}$$

where y_e s the value of the investigated variable (*MRR* or *Rz*) obtained experimentall and, y_m is the value of the investigated variable calculated using the prediction model (18) or (19).

A graphical representation of the deviation of the measured and calculated MSC values by models (18) and (19) is shown in Figure 5.

In the next stage, it was necessary to perform a cumulative analysis of the usefulness of the developed MSC models describing the influence of input factors on *MRR* and *Rz* for μ-WEDM. A cumulative analysis of the quality of the fit of the MSC models for the prediction of the output parameters *MRR* and *Rz* of the μ-WEDM for maraging steel MS1 is carried out and presented in the following Table 6.

Table 6. Cumulative analysis of the quality of the fit of the MSC models for prediction of the output parameters *MRR* and *Rz* of the electrical discharge process.

Parameter for *MRR*	Value
RSquare	0.998175
RSquare Adj	0.997846
Root Mean Square Error	0.002675
Mean of Response	0.136644
Observations (or Sum Wgts)	73
Parameter for *Rz*	**Value**
RSquare	0.998773
RSquare Adj	0.998528
Root Mean Square Error	0.164369
Mean of Response	9.332192
Observations (or Sum Wgts)	73

To express the suitability of the regression model application, appropriate model fit diagnostic tools need to be implemented to assess the specified MSC model in terms of graph fit, lack of fit, and likelihood of residuals. As can be observed from the results presented in Table 6, the variability index denoted as RSquare of the established MSC model for the prediction of the output quantitative performance parameters *MRR* in the μ-WEDM has a value of 99.8175% and for the prediction of the output qualitative parameters *Rz* has a value of 99.8773%. The adjusted index of determination denoted as RSquare Adj, which indicates the overall level of model variability, reaches a value of 99.7846% in the case of the MSC model for *MRR* and 99.8528% in the case of the MSC model for *Rz*. The average error of the proposed MSC model is 0.002675 $mm^3 \cdot min^{-1}$ for *MRR* and 0.164369 μm for *Rz*. The average value of the output quantitative performance parameter *MRR* in the μ-WEDM is 0.136644 $mm^3 \cdot min^{-1}$ and the average value of the output qualitative parameter *Rz* of the machined area is 9.332192 μm. However, the R^2 index alone is not a sufficient indicator to investigate the validity of the established MSC model for the prediction of the *MRR* and *Rz* parameters, mainly because an advanced analysis of the measured data has been performed. Based on the indices presented in Table 6, it can be concluded that the achieved value of the adjusted index of determination is satisfactory in terms of performance, quality, and the functionality of the developed MSC models. Therefore, the given models are suitable for the further optimization process.

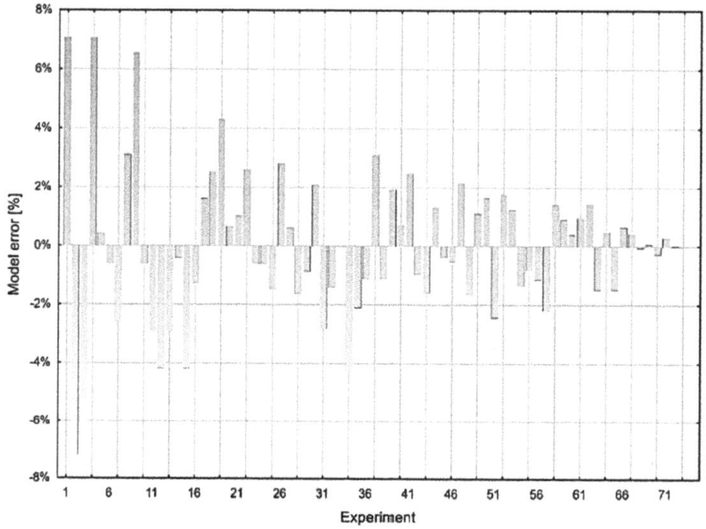

(**a**)

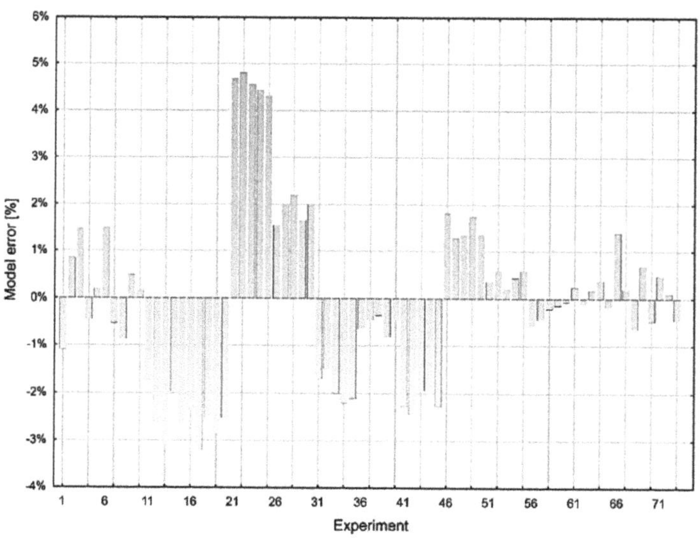

(**b**)

Figure 5. Graphical analysis of the predictive MSC models. (**a**) MSC model deviations for *MRR*. (**b**) MSC model deviations for *Rz*.

To assess the suitability of the application of MSC models for the prediction of *MRR* and *Rz* parameters, the parametric statistical method analysis of variance (ANOVA) is also a suitable tool. The results of this analysis are presented in Table 7 below.

Table 7. Results of evaluation of MSC models for prediction of *MRR* and *Rz* parameters by ANOVA.

Source for *MRR*	DF	Sum of Squares	Mean Square	*F*-Ratio	*Prob* > *F*
Model	11	0.2387703	0.021706	3033.879	0.0001 *
Error	61	0.0004364	7.16×10^{-6}		
C. Total	72	0.2392067			
Source for *Rz*	**DF**	**Sum of Squares**	**Mean Square**	***F*-Ratio**	***Prob* > *F***
Model	12	1319.781	109.982	4070.833	0.0001 *
Error	60	1.621	0.027		
C. Total	72	1321.402			

*—significant at the level of $\alpha = 0.05$.

From the results of the evaluation of the MSC models for the prediction of the output parameters of the *MRR* and *Rz* μ-WEDM by ANOVA presented in Table 7, it can be observed that the variability due to random errors is significantly smaller than the variability of the values determined and explained by the model. If the obtained value (*Prob* > *F*) is less than the significance level α, it can be said that there is at least one non-zero term in the proposed MSC model that significantly affects the value of the variable under study. Based on the application of the Fisher–Snedecor test criterion, it was found that the achieved value (*Prob* > *F*) for the MSC model predicting the output performance parameter *MRR* of the μ-WEDM was 0.0001 at the significance level $\alpha = 5.0\%$ that of the MSC model predicting the output qualitative parameter of the machined surface *Rz* was 0.0001 at the selected significance level $\alpha = 5.0\%$. This indicates the adequacy of the determined MSC models, which is also implied by the verification of null statistical hypothesis. The latter is further confirmed by the fact that none of the input factors of the MSC model significantly affects the resulting value of the variable under study.

In terms of the complexity of assessing the validity of the MSC models for the prediction of the output parameters of the electrical discharge process *MRR* and *Rz* in the machining of maraging steel MS1 by μ-WEDM technology, it was still appropriate to apply the ANOVA lack of goodness-of-fit test. This is to reveal the predictive power of the established MSC models. The variance of the residuals and the variance of the measured data should be applied to diagnose whether the proposed MSC model fits the observed dependence well. The results of the ANOVA lack of goodness-of-fit test are presented in Table 8 below.

Table 8. Results of the lack of goodness-of-fit test of MSC models for the prediction of the output parameters *MRR* and *Rz* in μ-WEDM by ANOVA.

Source	DF	Sum of Squares	Mean Square	*F* Ratio	*Prob* > *F*
Lack of Fit	7	0.0001104	0.000016	2.6133	0.0513
Pure Error	54	0.000326	6.04×10^{-6}		
Total Error	61	0.0004364			
Source	**DF**	**Sum of Squares**	**Mean Square**	***F* Ratio**	***Prob* > *F***
Lack of Fit	6	1.588221	0.264703	435.7923	0.0677
Pure Error	54	0.0328	0.000607		
Total Error	60	1.621021			

From the results presented in Table 8, it can be observed that the residual variability was compared to the within-group variability of the measured data. Thus, it can be argued that the null statistical hypothesis (H_0) and the variance of the residuals are less than the within-group variance. At the same time, the alternative hypothesis (H_1) was tested, i.e., the hypothesis that the variance of the residuals is greater than the within-group variance. Then, at the chosen significance level $\alpha = 5.0\%$, the value of Fisher's criterion converted to a probability scale (*Prob* > *F*) for the MSC model predicting the output performance parameter *MRR* is 0.0513 and that of the MSC model predicting the output qualitative parameter *Rz* is 0.0677. Based on this, it can be argued that there is sufficient evidence to reject the null statistical hypothesis (H_0). As a result, the variance of the residuals is less than or equal to the within-group variance and therefore the model is statistically significant. When a sufficiently low model error is obtained, the model shows an excellent fit to the real data. Based on the obtained results of the lack of agreement test of the proposed MSC models by ANOVA for the prediction of the output quantitative performance parameters of the *MRR* and the output qualitative parameters *Rz* of the machined area in the machining of maraging steel MS1 by µ-WEDM technology, it can be claimed that the established MSC models are significantly valid.

3.3. Optimization of Process Efficiency in µ-WEDM Maraging Steel MS1

Based on the empirically designed MSC models predicting the behaviour of the output quantitative process parameter *MRR* and the output qualitative parameter *Rz* of the machined area at µ-WEDM of maraging steel MS1, verified by numerical and statistical analysis, it is possible to proceed with the optimization of the performance of the µ-WEDM process. In the context of techno-economic optimization of µ-WEDM, the MSC models defined through relations (18) and (19) can be used to formulate the optimization criterion. In practice, the performance of the electrical discharge process at µ-WEDM of maraging steel MS1 is defined as the amount of material removed per unit of time while achieving the desired quality level of the machined surface in terms of its roughness. Therefore, when optimizing this process, we are looking to save machining time while maximizing the quality of the machined surface, which is defined by the lowest possible value of the machined surface roughness parameter *Rz*. At the same time, the optimal criteria must take into account the aspect of the economic efficiency of the process, with machining time being the decisive indicator of economic efficiency. By minimizing the machining time to achieve the desired value of the machined surface roughness parameter *Rz* by setting the optimum combination of values of the input factors, which are represented by the MTP, we can maximize the economic benefit. Therefore, in the optimization procedure, our aim was to minimize the machining time for a predetermined quality level of the machined surface in terms of the roughness parameter *Rz*. Since the empirically determined MSC models (18) and (19) are parametrically nonlinear, it can be expected that the objective function is also nonlinear. Therefore, nonlinear programming was chosen to perform the optimization procedure using the MATLAB software system.

The formation of a mathematical model is the first step towards the optimal solution, followed by the formulation of the optimization problem, which means the formulation of the objective function and the determination of the limiting constraints on the process to avoid unrealistic solutions during the optimization procedure. The choice of the appropriate optimization method and suitable software support is up to the user, as there is no one-size-fits-all optimization method suitable for any optimization problem. Our investigated process is limited by several constraints and exhibits nonlinearity in functional dependencies, so it is necessary to formulate the optimization problem with constraints and select from the optimization methods of nonlinear programming.

The optimization problem of mathematical programming aims at the extremization of the objective function $f_0(x)$, i.e., to find the minimum/maximum of the objective function $f_0(x)$ while solving the minimization/maximization problem. The validity of the relationship $\min_{x \in \mathbb{R}^n} f(x) = -\max_{x \in \mathbb{R}^n}(-f(x))$ allows us to transform each maximization into a minimization optimization problem. The optimization problem with equality and inequality constraints is generally focused on the minimization of an objective function $f_0(x)$ in a feasible region K and is formulated in the following shape:

$$Min\{f_0(x) \ \mid x \in X, f_i(x) \leq 0, \ i \in I, \ h_j(x) \ = \ 0, \ j \in J\} \tag{21}$$

where I and J are index sets, OP (3) is called a mathematical programming problem. The NLP problem arises when at least one of the functions $f_0, \ f_i, \ i \in I, \ h_j, \ j \in J$ is nonlinear in the optimization process (21). Clearly, the optimal solution of the NLP (21) is such a vector $x^* \in X$, which satisfies the condition $\forall x \in X : f_0(x^*) \leq f_0(x)$, i.e., the objective function attains the smallest value.

From the point of view of the exact execution of the optimization of the electrical discharge process in the machining of Maraging Steel MS1, it is necessary to define constraints. These constraints need to be defined within the framework of the experiments carried out and take into account the MTP intervals, which are shown in Table 1. Based on the definition of the specific values of the input parameters of the μ-WEDM defined by the MTP, it is then possible to define the boundary conditions in the following form (22):

$$\begin{array}{c} 2.0 \leq \text{Maximum peak current } I \text{ (A)} \leq 8.0 \\ 5.0 \leq \text{Pulse-on time duration } t_{on} \text{ (μs)} \leq 40.0 \\ 3.0 \leq \text{Pulse-off time duration } t_{off} \text{ (μs)} \leq 15.0 \\ 70.0 \leq \text{Voltage of discharge } U \text{ (V)} \leq 90.0 \end{array} \tag{22}$$

The optimization of the performance of the electrical discharge process in the machining of Maraging Steel MS1 by μ-WEDM technology involves the optimization criteria, which are the maximization of the objective function defined by the MSC model (18) and at the same time the minimization of the objective function defined by the MSC model (19) with the application of the optimization constraints defined by the functions (22). Subsequently, by applying nonlinear programming in MATLAB 2019a software, the corresponding script was created. The task of the performed nonlinear optimization was to find the local extreme of the objective function for a given optimization problem.

The objective function (18) modelling the performance of the *MRR*, where x_1—maximum peak current I [A]; x_2—pulse-on time duration t_{on} [μs]; x_3—pulse-off time duration time t_{off} [μs]; x_4—voltage U [V], can be written in this case as an optimization problem in general form:

$$MRR \ = \ f_0(x_1, \ x_2, \ x_3, x_4) \ \ \rightarrow \ \ \max \tag{23}$$

or in natural scale in abbreviated form:

$$MRR \ = \ f_0\Big(I\ [A]\ (x_1),\ t_{on}\ [\mu s]\ (x_2), t_{off}\ [\mu s]\ \ (x_3),\ U\ [V]\ \ (x_4)\Big) \ \ \rightarrow \ \ \max \tag{24}$$

This maximization problem was transformed into a minimization problem since this is the method by which software programs work in the MATLAB environment, where each optimization problem must be rewritten in a suitable format:

$$\min f(x) \begin{cases} c(x) \leq 0 \\ ceq(x) \ = \ 0 \\ \mathbf{A} \cdot \mathbf{x} < \mathbf{b} \\ \mathbf{Aeq} \cdot \mathbf{x} \ = \ \mathbf{beq} \\ \mathbf{lb} \leq \mathbf{x} \leq \mathbf{ub} \end{cases} \tag{25}$$

where **x**, **b**, **beq**, and **lb** (lower bound) and **ub** (upper bound) are vectors, **A** and **Aeq** are matrices with constant coefficients, *c(x)* and *ceq(x)* are vector functions, and *f(x)* is a scalar function. The functions *f(x)*, *c(x)* and *ceq(x)* are nonlinear.

In the case of optimization of the output qualitative roughness parameter *Rz*, the objective function (19) modelling the values of the resulting roughness *Rz*, where x_1—maximum peak current I (A), x_2 is pulse-on time duration t_{on} (μs), x_3 pulse-off time duration time t_{off} (μs), and x_4 is voltage U (V) can be written as an optimization problem in a general form:

$$Rz = f_0(x_1, x_2, x_3, x_4) \rightarrow \min \quad (26)$$

or in natural scale in abbreviated form:

$$Rz = f_0\left(I\,[A]\,(x_1),\ t_{on}\,[\mu s]\,(x_2), t_{off}\,[\mu s]\,(x_3),\ U\,[V]\,(x_4)\right) \rightarrow \min \quad (27)$$

Subsequently, the objective functions (18) and (19), respecting the optimization constraints defined by relations (22), were rewritten into a form suitable for optimization in the MATLAB software environment. The performed optimization obtained combinations of input parameters that represent the MTP and the optimum time of the maraging steel MS1 μ-WEDM in achieving the desired quality level of the machined surface in terms of the surface roughness parameter *Rz*. To better illustrate the performed optimization of the performance of the machining maraging steel MS1 by μ-WEDM technology, a graphical dependence was constructed, which is presented in the following Figure 6.

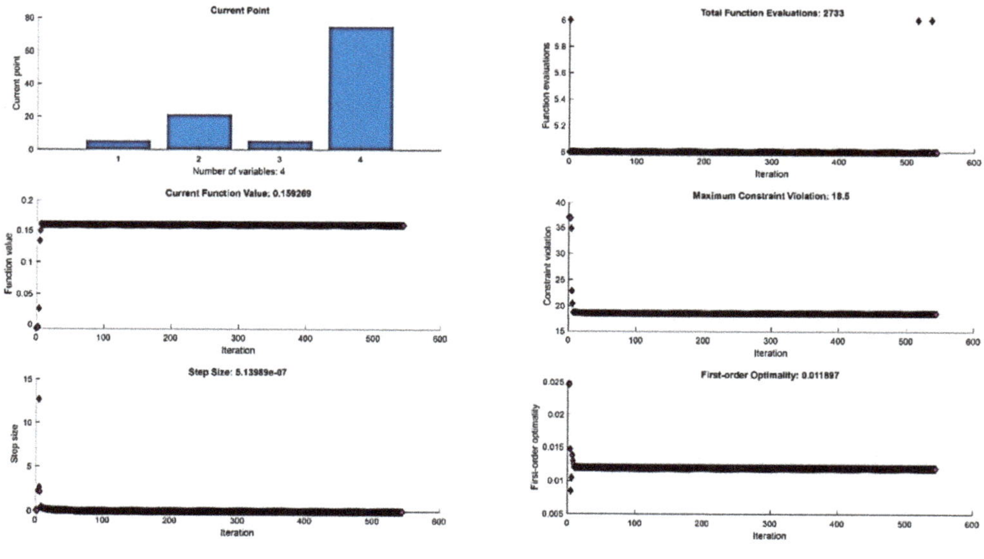

Figure 6. Graphical representation of the outputs of the performed optimization of the performance parameter of the μ-WEDM for maraging steel MS1.

The achieved value of the local maximum of function (18) is $MRR = 0.159$ mm^3·min^{-1}, and this value is reached by the function at: $I = 5.5$ A, $t_{on} = 21.5$ μs, $t_{off} = 5.5$ μs, and $U = 75$ V.

To better illustrate the optimization performed on the quality of the machined maraging steel MS1 by μ-WEDM technology, a graphical dependence was constructed and is presented in the following Figure 7.

The achieved value of the local minimum of function (19) is $Rz = 1.051$ μm, while this value is reached by the function at: $I = 3.014$ A, $t_{on} = 3.0$ μs, $t_{off} = 3.0$ μs and $U = 70$ V.

As can be observed from the results of the analysis of the recorded data and the outputs of the optimization process for μ-WEDM of maraging steel MS1, the performance of the *MRR* μ-WEDM and the quality of the machined surface, in terms of the roughness parameter *Rz*, are most influenced by the maximum discharge current *I* and the pulse duration t_{on}. The results indicate that the optimum performance of the μ-WEDM can be achieved by increasing the maximum discharge current while maintaining a constant value of the electrical discharge voltage and, at the same time, setting the values of the pulse duration close to the mean value and the duration of the pause between pulse discharges close to the lower limit of the interval defined in constraints (8). At the same time, the results of the performance optimization process of the electrical discharge machining of maraging steel MS1 were verified in real operating conditions, showing a close agreement. Although based on the research result, it is clear that the maximum discharge current has the greatest influence on the performance of the μ-WEDM, increasing its value has limit constraints. It is necessary to consider the value of the critical discharge current *I*, at which the wire tool electrode is destroyed, which reduces the efficiency and performance of the μ-WEDM of maraging steel MS1 process. Therefore, its value was limited to 10.0 A based on the experimental validation results. If it is exceeded, the wire tool electrode is destroyed and the quality of the machined surface in terms of the *Rz* parameter is no longer satisfactory.

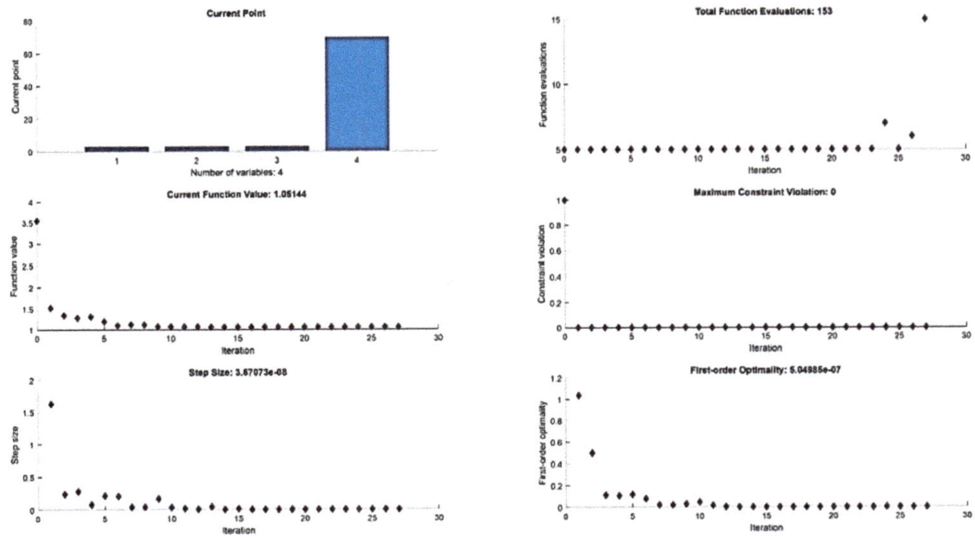

Figure 7. Graphical representation of the outputs of the performed optimization of the quality parameter of the machined surface after μ-WEDM of maraging steel MS1.

4. Conclusions

A typical feature of contemporary industrial production is the drive to increase the overall efficiency of the production process. The aim is to achieve maximum machining performance while maintaining a high quality of the machined surface. Achieving this is often problematic because, in general, as cutting power increases, the quality of the machined area decreases. This is to be aided by the optimization of the μ-WEDM, which identifies the appropriate combination of input MTP values. This combination of values of the input parameters of the μ-WEDM is then intended to guarantee its high efficiency. Therefore, the objective of the performed optimization of the performance of the electrical discharge process in the machining of maraging steel MS1 was to maximize the *MRR* while maintaining the quality level of the machined surface as high as possible. The optimization of the output parameters of the μ-WEDM in relation to the MTP is described in detail in the paper. An inappropriate combination results in a very low cutting performance and poor quality of the machined surface not only in terms of geometric accuracy but also in terms of surface roughness parameters. The developed MSC models predicting the output quantitative parameter *MRR* and the qualitative parameter *Rz*, provided the basis for the subsequent optimization of the μ-WEDM. In addition, based on the performed experimental measurements, the original dependencies of the influence of the MTP process input parameters on *MRR* and *Rz* for μ-WEDM maraging steel MS1 were obtained. At the same time, several significant facts were observed from the results of the experimental measurements.

The results of the experiment are summarized in the following points:

- It was found that in terms of the four considered input factors of the electroerosive process, maximum peak current *I* with a weight of 37.513%, pulse-on time duration t_{on} with a weight of 11.306%, pulse-off time duration t_{off} with a weight of 9.51%, and voltage of discharge *U* with a weight of 5.18% have the main influence on *MRR*;
- It was also found that in terms of the four considered input factors of the electroerosive process, maximum peak current *I* with a weight of 41.466%, pulse-on time duration t_{on} with a weight of 20.814%, pulse-off time duration t_{off} with a weight of 12.76%, and voltage of discharge *U* with a weight of 7.42% have the main influence on *Rz*;
- It was found that with increasing values of the input parameters of the electrical discharge process *I* and t_{on}, the cutting performance of *MRR* increases but the quality of the machined surface decreases in terms of the surface roughness parameter *Rz*. The highest value of the *MRR* parameter = 0.190 $mm^3 \cdot min^{-1}$ was obtained for the combination of MTP: I = 8.0 A, t_{on} = 40 µs, t_{off} = 3 µs, and U = 70 V;
- At the same time, it was found that with increasing values of the input parameters of the electrical discharge process t_{off} and *U*, the cutting performance of the *MRR* decreases but the quality of the machined surface increases in terms of the surface roughness parameter *Rz*. The lowest value of the *Rz* parameter = 0.09 µm was obtained for the combination of MTP: I = 2.5 A, t_{on} = 3 µs, t_{off} = 15 µs, and U = 90 V;
- It was found that the aforementioned pairs of MTP input parameters in the electrical discharge process behave oppositely in relation to *MRR* and *Rz*;
- For the given reasons, it was necessary to search for an appropriate ratio of the MTP input parameters in the electrical discharge process to achieve the optimum value of the process output performance parameter *MRR* and the quality parameter of the machined surface *Rz*;
- Optimization was performed with respect to maximizing the output parameter of the *MRR* and minimizing the quality parameter of the machined surface in terms of the surface roughness parameter *Rz* for µ-WEDM maraging steel MS1. Through the optimization, a local maximum of 0.159 $mm^3 \cdot min^{-1}$ of the *MRR* parameter can be achieved at with MTP settings of I = 5.5 A, t_{on} = 21.5 µs, t_{off} = 5.5 µs, and U = 75 V. Conversely, through optimization a local minimum of 1.051 µm of the *Rz* parameter can be achieved at MTP settings of I = 3.014 A, t_{on} = 3.0 µs, t_{off} = 3.0 µs, and U = 70 V;
- The performed optimization of the electrical discharge process can generally achieve an increase in the overall efficiency of µ-WEDM in the machining of maraging steel MS1.

Further scientific research in this area needs to be oriented towards a more comprehensive approach to optimizing the electrical discharge process as well as other machined materials using different wire tool electrodes to take into account the critical values of all important factors. To perform µ-WEDM of maraging steel MS1 under optimum machining process conditions, it is necessary to consider the value of the critical discharge current and pulse duration, as exceeding them results in the destruction of the wire tool electrode, which results in a decrease in the overall efficiency of the electrical discharge process. Therefore, in order to increase the efficiency of the actual electrical discharge process in practice, future experimental and research activities will focus on data analysis to determine the critical values of these defined MTPs.

Author Contributions: Conceptualization, L'.S.; methodology, A.V.; validation, L'.S.; formal analysis, M.G.; investigation, L'.S.; data curation, M.G.; writing—original draft preparation, L'.S.; visualization, P.K.; project administration, L'.S.; funding acquisition, L'.S. All authors have read and agreed to the published version of the manuscript.

Funding: This research was funded by the Scientific Grant Agency of the Ministry of Education, Science, Research and Sport of the Slovak Republic, grant number VEGA 1/0205/19.

Institutional Review Board Statement: Not applicable.

Informed Consent Statement: Not applicable.

Data Availability Statement: All data are published with the paper.

Acknowledgments: The authors would like to thank the grant agency for supporting this research work through the project VEGA 1/0205/19.

Conflicts of Interest: The authors declare no conflict of interest.

Nomenclature

CDA	Confirmatory Data Analysis
DOE	Design of Experiments
DMLS	Direct Metal Laser Sintering
D	Total required functionality
EDA	Exploratory Data Analysis
I	Maximum peak current (A)
IPM	Interior Point Method
MSC	Mathematical-Statistical Computational
MTP	Main Technological Parameters
MRR	Material Removal Rate
NLP	Non-Linear Programming
QNM	Quasi-Newton Methods
Rz	Ten-point Mean Roughness (μm)
SDM	Steepest Descent Method
SQP	Sequential Quadratic Programming
t_{on}	Pulse-on time duration (μs)
t_{off}	Pulse-off time duration (μs)
U	Voltage of discharge (V)
y	Desired value function
$y_{min/max}$	Lower/upper response limit values
μ-WEDM	micro-Wire Electrical Discharge Machining

References

1. Ali, M.Y.; Banu, A.; Salehan, M.; Adesta, Y.E.T.; Hazza, M.; Shaffiq, M. Dimensional Accuracy in Dry Micro Wire Electrical Discharge Machining. *J. Mech. Eng. SCIE* **2018**, *12*, 3321–3329.
2. Mičietová, A.; Neslušan, M.; Čilliková, M. Influence of surface geometry and structure after non-conventional methods of parting on the following milling operations. *Manuf. Technol.* **2013**, *13*, 199–204.
3. Wang, J.; Sánchez, J.A.; Izquierdo, B.; Ayesta, I. Experimental and numerical study of crater volume in wire electrical discharge machining. *Materials* **2020**, *13*, 577.
4. Daneshjo, N.; Malega, P.; Drabik, P. Techniques for Production Quality Control in the Global Company. *Adv. Sci. Technol. Res. J.* **2021**, *15*, 174–183.
5. Dzionk, S.; Siemiatkowski, M.S. Studying the effect of working conditions on WEDM machining performance of super alloy Inconel 617. *Machines* **2020**, *8*, 54.
6. Firouzabadi, H.A.; Parvizian, J.; Abdullah, A. Improving accuracy of curved corners in wire EDM successive cutting. *Int. J. Adv. Manuf. Technol.* **2015**, *76*, 447–459.
7. Islam, M.N.; Rafai, N.H.; Subramanian, S.S. An Investigation into Dimensional Accuracy Achievable in Wire-cut Electrical Discharge Machining. In Proceedings of the World Congress on Engineering, WCE 2010, London, UK, 30 June–2 July 2010; pp. 1–6.
8. Trishch, R.; Nechuiviter, O.; Dyadyura, K.; Vasilevskyi, O.; Tsykhanovska, I.; Yakovlev, M. Qualimetric method of assessing risks of low quality products. *MM Sci. J.* **2021**, *10*, 4769–4774.
9. Aggarwal, V.; Khangura, S.S.; Garg, R.K. Parametric modeling and optimization for wire electrical discharge machining of Inconel 718 using response surface methodology. *Int. J. Adv. Manuf. Technol.* **2015**, *79*, 31–47.
10. Evin, E.; Tomáš, M.; Kmec, J. Optimization of electro-discharge texturing parameters for steel sheets' finishing rollers. *Materials* **2020**, *13*, 1223.

11. Chen, Z.; Huang, Y.; Zhang, Z.; Li, H.; Zhang, G. An analysis and optimization of the geometrical inaccuracy in WEDM rough corner cutting. *Int. J. Adv. Manuf. Technol.* **2014**, *74*, 917–929.
12. Hamed, S.; Al-Juboori, L.A.; Najm, V.N.; Saleh, A.M. Analysis the impact of WEDM parameters on surface microstructure using response surface methodology. In Proceedings of the ASET 2020, Advances in Science and Engineering Technology International Conferences, Dubai, United Arab Emirates, 4–9 February 2020; p. 9118208.
13. Kiyak, M. Investigation of effects of cutting parameters on surface quality and hardness in the wire-EDM process. *Int. J. Adv. Manuf. Tech.* **2021**, *119*, 647–655. [CrossRef]
14. Mouralova, K.; Zahradnicek, R.; Houska, P. Evaluation of surface quality of X210Cr12 steel for forming tools machined by WEDM. *MM Sci. J.* **2016**, *5*, 1366–1369. [CrossRef]
15. Pi, V.N.; Tam, D.T.; Cuong, N.M.; Tran, T.H. Multi-objective optimization of PMEDM process parameters for processing cylindrical shaped parts using taguchi method and grey relational analysis. *Int. J. Mech. Prod. Eng. Res. Develop.* **2020**, *10*, 669–678.
16. Raksiri, C.; Chatchaikulsiri, P. CNC Wire-Cut Parameter Optimized Determination of the Stair Shape Workpiece. *Int. J. Mech. Mechatron. Eng.* **2010**, *4*, 924–929.
17. Zhu, Z.; Guo, D.; Xu, J.; Lin, J.; Lei, J.; Xu, B.; Wu, X.; Wang, X. Processing Characteristics of Micro Electrical Discharge Machining for Surface Modification of TiNi Shape Memory Alloys Using a TiC Powder Dielectric. *Micromachines* **2020**, *11*, 1018. [CrossRef]
18. Meshram, D.B.; Puri, Y.M. Optimized curved electrical discharge machining-based curvature channel. *J. Braz. Soc. Mech. Sci.* **2020**, *42*, 82. [CrossRef]
19. Zhu, S.; Chen, W.; Zhan, X.; Ding, L.; Zhou, J. Parameter optimisation of laser cladding repair for an Invar alloy mould. *Proc. Inst. Mech. Eng. Part B J. Eng. Manuf.* **2019**, *233*, 1859–1871. [CrossRef]
20. Pradhan, B.B.; Masanta, M.; Sarkar, B.R.; Bhattacharyya, B. Investigation of electro-discharge micro-machining of titanium super alloy. *Int. J. Adv. Manuf. Technol.* **2009**, *41*, 1094–1106. [CrossRef]
21. Meena, V.K.; Azad, M.S. Grey relational analysis of micro-EDM machining of Ti-6Al-4V alloy. *Mater. Manuf. Process.* **2012**, *27*, 973–977. [CrossRef]
22. Somashekhar, K.P.; Ramachandran, N.; Mathew, J. Optimization of material removal rate in micro-EDM using artificial neural network and genetic algorithms. *Mater. Manuf. Process.* **2010**, *25*, 467–475. [CrossRef]
23. Jabbaripour, B.; Sadeghi, M.H.; Faridvand, S.; Shabgard, M.R. Investigating the effects of EDM parameters on surface integrity, MRR and TWR in machining of Tie6ALe4V. *Mach. Sci. Technol.* **2012**, *16*, 419–444. [CrossRef]
24. Kuruvila, N.; Ravindra, H.V. Parametric influence and optimization of wire EDM of Hot die steel. *Mach. Sci. Technol.* **2011**, *15*, 47–75. [CrossRef]
25. Mukherjee, R.; Chakraborty, S. Selection of EDM process parameters using biogeography-based optimization algorithm. *Mater. Manuf. Process.* **2012**, *27*, 954–962. [CrossRef]
26. Lin, Y.C.; Chen, Y.F.; Lin, C.T.; Tzeng, H.J. Electrical discharge machining (EDM) characteristics associated with electrical discharge energy on machining of cemented tungsten carbide. *Mater. Manuf. Process.* **2008**, *23*, 391–399. [CrossRef]
27. Strasky, J.; Janecek, M.; Harcuba, P. Electric Discharge Machining of Tie6Ale4V Alloy for Biomedical Use. In *WDS'11. Proceedings of Contributed Papers*; Part III; MatfyzPress: Prague, Czechia, 2011; pp. 127–131.
28. Baron, P.; Kaščák, J. Comparison of selected materials intended for the manufacture of plastic holder for reverse parking sensor with the use of computer simulation tools. *MM Sci. J.* **2019**, *12*, 3463–3467. [CrossRef]
29. Panda, A.N.T.O.N.; Anisimov, V.M.; Anisimov, V.V.; Diadiura, K.O.; Pandova, I.V.E.T.A. Increasing of wear resistance of linear block-polyurethanes by thermal processing methods. *MM Sci. J.* **2021**, *10*, 731–4735. [CrossRef]
30. Simkulet, V.; Mitaľová, Z.; Lehocká, D.; Kočiško, M.; Manduľák, D. *Evaluation of Fracture Surface Samples by Impact Energy Test Prepared after DMLS Additive Manufacturing Technology*; SAS: Košice, Slovakia, 2017; pp. 82–83.
31. Pollak, M.; Kocisko, M.; Basistova, A.; Hlavata, S. Production of fiber as an input material for the 3D printing process. *MM Sci. J.* **2021**, *6*, 4414–4419. [CrossRef]
32. Mascenik, J.; Pavlenko, S. Determination of stress and deformation during laser welding of aluminium alloys with the PC support. *MM Sci. J.* **2020**, *2020*, 4104–4107. [CrossRef]
33. Simkulet, V.; Duplakova, D.; Kovalcikova, A.; Hatala, M.; Botko, F.; Mitalova, Z.; Vandzura, R. Evaluation of tribological characteristics of material prepared by DMLS technology. *MM Sci. J.* **2021**, *10*, 4941–4945. [CrossRef]
34. Rimár, M.; Abraham, M.; Fedák, M.; Kulikov, A.; Oravec, P.; Váhovský, J. Methods of increasing the efficiency of cogeneration based energy equipment. *MM Sci. J.* **2019**, *6*, 2935–2938. [CrossRef]
35. Sarafan, S.; Wanjara, P.; Gholipour, J.; Bernier, F.; Osman, M.; Sikan, F.; Molavi-Zarandi, M.; Soost, J.; Brochu, M. Evaluation of Maraging Steel Produced Using Hybrid Additive/Subtractive Manufacturing. *J. Manuf. Mater. Process.* **2021**, *5*, 107. [CrossRef]
36. Yan, M.T.; Wang, P.W.; Lai, J.C. Improvement of part straightness accuracy in rough cutting of wire EDM through a mechatronic system design. *Int. J. Adv. Manuf. Technol.* **2016**, *84*, 2623–2635. [CrossRef]
37. Zhang, W.; Wang, X. Simulation of the inventory cost for rotable spare with fleet size impact. *Acad. J. Manu. Eng.* **2017**, *15*, 124–132.
38. Grigoriev, S.N.; Pivkin, P.M.; Kozochkin, M.P.; Volosova, M.A.; Okunkova, A.A.; Porvatov, A.N.; Zelensky, A.A.; Nadykto, A.B. Physicomechanical nature of acoustic emission preceding wire breakage during wire electrical discharge machining (WEDM) of advanced cutting tool materials. *Metals* **2021**, *11*, 1865. [CrossRef]
39. Glowacz, A. Fault diagnosis of electric impact drills using thermal imaging. *Measurement* **2021**, *171*, 108815. [CrossRef]

40. Yaman, S.; Cakir, O. Investigation of the effects of EDM parameters on surface roughness. *J. Adv. Manu. Eng.* **2020**, *1*, 46–55.
41. Ivanov, V.; Dehtiarov, I.; Evtuhov, A.; Pavlenko, I.; Ruban, A. Multiaxis machining of fork-type parts: Fixture design and numerical simulation. *Int. Conf. New Technol. Dev. Appl.* **2021**, *233*, 42–152.
42. Kascak, J.; Gaspar, S.; Pasko, J.; Husar, J.; Knapcikova, L. Polylactic acid and its cellulose based composite as a significant tool for the production of optimized models modified for additive manufacturing. *Sustainability* **2021**, *13*, 1256. [CrossRef]
43. Mascenik, J.; Pavlenko, S. Controlled testing of belt transmissions at different loads. *MM Sci. J.* **2021**, *12*, 5497–5501. [CrossRef]
44. Van, D.N.; Van, B.P.; Huu, P.N. Application of Deng's similarity-based analytic hierarchy process approach in parametric optimization of the electrical discharge machining process of SDK11 die steel. *Trans. Can. Soc. Mech. Eng.* **2020**, *44*, 294–310. [CrossRef]
45. Mouralova, K.; Benes, L.; Zahradnicek, R.; Bednar, J.; Zadera, A.; Fries, J.; Kana, V. WEDM Used for machining high entropy alloys. *Materials* **2020**, *13*, 4823. [CrossRef]
46. Rani, S.S.; Sundari, V.K.; Jose, P.S.H.; Sivaranjani, S.; Stalin, B.; Pritima, D. Enrichment of material subtraction rate on Eglin steel using electrical discharge machining process through modification of electrical circuits. *Mater. Today* **2020**, *33*, 4428–4430. [CrossRef]
47. Mižáková, J.; Piteľ, J. An analytical dynamic model of heat transfer from the heating body to the heated room. *MATEC Web Conf.* **2017**, *125*, 02047. [CrossRef]
48. Świercz, R.; Oniszczuk-Świercz, D. Experimental Investigation of Surface Layer Properties of High Thermal Conductivity Tool Steel after Electrical Discharge Machining. *Metals* **2017**, *7*, 550. [CrossRef]
49. Sanchez, J.A.; Rodil, J.L.; Herrero, A.; De Lacalle, L.N.L.; Lamikiz, A. On the influence of cutting speed limitation on the accuracy of wire-EDM corner-cutting. *J. Mater. Process. Technol.* **2007**, *182*, 574–579. [CrossRef]
50. Ranjan, J.; Patra, K.; Szalay, T.; Mia, M.; Gupta, M.K.; Song, Q.; Krolczyk, G.; Chudy, R.; Pashnyov, V.A.; Pimenov, D.Y. artificial intelligence-based hole quality prediction in micro-drilling using multiple sensors. *Sensors* **2020**, *20*, 885. [CrossRef]
51. Oniszczuk-Swiercz, D.; Swiercz, R.; Chmielewski, T.; Salacinski, T. Experimental investigation of influence WEDM parameters on surface roughness and flatness deviation. *METAL* **2020**, *29*, 611–617.
52. Maraging Steel MS1 Properties. Available online: https://www.scribd.com/document/48576364/z3T-Maraging-Steel-1-2709-EOS-MS1 (accessed on 20 July 2022).
53. Sambaran, M.; Devesh, R. EDM of titanium foam: Electrode wear rate, oversize, and MRR. *Mater. Manuf. Process.* **2021**, *37*, 825–837.
54. Rashedul, I.M.; Zhang, Y.; Zhou, K.; Wang, G.; Xi, T.; Ji, L. Influence of different tool electrode materials on electrochemical discharge machining performances. *Micromachines* **2021**, *12*, 1077. [CrossRef]
55. Ťavodová, M. Research state heat affected zone of the material after wire EDM. *Acta Fac. Tech.* **2014**, *19*, 145–152.
56. Židek, K.; Piteľ, J.; Adámek, M.; Lazorík, P.; Hošovský, A. Digital Twin of Experimental Smart Manufacturing Assembly System for Industry 4.0 Concept. *Sustainability* **2020**, *12*, 3658. [CrossRef]
57. Sun, S.; Przystupa, K.; Wei, M.; Yu, H.; Ye, Z.; Kochan, O. Fast bearing fault diagnosis of rolling element using Levy Moth-Flame optimization algorithm and Naive Bayes. *Eksploat. Niezawodn.—Maint. Reliab.* **2020**, *22*, 730–740. [CrossRef]
58. Iyyappan, S.; Sudhakarapandian, G.R.; Sakthivel, M. Influence of silicon carbide mixed used engine oil dielectric fluid on EDM characteristics of AA7075/SiCp/B4Cp hybrid composites. *Mater. Res. Express* **2021**, *8*, 086514. [CrossRef]
59. Grigoriev, S.N.; Kozochkin, M.P.; Porvatov, A.N.; Volosova, M.A.; Okunkova, A.A. Electrical discharge machining of ceramic nanocomposites: Sublimation phenomena and adaptive control. *Heliyon* **2019**, *5*, e02629. [CrossRef] [PubMed]
60. Hašová, S.; Straka, Ľ. Design and verification of software for simulation of selected quality indicators of machined surface after WEDM. *Acad. J. Manu. Eng.* **2016**, *14*, 13–20.
61. Rouniyar, A.K.; Shandilya, P. Fabrication and experimental investigation of magnetic field assisted powder mixed electrical discharge machining on machining of aluminum 6061 alloy. *Proc. Inst. Mech. Eng. Part B J. Eng. Manuf.* **2019**, *233*, 2283–2291. [CrossRef]

micromachines

Article

Analysis of Wire-Cut Electro Discharge Machining of Polymer Composite Materials

Timur Rizovich Ablyaz [1,*], Evgeny Sergeevich Shlykov [1], Karim Ravilevich Muratov [1] and Sarabjeet Singh Sidhu [2,†]

1 Department of Mechanical Engineering, Perm National Research Polytechnic University, 614000 Perm, Russia; kruspert@mail.ru (E.S.S.); karimur_80@mail.ru (K.R.M.)
2 Department of Mechanical Engineering, Sardar Beant Singh State University, Gurdaspur 143521, India; sarabjeetsidhu@yahoo.com
* Correspondence: lowrider11-13-11@mail.ru
† Former address: Beant College of Engineering and Technology, Gurdaspur 143521, India.

Abstract: This study presents the analysis of wire-cut electro-discharge machining (WIRE-EDM) of polymer composite material (PCM). The conductivity of the workpiece is improved by using 1 mm thick titanium plates (layers) sandwiched on the PCM. Input process parameters selected are variable voltage (50–100 V), pulse duration (5–15 μs), and pause time (10–50 μs), while the cut-width (kerf) is recognized as an output parameter. Experimentation was carried out by following the central composition design (CCD) design matrix. Analysis of variance was applied to investigate the effect of process parameters on the cut-width of the PCM parts and develop the theoretical model. The results demonstrated that voltage and pulse duration significantly affect the cut-width accuracy of PCM. Furthermore, the theoretical model of machining is developed and illustrates the efficacy within the acceptable range. Finally, it is concluded that the model is an excellent way to successfully estimate the correction factors to machine complex-shaped PCM parts.

Keywords: wire electro-discharge machining; polymer composite materials; processing precision; interelectrode gap

Citation: Ablyaz, T.R.; Shlykov, E.S.; Muratov, K.R.; Sidhu, S.S. Analysis of Wire-Cut Electro Discharge Machining of Polymer Composite Materials. *Micromachines* **2021**, *12*, 571. https://doi.org/10.3390/mi12050571

Academic Editors: Irene Fassi and Francesco Modica

Received: 26 April 2021
Accepted: 13 May 2021
Published: 18 May 2021

1. Introduction

Recently, replacing metallic machine parts with composite material has been seen as a potential alternative to various issues, including high metal costs, rusting, and the weight of the components. In the modern machining industry, composite materials that possess similar or even enhanced physical and mechanical properties compared to metals are highly encouraged [1–3]. Polymer composite materials (PCMs) are recognized as a group of difficult-to-machine materials [4]. The development of light-weight PCMs plays a significant role in aviation and many critical industrial applications. These materials are economically efficient and reduce the CO_2 emission load [1]. The binders used in PCM have good strengths and are heat resistant, resulting in high elastic strength and operational stability. In contrast, the matrix phase in the PCM is the ductile phase that transfers the external load stress to the filler phase. The filler/reinforcement used in a PCM determines its mechanical properties, such as strength, stiffness, and deformability. The filler used may be carbon/ceramic fibers. These fibers have good physical and mechanical properties. These fibers are converted into fabrics by weaving [5–7]. A typical PCM is shown in Figure 1a, and the weaving pattern of the fibers forms the reinforcement phase and the possible defects in conventional machining.

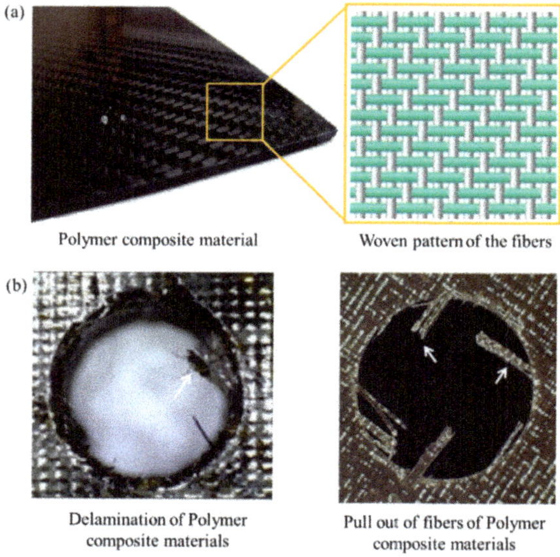

Figure 1. (**a**) Polymer composite materials and woven patterns of fibers of the reinforcement phase; and (**b**) delamination and pull out of fibers during the drilling of polymer composite materials.

PCMs possess many limitations during their machining due to mismatch of properties in the filler and matrix phases [8,9]. PCMs tend to delaminate during machining due to layering, structural heterogeneity, high hardness of the filler material, and low plasticity of the binder. The machining of PCMs also results in high cutting forces and vibrations, which causes pullout of fibers and other detrimental effects at the machined zone. Delamination and pull out of fibers can be observed in Figure 1b, which diminishes the quality of the fabricated feature. Moreover, during PCM machining, a high tool wear rate causes a decrease in productivity and enhances manufacturing costs. To avoid such problems, various non-conventional machining methods have been developed for processing these materials (Figure 2). Despite having many advantages over conventional machining methods, non-conventional processes such as electrochemical and chemical processing are hazardous to the environment [10,11]. On the other hand, techniques such as laser treatment [12–15], plasma treatment [16,17], electron beam treatment [18,19], ultrasonic treatment [20–22], and water jet treatment [23–27] are considered advantageous only where accuracy is not the primary concern. Moreover, these machining methods have disadvantages such as thermal destruction of the matrix phase and lack of accuracy, limiting their application for machining small-sized components.

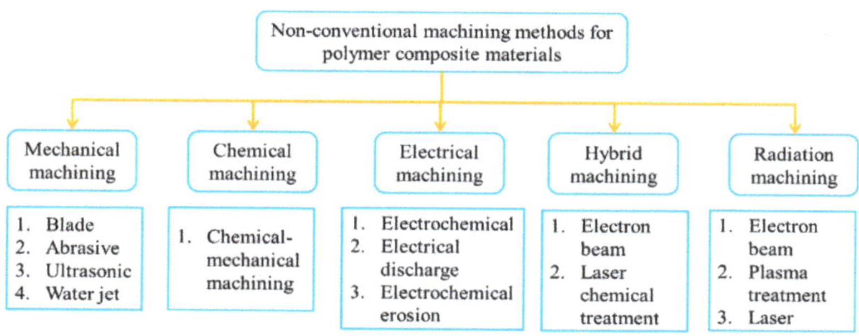

Figure 2. Non-conventional machining methods of polymer composite materials.

Among the various non-conventional machining techniques, Wire-cut Electrical Discharge machining (WIRE-EDM) has been proven to be a potential candidate to prepare small components with high accuracy [28–30]. In this process, thermal-electrical energy is involved in transforming the electrical energy to heat energy sufficient to melt the target zone, and an accurate curved profile can be obtained on metallic as well as polymer composite materials. A schematic diagram of the WIRE-EDM process is shown in Figure 3. In this process, the wire electrode moves vertically (mostly) over sapphire or diamond guides, which are controlled by Computer Numerical Control (CNC) program. A steady stream of deionized water or other fluid is used as a dielectric medium, flushes out debris, and cools the workpiece and the wire electrode. The dielectric fluid gets ionized, thereby producing a spark between the wire electrode tool (ET) and the workpiece electrode (WE) (Figure 3). The ionization of dielectric fluid depends on many factors of WIRE-EDM, such as properties of the working fluid, degree of contamination of the working fluid with erosion, the material of the electrodes, and dielectric flow pressure. The amount of thermal energy generated within the electrodes affects the amount of material removed from the surfaces of the ET and WE differently. This unevenness in material removal depends on the thermophysical properties of the ET and WE, and the process parameters of WIRE-EDM [29–31]. By varying these factors, electrode erosion can be precisely controlled. The spark energy within the electrodes depends on the voltage, the pulse formation time, the state of the working fluid, and the size of the interelectrode gap. Thus, the accuracy of the WIRE-EDM of PCMs is influenced by the size of the inter-electrode gap. The error of the inter-electrode gap depends on the inhomogeneity in the structures/properties of ET and the WE, and also the properties of the working fluid. The WIRE-EDM of PCMs results in the formation of dimples/craters of different sizes on the surface of the workpiece. These randomly formed dimples/craters are a factor that complicates the prediction of the inter-electrode gap [31].

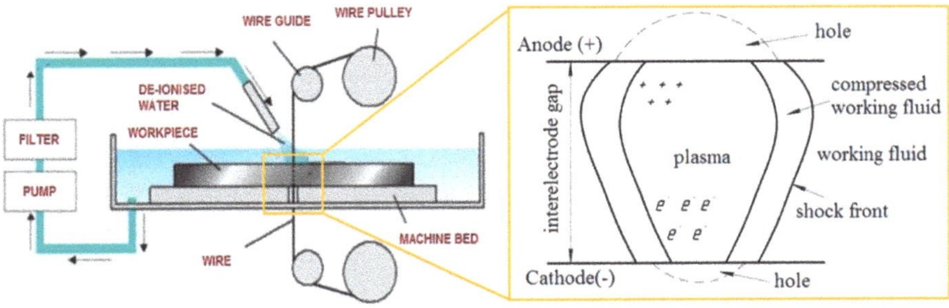

Figure 3. Schematic diagram and working principle of the wire-cut electro-discharge machining (WIRE-EDM) process [31].

The application of WIRE-EDM on PCMs has been investigated by various researchers [8,9]. However, it is a known fact that the conductivity of PCMs is limited. Thus, during the WIRE-EDM process, the resin of the PCM was destroyed at the edges of the holes. This was due to high temperatures and ineffective cooling at the machining zone.

Abdallah et al. [32] used WIRE-EDM to study the effects of gap voltage, current, pulse-on time, and pulse-off time on the material removal rate (MRR), top and bottom cut-width (kerf), and workpiece edge damage in unidirectional carbon fiber reinforced polymer (CFRP) composites. Current and pulse-off time were found to be statistically important parameters in terms of MRR, with current being the only factor affecting cut-width on the top surface. Recently, Dutta et al. [33] investigated a modified version of WIRE-EDM for CFRP composite cutting by using H13 steel plates as sandwich, assisting the electrodes to trigger the electrical spark during CFRP composite WIRE-EDM. Using metal plates (H13 steel) as assisting electrodes, problems such as incomplete cuts and deviations in the machining direction during CFRP WIRE-EDM were controlled. The results showed that increasing the current (from 2A to 12A) reduced the cutting time (by 60.95%) while keeping all other parameters constant.

Likewise, in similar studies [34–48] related to the WIRE-EDM of PCMs, it was observed that the quality and accuracy of holes in a low-conductive material can be regulated by applying a conductive layer above the non-conductive PCM. Also, the development of theoretical models of the WIRE-EDM of PCMs provides a guide to obtain the accuracy required in the process [37].

A schematic of WIRE-EDM process is represented in Figure 4. Herein, the size of the ET (2R), and the value of interelectrode gap(overcut)S are taken into account for accurate machining of the product. The correction in machining can be done in reference to the center of the ET(wire)in a CNC controller.

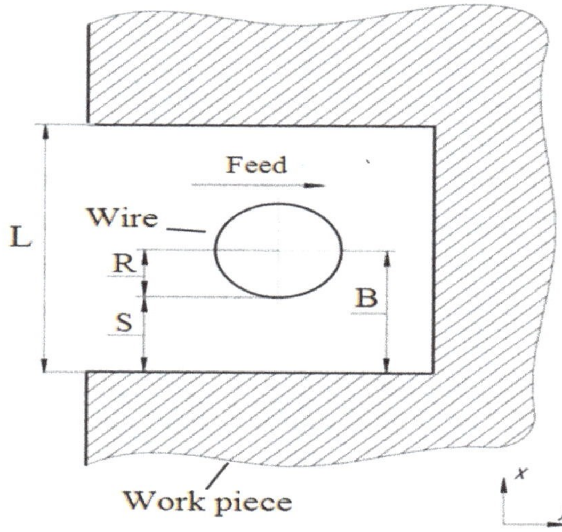

Figure 4. WIRE-EDM processing in the XY plane: R—radius of the electrode tool (ET); S—interelectrode gap;(B)—reference for correction; L—cut-width (Kerf).

The material removal rate from the workpiece during a single pulse is estimated by the following equation [49–51]:

$$MRR = \frac{m}{t_{on}} \quad (1)$$

where t_{on} is the duration of a single pulse (μs), and m is the weight lost during the EDM process (kg). Data in the literature [31,51] indicate that *MRR* depends on the value of the interelectrode gap S (m), the feed rate V (m/s), the physical and mechanical properties of the processed material, and the workpiece thickness h (m). *MRR* is calculated using Equation (2):

$$MRR = 2(R + S)hV\rho \quad (2)$$

where R is the radius of the ET (m), S is the interelectrode gap (m); ρ is the density of the processed material (kg/m^3), h is the thickness of workpiece (m), and V is the feed rate.

The spark energy, W (J), is released in the interelectrode gap and is distributed between the ET and the workpiece. The material is removed from the workpiece by the mean of the spark energy [31]. The pulse energy is calculated as:

$$W = \int_0^{t_p} UIdt_{on} \quad (3)$$

where U is the voltage, (V); I is the current strength, (A); t_{on} is the pulse duration, (μs); and tp is the pulse width (μs). A correction factor is introduced in Equation (3) to improve the accuracy of the calculations.

$$W = (\eta_u)UIt_{on} \tag{4}$$

The coefficient for a fraction of the energy (η_u) utilized in the machining process is represented in Equation (5) [31,50]:

$$\eta_u = (1 - K_1)(1 - K_2) \tag{5}$$

where K_1 is the amount of energy lost during the heating and evaporation of the dielectric fluid; K_2 is amount of energy lost in the ET. The amount of heat Q (J) transferred to the WE for heating and melting is determined by the formula:

$$Q = m(C_1\Delta T_1 + \lambda + C_2\Delta T_2 + r) \tag{6}$$

where m is the mass of the workpiece (kg); C_1 is the specific heat capacity of the material in the solid state (J/kg K) C_2 is the specific heat capacity of the material in the liquid state (J/kg K); ΔT is the temperature difference between the initial and final heating points (K); λ is the specific heat of fusion of the material (J/kg); and r is the specific heat of vaporization (J/kg).

Taking into the account the equal coefficients of energy loss in Equation (4) and Equation (6) (i.e., W and Q) and $Z = C_1\Delta T_1 + \lambda + C_2\Delta T_2 + r$. Equation (1) is represented as:

$$MRR = \frac{Q\eta_u \mathrm{UI}}{ZW} = \frac{\eta_u UI}{Z} \tag{7}$$

thus equating expressions (2) and (7). The value of the interelectrode gap S is calculated as:

$$S = \frac{\eta_u UI}{2ZhV\rho} - R \tag{8}$$

The value of B (as shown in Figure 4) can be calculated, and thus the cut width (L) is obtained. This theoretical model can be used to estimate the process parameter affecting the cut width (L), and can thus suggest the amount of correction required for accurate machining.

The PCM is made up of both electrically conductive carbon fiber and non-conductive epoxy resin. As a result, machining such composites with prominent machining techniques i.e., WIRE-EDM or EDM is a difficult job. In the literature, research on the WIRE-EDM of PCM is limited to where the effect of WIRE-EDM on PCM is explored for higher machining rate, electrode wear, and performance in the linear cut.

In this work, the authors investigated the performance of WIRE-EDM on a patent carbon fiber-reinforced PCM possibly adopted in the aviation industry. The voltage, pulse duration, and pause time were selected as process parameters. These parameters were statistically evaluated, and the level of the significance of factors affecting the cut width was determined using analysis of variance. Finally, the experimental values obtained for cut width were modeled mathematically in terms of significant factors using response surface methodology.

Purpose of Study

- To assess the influence of key process parameters on the cut width (kerf) and surface quality of PCM sandwiched in Titanium alloy.
- To develop a regression model using response surface methodology, which is further examined with the experimental results for non-linear machining cut-width on the selected PCM.
- To determine the trajectory of ET to machine PCM in the form of a complex shaped part, such as a gear.

2. Material and Methods

2.1. Material

In this study, a polymer composite material (VKU-39, pl. refer https://viam.ru, 11 May 2021) used in the aviation industry was chosen. The workpiece is a laminated fibrous polymer composite made of carbon fiber twill as reinforcement/filler, with epoxy as a binder material. The property of the selected PCM is as shown in Table 1.

Table 1. Properties of polymer composite material (PCM).

Property	Average Value
Filler material	Carbon fabrics twill (Porcher-3692)
Heat resistance °C	170°
Monolayer thickness, mm	0.2
Tensile strength, MPA	945
Tensile modulus, GPA	69
Compressive strength, MPA	610
Compression modulus, GPA	54
Density of carbon fiber reinforced polymer (CFRP), kg/m^3	1550
Weaving type	Twill, at an angle of 90°
Electrical conductivity, S/m	10^{-7}

A PCM plate of thickness 2 mm was used for the WIRE-EDM experiments. To improve the conductivity of the PCM, a conductive layer of titanium (1 mm) was applied on both sides (Figure 5). The study was carried out on a wire-cut EDM machine, "Electronica EcoCut." The electrode tool (ET) used was a brass wire with a diameter of 0.25 mm. Distilled water was used as a dielectric medium, and was sprayed on the ET (Figure 5) instead of immersing the PCMs in a water bath, as PCM immersion in a water bath causes defects such as swelling and filling of water.

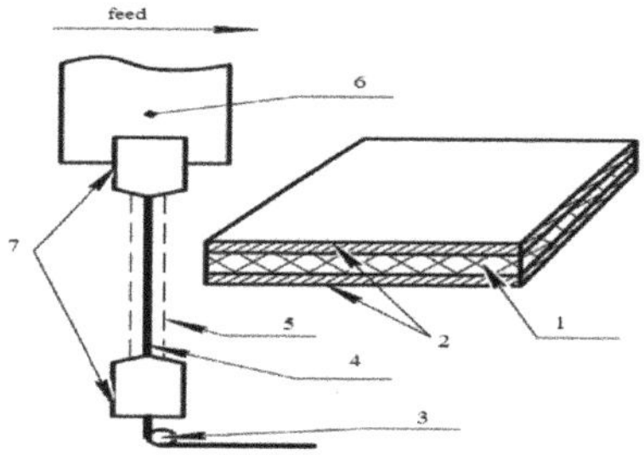

Figure 5. Schematic of the WIRE-EDM of polymer composite material (PCM) with conductive layers used in the present study (1: PCM, 2: conductive Ti-layers, 3: tension roller, 4: ET wire, 5: flushing, 6: gearbox, 7: upper and lower diamond wire guides).

2.2. Method

The process parameters selected were U—voltage, V; Ton—pulse duration, μs; and Toff—the pause time in pulse duration. The experimental runs were carried out by employing the orthogonal central composition design (CCD) matrix, where α denotes the distance between the star point and central point i.e., value of α = 1.215. The design matrix in the study was obtained with the assistance of Design-Expert software. Orthogonality in the design assisted in estimating the independent regression coefficients [49,52]. The process parameters are presented in Table 2. The output parameter was the value of the EDM cut width (L). The cut width was the wire diameter of the ET and the size of the side clearance (overcut).

Table 2. Process parameters of wire-electrical discharge machining (EDM).

Factors, Units	Units	Lower Level(−1)	Upper Level(+1)	Average Level	Lower "Star" Point	Top "Star" Point
U	V	50	100	75	40	110
Ton	μs	5	15	10	2	20
Toff	μs	10	50	30	5	60

The experiment design matrix and response (cut width) is presented in Table 3. Each experimental run was replicated thrice for more accuracy. A pictorial view of the machining process is shown in Figure 6.

Table 3. Experimental central composition design (CCD) matrix with the responses the experiment obtained.

Exp. No.	Process Parameters			Response
	Voltage U (V)	Pulse Duration Ton (μs)	Pause Time Toff (μs)	Cut-Width L (μm)
1	100	15	50	337
2	50	15	10	286
3	100	5	50	280
4	75	10	60	323
5	75	10	5	327
6	50	5	10	275
7	50	5	50	284
8	75	20	30	355
9	100	15	10	342
10	50	15	50	276
11	75	2	30	273
12	110	10	30	315
13	75	10	30	294
14	100	5	10	255
15	40	10	30	283

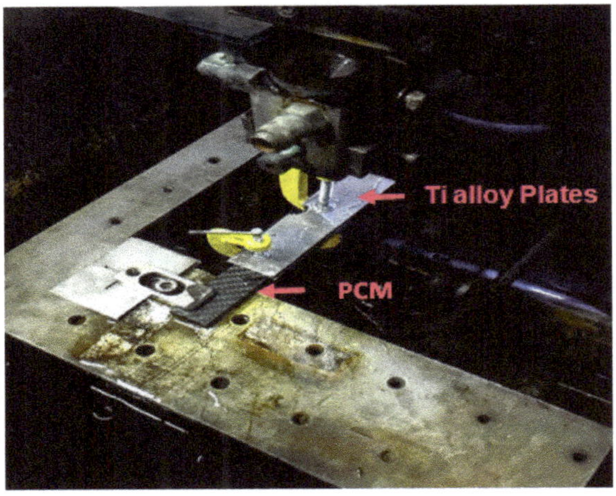

Figure 6. Pictorial view of PCM machining setup.

3. Results and Discussion

Analysis of variance (ANOVA) is a commonly used statistical technique to test the significance of a model and the contribution of each process parameter on the experimental response. The mathematical model was predicted using Design Expert software and is summarized in Table 4.

From Table 4, it is observed that the value of adjusted R-square and predicted R-square values were higher for the 2-way interaction model. Thus, this model is suggested for further analysis.

Statistical analysis of the experimental data (Table 5) revealed the significance of the process parameters, namely voltage, pulse duration, and pause time, on the measured response i.e., cut width. The results are summarized in Table 5, with a 95% confidence level.

Table 4. Mathematical model analysis.

Model	*p*-Value	Adjusted R^2	Predicted R^2	
Linear	0.0127	0.5061	0.2412	-
2-way interaction	0.0663	**0.7100**	**0.3711**	Suggested Model
Quadratic	0.4820	0.7049	−0.0121	-

Table 5. Analysis of variance (ANOVA) of process parameters for 2-way interaction model.

ANOVA for Response Surface						
Source	**Sum of Squares**	**df**	**Mean Square**	**F-Value**	***p*-Value**	
Model	10,655.21	6	1775.87	6.71	0.0085	Significant
U-Voltage	1593.02	1	1593.02	6.02	0.0397	-
Ton	6189.06	1	6189.06	23.40	0.0013	-
Toff	33.21	1	33.21	0.1256	0.7322	-
U × Ton	2485.13	1	2485.13	9.40	0.0155	-
U × Toff	55.13	1	55.13	0.2084	0.6602	-
Ton × Ton	300.12	1	300.12	1.13	0.3179	-
Residual	2116.12	8	264.52	-	-	-
Total	12,771.33	14	1775.87	-	-	-

The 2-way interaction model F-value of 6.71 implies that the model is significant, and only 0.85% chance of noise exists in this model. From Table 5, the *p*-value less than 0.05 indicated that the model terms of voltage, Ton, and their interaction (i.e., U × T-on) were significant, whereas Toff, U × Toff and Ton × Ton were insignificant. Eliminating insignificant terms results in the improvement in the model, as presented in Table 6.

Table 6. ANOVA for the reduced 2-way interaction model.

ANOVA for Response Surface						
Source	**Sum of Squares**	**df**	**Mean Square**	**F-Value**	***p*-Value**	
Model	10,266.75	3	3422.25	15.03	0.0003	Significant
U-Voltage	1593.02	1	1593.02	7.00	0.0228	-
Ton	6188.60	1	6188.60	27.18	0.0003	-
U × Ton	2485.13	1	2485.13	10.91	0.0070	-
Residual	2504.58	11	227.69	-	-	-
Total	12,771.33	14	3422.25	-	-	-

The model regression statistics for the selected model demonstrated a high R^2 value (i.e., 0.80), which is acceptable. The predicted R^2 of 0.6677 was in reasonable agreement with the adjusted R^2 of 0.7504; i.e., the difference was less than 0.2.

In order to check the adequacy of model, the predicted value and the actual experimental values were compared along a 45° line, as shown in Figure 7. This implies that the proposed model is adequate and there is no violation of the independent or constant variance assumptions.

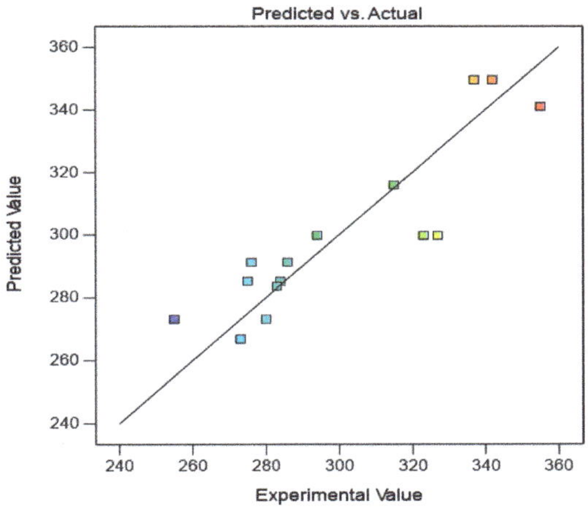

Figure 7. Comparison of model-predicted values versus actual experiment values for cut width.

The final equation obtained for the prediction of cut-width (L) is represented as Equation (9):

$$L = 329.6 - 0.9475 \times U - 6.45 \times Ton + 0.141 \times U \times Ton \tag{9}$$

The regression equation analysis (9) shows the combination (interaction) of input process parameters (factors) affecting the value of the cut width.

Figures 8 and 9 presents the response surface describing the dependence of the cut width on the voltage and pulse duration.

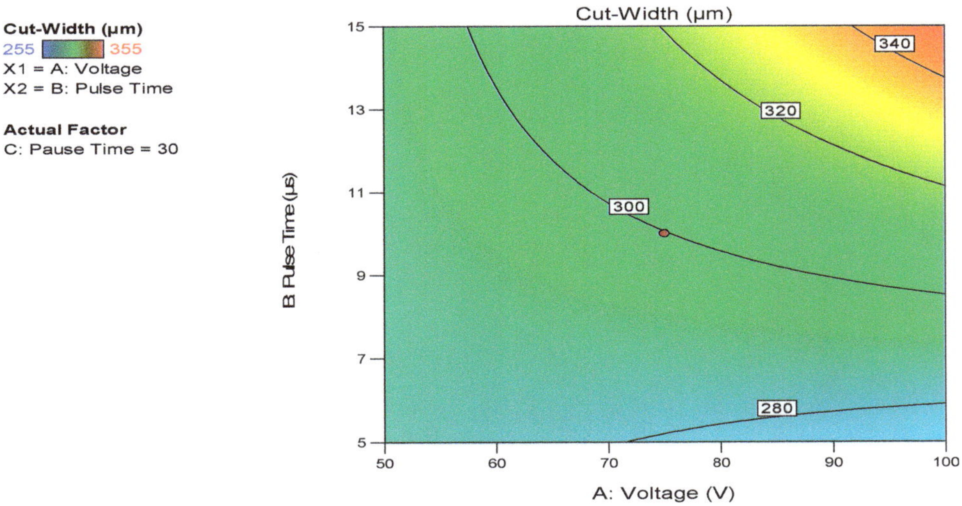

Figure 8. Contour plot for cut-width dependance on voltage and pulse duration for pause time Toff = 30 μs.

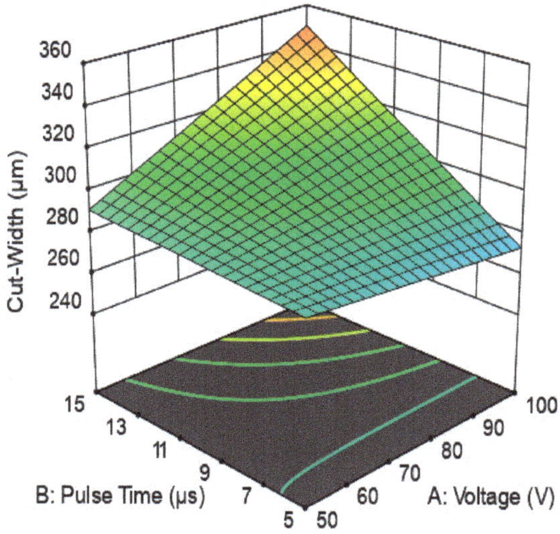

Figure 9. Response surface 3D-plot for cut-width dependence on voltage and pulse duration for pause time Toff = 30µs.

Examination of the Predicted Model forWIRE-EDM onComplex-Shaped PCM Parts

It is depicted from the plots (Figures 8 and 9) that cut width is directly proportional to the voltage and pulse duration. Thus, for machining complex PCM parts, a high value for WIRE-EDM i.e., U = 100 V, Ton = 15 µs, Toff = 30 µs was selected. The obtained cut width path is presented in Figure 10. The cut width (Figure 10) was measured at various WIRE-EDM zones, i.e., at the entrance to the PCM, at the corner, and the end of processing. The average value of the cut width was calculated as L = 330 µm.

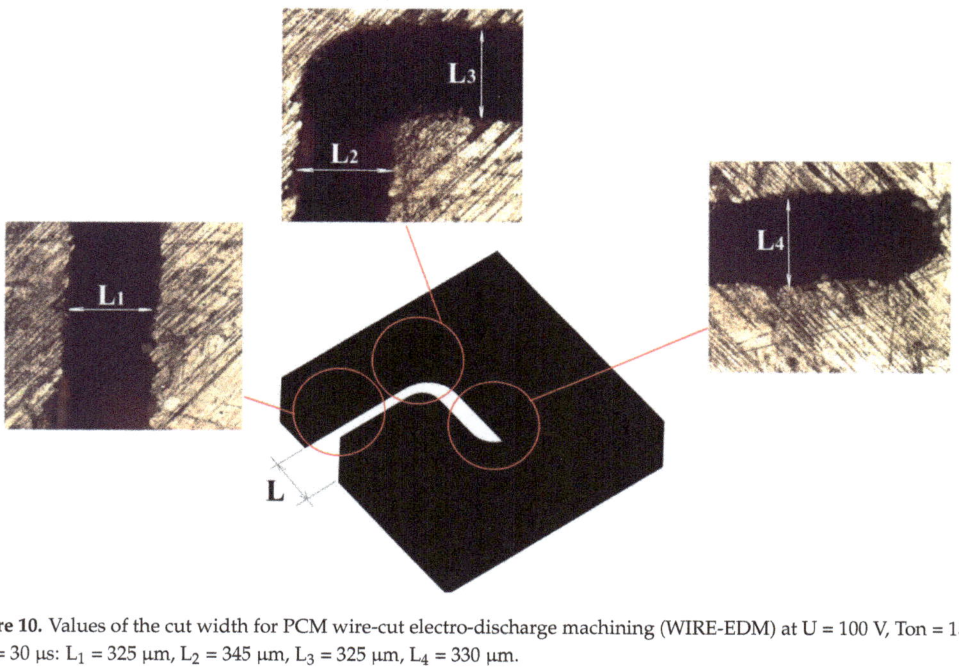

Figure 10. Values of the cut width for PCM wire-cut electro-discharge machining (WIRE-EDM) at U = 100 V, Ton = 15 µs, Toff = 30 µs: L_1 = 325 µm, L_2 = 345 µm, L_3 = 325 µm, L_4 = 330 µm.

Figure 11a,b reveals the surface the cutting edge achieved after applying the conductive Ti-alloy layer (plates) to the surface of PCM. It is found that this method of machining results in attaining a defect-free smooth surface on both sides of the processed PCM sheet at the entrance, at the corner, and at the end.

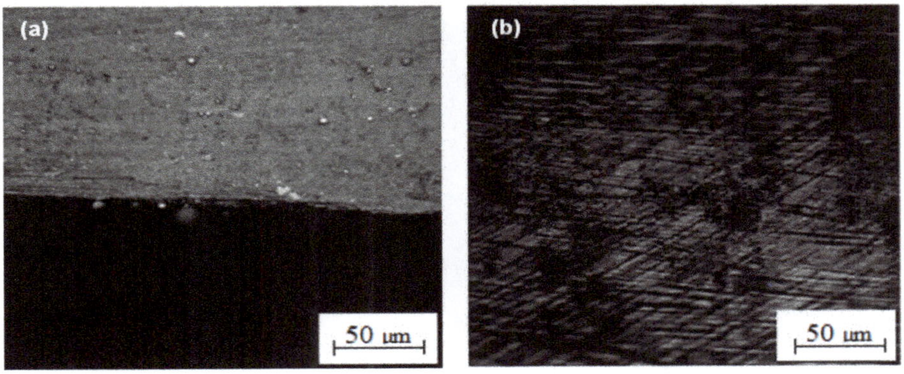

Figure 11. Machined PCM surfaces after application of a conductive Ti-layer on the surface of PCM; (**a**)—sheet surface at the processing zone (at the edge), (**b**)—along the cross-section of sheet width.

Thus, the Ti layer sandwich method for PCM enhanced the quality of the machined surface. Additionally, it has the potential to obtain a smooth, defect-free surface within the processed slot without causing any damage to the fibers/binder of the PCM material.

The percentage error between the values of the cut-width was calculated using Equation (10). The expected value from Equation (9) was calculated as 349.6 µm, and the experimental value (actual value) measured was 330 µm (Figure 10). Thus, the percentage error obtained is calculated as 5.906%, which is acceptable and shows the competency of the model.

$$\%\ error = \frac{|Expected\ value - Actual\ value|}{Expected\ value} \tag{10}$$

The regression Equation (9) was further examined to accurately machine a gear-shaped PCM part. An "Electronica EcoCut" CNC WIRE-EDM machine was programmed. The program was used to machine a PCM workpiece into the gear shape. The machining parameters were carried out at Ton = 15µs, Toff = 30 µs, and U = 100 V. Based on this regression model (Equation (9)), the trajectory correction value was calculated as B = 0.165 mm. When machining the PCM, the offset was added into the control program using the command "G41 B" = 0.165. In Figure 12, the trajectory of the ET and the finished product "gear" are presented.

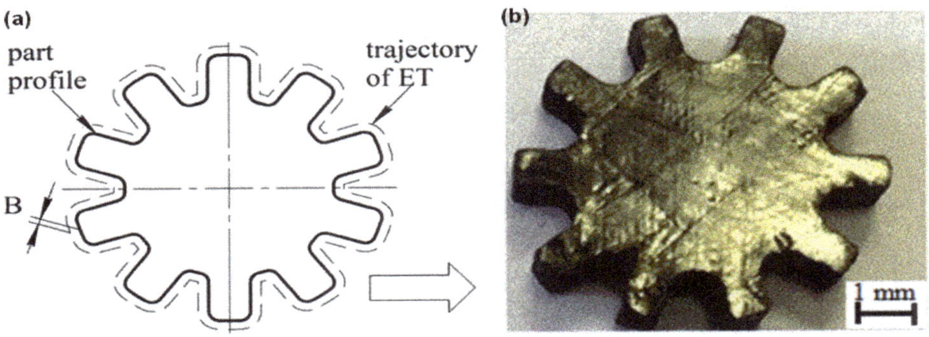

Figure 12. (**a**) Trajectory of the ET, and (**b**) finished product.

4. Conclusions

The experimental work reveals the dependence between the WIRE-EDM cut width and processing parameters such as voltage, pulse-on time, pulse-off time. These results can be expedited to adjust the size of the ET and ensure precision in the WIRE-EDM of PCM (VKU-39) workpieces. It is ascertained that voltage and pulse duration and their interaction are the significant factors affecting the process parameters for machining the PCM workpiece. Furthermore, a2-way interaction model is developed to estimate the cut-width, which shows excellent adequacy with the experimental values obtained. Based on the developed model, the cut-width correction factor for the trajectory of ET was estimated in a WIRE-EDM CNC machine for the accurate machining of a complex-shaped PCM product. Consequently, it is suggested that the proposed model successfully facilitates the forecasting of WIRE-EDM accuracy.

Author Contributions: Conceptualization, E.S.S. and T.R.A.; methodology, K.R.M. and T.R.A.; validation, S.S.S.; formal analysis, T.R.A., E.S.S. and S.S.S.; investigation, E.S.S. and T.R.A.; resources, E.S.S., T.R.A. and K.R.M.; data curation, E.S.S. and K.R.M.; writing—original draft preparation, E.S.S. and T.R.A.; writing—review and editing, K.R.M. and T.R.A.; visualization, E.S.S. and T.R.A.; supervision, S.S.S. and T.R.A.; project administration, K.R.M. and T.R.A.; funding acquisition, T.R.A., E.S.S. and K.R.M. All authors have read and agreed to the published version of the manuscript.

Funding: This work was supported by the Russian President in encouragement of young Russian scientists (grant MK-566.2021.4).

Conflicts of Interest: The authors declare no conflict of interest. The funders had no role in the design of the study; in the collection, analyses, or interpretation of data; in the writing of the manuscript, or in the decision to publish the results.

References

1. Sarde, B.; Patil, Y.D. Recent Research Status on Polymer Composite Used in Concrete—An Overview. *Mater. Today Proc.* **2019**, *18*, 3780–3790. [CrossRef]
2. Yahaya, R.; Sapuan, S.; Jawaid, M.; Leman, Z.; Zainudin, E. Mechanical performance of woven kenaf-Kevlar hybrid composites. *J. Reinf. Plast. Compos.* **2014**, *33*, 2242–2254. [CrossRef]
3. Hsissou, R.; Benhiba, F.; Echihi, S.; Benkhaya, S.; Hilali, M.; Berisha, A.; Briche, S.; Zarrouk, A.; Nouneh, K.; Elharfi, A. New epoxy composite polymers as a potential anticorrosive coatings for carbon steel in 3.5% NaCl solution: Experimental and computational approaches. *Chem. Data Collect.* **2021**, *31*, 100619. [CrossRef]
4. Valino, A.D.; Dizon, J.R.C.; Espera, A.H.; Chen, Q.; Messman, J.; Advincula, R.C. Advances in 3D printing of thermoplastic polymer composites and nanocomposites. *Prog. Polym. Sci.* **2019**, *98*, 101162. [CrossRef]
5. Hsissou, R.; Bekhta, A.; Dagdag, O.; El Bachiri, A.; Rafik, M.; Elharfi, A. Rheological properties of composite polymers and hybrid nanocomposites. *Heliyon* **2020**, *6*, e04187. [CrossRef]
6. Mehra, N.; Mu, L.; Ji, T.; Yang, X.; Kong, J.; Gu, J.; Zhu, J. Thermal transport in polymeric materials and across composite interfaces. *Appl. Mater. Today* **2018**, *12*, 92–130. [CrossRef]
7. Thomason, J. A review of the analysis and characterisation of polymeric glass fibre sizings. *Polym. Test.* **2020**, *85*, 106421. [CrossRef]
8. Shlykov, E.S.; Ablyaz, T.R.; Oglezneva, S.A. Electrical Discharge Machining of Polymer Composites. *Russ. Eng. Res.* **2020**, *40*, 878–879. [CrossRef]
9. Ablyaz, T.R.; Muratov, K.R.; Shlykov, E.S.; Shipunov, G.S.; Shakirzyanov, T.V. Electric-Discharge Machining of Polymer Composites. *Russ. Eng. Res.* **2019**, *39*, 898–900. [CrossRef]
10. Yao, Y.; Wang, B.; Wang, J.; Jin, H.; Zhang, Y.; Dong, S. Chemical machining of Zerodur material with atmospheric pressure plasma jet. *CIRP Ann.* **2010**, *59*, 337–340. [CrossRef]
11. Shimpi, J.R.; Sidhaye, D.S.; Prasad, B.L. Digestive Ripening: A Fine Chemical Machining Process on the Nanoscale. *Langmuir* **2017**, *33*, 9491–9507. [CrossRef] [PubMed]
12. Romoli, L.; Tantussi, G.; Dini, G. Experimental approach to the laser machining of PMMA substrates for the fabrication of microfluidic devices. *Opt. Lasers Eng.* **2011**, *49*, 419–427. [CrossRef]
13. Li, Z.; Zheng, H.; Lim, G.; Chu, P.; Li, L. Study on UV laser machining quality of carbon fibre reinforced composites. *Compos. Part A Appl. Sci. Manuf.* **2010**, *41*, 1403–1408. [CrossRef]
14. Samant, A.N.; Dahotre, N.B. Laser machining of structural ceramics—A review. *J. Eur. Ceram. Soc.* **2009**, *29*, 969–993. [CrossRef]
15. Pham, D.T.; Dimov, S.S.; Ji, C.; Petkov, P.V.; Dobrev, T. Laser milling as a 'rapid' micromanufacturing process. *Proc. Inst. Mech. Eng. Part B J. Eng. Manuf.* **2004**, *218*, 1–7. [CrossRef]

16. Shi, B.; Dai, Y.; Xie, X.; Li, S.; Zhou, L. Arc-Enhanced Plasma Machining Technology for High Efficiency Machining of Silicon Carbide. *Plasma Chem. Plasma Process.* **2016**, *36*, 891–900. [CrossRef]
17. Malhotra, R.; Saxena, I.; Ehmann, K.F.; Cao, J. Laser-induced plasma micro-machining (LIPMM) for enhanced productivity and flexibility in laser-based micro-machining processes. *CIRP Ann.* **2013**, *62*, 211–214. [CrossRef]
18. Spinney, P.S.; Howitt, D.G.; Smith, R.L.; Collins, S.D. Nanopore formation by low-energy focused electron beam machining. *Nanotechnology* **2010**, *21*, 375301. [CrossRef]
19. Parthasarathy, J.; Starly, B.; Raman, S.; Christensen, A. Mechanical evaluation of porous titanium (Ti6Al4V) structures with electron beam melting (EBM). *J. Mech. Behav. Biomed. Mater.* **2010**, *3*, 249–259. [CrossRef]
20. Koike, M.; Martinez, K.; Guo, L.; Chahine, G.; Kovacevic, R.; Okabe, T. Evaluation of titanium alloy fabricated using electron beam melting system for dental applications. *J. Mater. Process. Technol.* **2011**, *211*, 1400–1408. [CrossRef]
21. Rajurkar, K.; Sundaram, M.; Malshe, A. Review of Electrochemical and Electrodischarge Machining. *Procedia CIRP* **2013**, *6*, 13–26. [CrossRef]
22. Tiwari, A.; Mandal, A.; Kumar, K. Multi-objective Optimization of Electro-chemical Machining by Non-dominated Sorting Genetic Algorithm. *Mater. Today Proc.* **2015**, *2*, 2569–2575. [CrossRef]
23. Alberdi, A.; Rivero, A.; de Lacalle, L.N.L.; Etxeberria, I.; Suárez, A. Effect of process parameter on the kerf geometry in abrasive water jet milling. *Int. J. Adv. Manuf. Technol.* **2010**, *51*, 467–480. [CrossRef]
24. Ting, H.; Abou-El-Hossein, K.; Chua, H. Review of micromachining of ceramics by etching. *Trans. Nonferrous Met. Soc. China* **2009**, *19*, S1–S16. [CrossRef]
25. Matsumura, T.; Muramatsu, T.; Fueki, S. Abrasive water jet machining of glass with stagnation effect. *CIRP Ann.* **2011**, *60*, 355–358. [CrossRef]
26. Gupta, V.; Pandey, P.; Garg, M.P.; Khanna, R.; Batra, N. Minimization of Kerf Taper Angle and Kerf Width Using Taguchi's Method in Abrasive Water Jet Machining of Marble. *Procedia Mater. Sci.* **2014**, *6*, 140–149. [CrossRef]
27. Deris, A.M.; Zain, A.M.; Sallehuddin, R. A note of hybrid GR-SVM for prediction of surface roughness in abrasive water jet machining: A response. *Meccanica* **2017**, *52*, 1993–1994. [CrossRef]
28. Panner Selvam, M.; Ranjith Kumar, P. Optimization Kerf Width and Surface Roughness in Wirecut Electrical Discharge Machining Using Brass Wire. *Mech. Mech. Eng.* **2017**, *21*, 37–55.
29. Maher, I.; Ling, L.H.; Sarhan, A.A.D.; Hamdi, M. Improve wire EDM performance at different machining parameters—ANFIS modeling. *IFAC-PapersOnLine* **2015**, *48*, 105–110. [CrossRef]
30. Wu, C.; Cao, S.; Zhao, Y.J.; Qi, H.; Liu, X.; Liu, G.; Guo, J.; Li, H.N. Preheating assisted wire EDM of semi-conductive CFRPs: Principle and anisotropy. *J. Mater. Process. Technol.* **2021**, *288*, 116915. [CrossRef]
31. Abbas, N.M.; Solomon, D.G.; Bahari, F. A review on current research trends in electrical discharge machining (EDM). *Int. J. Mach. Tools Manuf.* **2007**, *47*, 1214–1228. [CrossRef]
32. Abdallah, R.; Soo, S.L.; Hood, R. A feasibility study on wire electrical discharge machining of carbon fibre reinforced plastic composites. *Procedia CIRP* **2018**, *77*, 195–198. [CrossRef]
33. Dutta, H.; Debnath, K.; Sarma, D.K. Investigation on cutting of thin carbon fiber-reinforced polymer composite plate using sandwich electrode-assisted wire electrical-discharge machining. *Proc. Inst. Mech. Eng. Part E J. Process. Mech. Eng.* **2021**. [CrossRef]
34. Fukuzawa, Y.; Katougi, H.; Mohri, N.; Furutani, K.; Tani, T. Machining properties of ceramics with an electric discharge machine. In Proceedings of the XII ISEM, Aachen, Germany, 11–13 May 1998; pp. 445–454.
35. Mohri, N.; Fukuzawa, Y.; Tani, T.; Saito, N.; Furutani, K. Assisting Electrode Method for Machining Insulating Ceramics. *CIRP Ann.* **1996**, *45*, 201–204. [CrossRef]
36. Mohri, N.; Fukusima, Y.; Fukuzawa, Y.; Tani, T.; Saito, N. Layer Generation Process on Work-piece in Electrical Discharge Machining. *CIRP Ann.* **2003**, *52*, 157–160. [CrossRef]
37. Lauwers, B.; Kruth, J.; Liu, W.; Eeraerts, W.; Schacht, B.; Bleys, P. Investigation of material removal mechanisms in EDM of composite ceramic materials. *J. Mater. Process. Technol.* **2004**, *149*, 347–352. [CrossRef]
38. Puertas, I.; Luis, C. A study on the electrical discharge machining of conductive ceramics. *J. Mater. Process. Technol.* **2004**, *153–154*, 1033–1038. [CrossRef]
39. Kucukturk, G.; Cogun, C. A New Method for Machining of Electrically Nonconductive Workpieces Using Electric Discharge Machining Technique. *Mach. Sci. Technol.* **2010**, *14*, 189–207. [CrossRef]
40. Hösel, T.; Müller, C.; Reinecke, H. Spark erosive structuring of electrically nonconductive zirconia with an assisting electrode. *CIRP J. Manuf. Sci. Technol.* **2011**, *4*, 357–361. [CrossRef]
41. Wüthrich, R.; Fascio, V. Machining of non-conducting materials using electrochemical discharge phenomenon—An overview. *Int. J. Mach. Tools Manuf.* **2005**, *45*, 1095–1108. [CrossRef]
42. Schubert, A.; Zeidler, H.; Wolf, N.; Hackert, M. Micro Electro Discharge Machining of Electrically Nonconductive Ceramics. In Proceedings of the 14th International Esaform Conference on Material Forming: ESAFORM 2011, Belfast, UK, 27–29 April 2011; AIP Publishing: College Park, MD, USA, 2011; pp. 1303–1308.
43. Saleh, M.; Anwar, S.; El-Tamimi, A.; Mohammed, M.K.; Ahmad, S. Milling Microchannels in Monel 400 Alloy by Wire EDM: An Experimental Analysis. *Micromachines* **2020**, *11*, 469. [CrossRef] [PubMed]

44. Kumar, V.; Kumar, V.; Jangra, K.K. An experimental analysis and optimization of machining rate and surface characteristics in WEDM of Monel-400 using RSM and desirability approach. *J. Ind. Eng. Int.* **2015**, *11*, 297–307. [CrossRef]
45. Manjaiah, M.; Narendranath, S.; Basavarajappa, S. Wire Electro Discharge Machining Performance of TiNiCu Shape Memory Alloy. *Silicon* **2016**, *8*, 467–475. [CrossRef]
46. Ahmed, N.; Mughal, M.P.; Shoaib, W.; Raza, S.F.; Alahmari, A.M. WEDM of Copper for the Fabrication of Large Surface-Area Micro-Channels: A Prerequisite for the High Heat-Transfer Rate. *Micromachines* **2020**, *11*, 173. [CrossRef] [PubMed]
47. Korlos, A.; Tzetzis, D.; Mansour, G.; Sargis, D.; David, C. The delamination effect of drilling and electro-discharge machining on the tensile strength of woven composites as studied by X-ray computed tomography. *Int. J. Mach. Mach. Mater. (IJMMM)* **2016**, *18*, 426–448. [CrossRef]
48. Wang, R.; Wang, J.; Yuan, W. Analysis and Optimization of a Microchannel Heat Sink with V-Ribs Using Nanofluids for Micro Solar Cells. *Micromachines* **2019**, *10*, 620. [CrossRef]
49. Yamada, H.; Mohri, N.; Furutani, K.; Saito, N.; Magara, T. Transient Response of Wire Vibration System in Wire Electrical Discharge Machining. *J. Jpn. Soc. Precis. Eng.* **1997**, *63*, 1548–1552. [CrossRef]
50. Enache, S.; Opran, C. Dynamic Stability of the Technological Machining System in EDM. *Ann. CIRP* **1993**, *42*, 209–214. [CrossRef]
51. Yan, M.-T.; Lai, Y.-P. Surface quality improvement of wire-EDM using a fine-finish power supply. *Int. J. Mach. Tools Manuf.* **2007**, *47*, 1686–1694. [CrossRef]
52. Schneider, A.; Hommel, G.; Blettner, M. Linear Regression Analysis: Part 14 of a Series on Evaluation of Scientific Publications. *Dtsch. Aerzteblatt Online* **2010**, *107*, 776–782. [CrossRef]

micromachines

Article

Parametric Optimization of Electric Discharge Machining of Metal Matrix Composites Using Analytic Hierarchy Process

Sarabjeet Singh Sidhu [1], Timur Rizovich Ablyaz [2,*], Preetkanwal Singh Bains [3], Karim Ravilevich Muratov [2], Evgeny Sergeevich Shlykov [2] and Vladislav Vitalyevich Shiryaev [2]

1 Mechanical Engineering Department, Sardar Beant Singh State University (Formerly Known as Beant College of Engineering and Technology), Gurdaspur 143521, India; sarabjeetsidhu@yahoo.com
2 Perm National Research Polytechnic University, 614000 Perm, Russia; karimur_80@mail.ru (K.R.M.); kruspert@mail.ru (E.S.S.); vlad2117@gmail.com (V.V.S.)
3 Mechanical Engineering Department, IKG Punjab Technical University, Kapurthala 144603, India; preetbains84@gmail.com
* Correspondence: lowrider11-13-11@mail.ru

Abstract: The present study reports on the method used to obtain the reliable outcomes for different responses in electric discharge machining (EDM) of metal matrix composites (MMCs). The analytic hierarchy process (AHP), a multiple criteria decision-making technique, was used to achieve the target outcomes. The process parameters were varied to evaluate their effect on the material erosion rate (*MER*), surface roughness (SR), and residual stresses (σ) following Taguchi's experimental design. The process parameters, such as the electrode material (Cu, Gr, Cu-Gr), current, pulse duration, and dielectric medium, were selected for the analysis. The residual stresses induced due to the spark pulse temperature gradient between the electrode were of primary concern during machining. The optimum process parameters that affected the responses were selected using AHP to figure out the most suitable conditions for the machining of MMCs.

Keywords: analytical hierarchy process; residual stress; metal erosion rate; surface roughness

Citation: Sidhu, S.S.; Ablyaz, T.R.; Bains, P.S.; Muratov, K.R.; Shlykov, E.S.; Shiryaev, V.V. Parametric Optimization of Electric Discharge Machining of Metal Matrix Composites Using Analytic Hierarchy Process. *Micromachines* **2021**, *12*, 1289. https://doi.org/10.3390/mi12111289

Academic Editors: Francesco Modica and Irene Fassi

Received: 24 August 2021
Accepted: 19 October 2021
Published: 21 October 2021

Publisher's Note: MDPI stays neutral with regard to jurisdictional claims in published maps and institutional affiliations.

1. Introduction

Composite materials have superior properties, such as high strength; high modulus, low coefficient of thermal expansion; and resistance to fatigue, corrosion, and wear. Due to these prominent properties and their high strength-to-weight ratios, composites are extensively used in numerous advanced engineering applications. Composites with different reinforcements (such as fibers or particles) are being researched widely for their use in several applications, including manufacturing and biomedical industries. The composite materials, usually called metal matrix composites (MMCs), consist of a metal or alloy in the ductile phase to absorb and equally distribute the external load and develop a percolating network to position the reinforced fibers or particles. Alongside this, a brittle constituent, i.e., reinforcement, is embedded in the metal matrix [1–3].The combined properties of these constituents in MMCs allow for high strength and fractural properties to be attained, as well as high temperature resistance, making them suitable for applications in the automobile and aviation industries, such as braking systems, piston rods, piston pins, and brake discs [4]. These composites are also used as thermal management solutions for high energy density miniature electronic components, such as microprocessor lids, flip-chip lids, and microwave housing, and can replace high-cost materials such as titanium-based alloy [5–7]. With the presence of a soft matrix phase and hard reinforced particles in MMCs, precise machining with conventional methods is challenging in terms of avoiding degradation of the material properties.

Such difficulties can be overcome by adopting newer machining methods that can achieve the desired geometry with minimum damage to the material properties [8]. One such method for machining MMCs is electric discharge machining (EDM), which operates by generating controlled electric sparks to machine composite materials with complex geometries and provides a better surface quality with high dimensional accuracy. In this process, the tool electrode produces its replica in the workpiece material, producing a series of discrete electrical sparks that are generated within the dielectric medium. One of the reasons for the tremendous popularity of this process is its ability to machine complex internal contours, even in hard-to-cut materials, with negligible cutting forces [2,8]; however, in this process, the rapid change in temperature gradient of the machined surface results in sub-surface defects such as cracks, spalling, porosity, residual stresses, and metallurgical transformation [9].

For the effective ED machining of MMCs, a higher value of discharge current and shorter pulse-on time is generally recommended. EDM was highlighted by several researchers in the literature as an effective non-traditional machining technique used for shaping and machining of difficult-to-machine materials such as Al-SiC metal matrix composite [10]. In a similar study, a ZrB_2-40% Cu composite electrode was reported as an alternative electrode with a better material removal rate and tool wear rate than a conventionally used copper tool [11]; however, the diametric overcut and surface roughness were better with the copper tool than the composite tool electrodes [12].

The output responses, such as the material erosion rate (*MER*), tool wear rate (TWR), and surface roughness (SR), have been studied and widely reported for the EDM process [2,8]; however, one critical factor that significantly affects the machined component's life is the residual stresses induced while machining. These stresses are quantified using destructive and non-destructive routes [13–15]. Non-destructive testing (NDT) X-ray diffraction techniques have been successfully used to evaluate the residual stresses in the materials, and accordingly the effect of the process parameters [16].

Many optimization techniques have been used to analyze the effects of non-conventional machining process parameters on the output responses, such as the *MER*, TWR, SR, and residual stresses. For instance, grey relational analysis has been applied to optimize EDM process parameters on Al-10% SiC composites [17].The multi-regression method was used to establish the relationships between the input and output parameters of the wire EDM process [18].The lexicographic goal programming method was used for optimization of EDM process parameters while machining MMC [19].

So far, the previous studies have mostly reported on the optimization of *MER*, TWR, and SR, but very few studies have focused on optimizing the process parameters in order to minimize residual stresses in MMCs. EDM has been widely used for metals and alloys, although its application on MMCs and analyses of the resulting residual stresses have been limited. As such, the present study aims to establish the best process parameter settings for 65vol% SiC/A356.2 and hybrid 10vol% SiC-5vol% quartz/Al. Three output responses, namely the *MER*, SR, and residual stresses, are optimized using the analytic hierarchy process (AHP). The AHP is a decision-aiding tool that involves specifying a goal, measuring the relative importance (priorities), and choosing the relevant criteria [20,21]. One of the advantages of this tool is that it merges both qualitative and quantitative factors. The tool was formulated to exhibit the way the decision-maker thinks and determines the options based on weighted values. The tangible (objective) and non-tangible (subjective) factors can be efficiently coordinated and can provide reliable findings utilizing simple calculations. The AHP is also validated in various other fields, such as for issues linked to the economy, the stock industry, aircraft manufacture, transportation, and in the construction industry [22].

The objectives of this study are as follows:

- Analyze the influence of the EDM process parameters on the 65vol% SiC/A356.2 (sample I, procured from CPS System, Dallas, TX, USA) and 10vol% SiC-5vol% quartz/Al composites (sample II, produced by a controlled environmental stir casting process);

- Evaluate the outcomes, such as the *MER* (metal erosion rate), SR (surface roughness), and σ (residual stresses), utilizing the L18 Taguchi experimental design and optimizing the process parameters using AHP.

2. Material and Methods

2.1. Material

Two different variants of particulate-reinforced MMCs were used in this study. The material (65vol% SiC/A356.2 metal matrix composite) used in the study was procured in rectangular plates from CPS, Boston, MA, USA. The other specimen used in this experiment was a hybrid metal matrix composite with 10vol% SiC-5vol% quartz in aluminum, which was prepared using the in-house stir-casting method. The material composition was quantified using optical emission spectroscopy (Make: Arun Technology PolySpek-M spectrometer) as 0.384% Zn, 0.498% Cu, 0.424% Fe, 2.063% Si, and 0.354% Pb, with the balance% as Al.

2.2. Method

The experiments were conducted on an OSCARMAX (SD550 ZNC, Taiwan) die-sinking EDM machine using a conventional polarity with the selected tool electrodes. The workpieces were machined in EDM oil as a dielectric fluid and with a suspended powder form of copper (5 g/L) and graphite (5 g/L) in the dielectric medium. To ensure the uniform mixing of the suspended powder, a stirrer pedal fixed at 1400 rpm was used during machining. Three tool electrode materials, namely (i) electrolytic copper, (ii) fined-grained graphite (Poco-EDM 3), and (iii) a copper–graphite composite (50% Cu, Grade 673) were used for the experimental study. The electrodes were machined to a cylindrical shape with a diameter of 18mm.

The responses such as the *MER*, SR, and σ were measured after each experimental trial. The *MER* was measured in terms of weight loss per unit time using a digital weighing machine (Chyo-MJ-300). The surface roughness was measured at three different directions of the machined surface using a Mitutoyo (SJ-400) surface roughness analyzer. The developed residual stresses while machining are quantified by X-ray diffraction technique using a PANalytical'sX'PertPro MPD (Almelo, The Netherlands) diffractometer. The diffractometer used in this study was a horizontal, fixed, laboratory-based system. Table 1 shows the brief experimental conditions for the PANalytical'sX'PertPro X-ray stress analysis.

Table 1. Residual stress measurement conditions.

Factors	Conditions
Characteristic X-ray	Cu–Kα1 + 2
Measurement method	Ω-Diffractometer method
Diffraction plane, (hkl)	(422)
Tube voltage, KV	45
Tube current, mA	40
Diffraction angle (2θ)	40°–140°
Diffraction plane, (hkl)	(422)

2.3. Experimentation

On the basis of the pilot study, workpiece material, dielectric type, tool electrode material, pulse-on and pulse-off durations, and current, the machining parameters were selected. The other factors, such as the open-circuit voltage (~135 V) and flushing pressure (0.6 kg/cm^2), were maintained as constant during the experimental study. Table 2 shows the control factors and settings used for the experiments.

Table 2. Parameters with different levels.

Parameters (Symbol)	Levels		
	Level-1	Level-2	Level-3
Workpiece (WP)	65 vol% SiC/A356.2 (WP I)	10 vol% SiC-5 vol% quartz/Al (WP II)	—
Tool Electrode (TE)	Cu	Gr	Cu-Gr
Dielectric medium(D)	EDM oil (D)	PMEDM (Cu)	PMEDM (Gr)
Current (A) Amp	4	8	12
Pulse-on (T_{on}) μs	10	30	50
Pulse-off (T_{off}) μs	15	30	45

Because the chosen factors for the experiments involved a combination of two and three levels, the degree of freedom (dof) for2-level factors was 1 and the doffor3-level factors was 2; hence, the total dof required was 11 (i.e., 1 (one 2-level factor) + 5 × 2 (five 3-level factors) = 11). The mixed-level orthogonal array involving a combination of two- and three-level factors with at least 11 dof was designated experimental design (DOE)L18. This DOE methodology used orthogonally designed arrays that significantly reduced the required number of experimental trials to record the necessary data without compromising the output data quality [23]. L18 signifies the18 distinct orthogonal trial conditions performed randomly to remove any undesirable inclinations in the study. The orthogonal arrays contained the two-level factor in column 1, with the option of assigning three-level factors to the other columns. The conditions of the experimental trials after the assignment of factors to a selected array are listed in Table 3. As can be seen from this design matrix, the first column represents the workpiece materials used in the study; thus, the first nine trials represent 65 vol% SiC/A356.2 MMC, hereafter designated as WP I. The remaining nine trials (trials 10 to 18) represent 10 vol% SiC-5 vol% quartz in aluminum, hereafter referred to as WP II. Other factors are assigned to the remaining columns of the L18 array, as listed in Table 3. The 18 experimental trials with two repetitions are performed as per Taguchi's design in random order. The mean *MER*, SR, and σ are measured at the end of each trial and are presented in Table 3 under output responses. The *MER* is calculated by the weight difference of the workpiece before and after machining, as given by Equation (1):

$$MER = \frac{\left(w_i - w_f\right)1000}{T} \text{mg/min} \quad (1)$$

where Δw if the change in weight, i.e., the weights before and after machining (gm); t is the machining time in minutes.

The SR was measured using a Mitutoyo SJ-400 surface roughness tester in terms of the arithmetic average of the absolute value Ra (μm). Each sample was measured from three locations diametrically on the machined surface and was averaged for investigation.

Residual stresses were estimated with the help of the classical X-ray diffraction procedure. The peak diffracted from the (422) plane was selected to measure the shift at a maximum 2θ angle. The change in d-spacing due to strain in the sample was minimal; hence, the highest possible 2θ angle peak was chosen. The relations of the d-spacing (Δd) with the diffraction peak (Δθ) is given by Equation (2):

$$\frac{\Delta d}{\Delta \theta} = (-)(\Delta\theta)\cot\theta \quad (2)$$

The stresses were calculated using the classical $\sin^2\psi$ technique [14,16], with the assumption that the stress state is unidirectional. Equation (3) was used to calculate normal the residual stresses:

$$a^{+} = \frac{1}{2}\left(Ie_{\varphi\psi+} + Ie_{\varphi\psi-}\right) = \frac{1}{2} S_2 \sin^2\psi(\sigma_\varphi) + Ie\varphi_0{}^{\circ} \quad (3)$$

Here, parameter a^+ is the average of the lattice strain for positive ($_{\varphi\psi+}$) and negative ($_{\varphi\psi-}$) values and ψ is the sample alignment (herein, $\varphi = 0°$); $1/2S_2 = (1 + \nu)/E$, $1/2S_2$ are the X-ray elastic constants (XEC's) and their values are represented in Table 4.

Table 3. Experimental layout (L18).

Trial. No. (T(n)) Where n = 1–18	Process Parameters						Output Responses			
	WP	TE	T_{off} (μs)	T_{on} (μs)	D	A	σ_0 (MPa)	σ_n (MPa) Calculated from Equation (5)	*MER* (mg/min)	SR (μm)
1	WP I	1	15	10	1	4	63.3	66.7	2.64	2.94
2	WP I	1	30	30	2	8	74.6	55.4	14.275	2.05
3	WP I	1	45	45	3	12	82.8	47.2	23.17	5.67
4	WP I	2	15	45	2	8	36.3	93.7	23.38	2.09
5	WP I	2	30	10	3	12	63.6	66.4	18.97	4.12
6	WP I	2	45	30	1	4	110.3	19.7	3.04	3.00
7	WP I	3	15	30	1	12	61.4	68.6	22.240	5.01
8	WP I	3	30	45	2	4	78.5	51.5	9.860	2.06
9	WP I	3	45	10	3	8	129	1	9.460	5.06
10	WP II	1	15	30	3	8	70.4	162.1	20.90	6.69
11	WP II	1	30	45	1	12	104	128.5	60.67	10.46
12	WP II	1	45	10	2	4	78.1	154.4	10.860	4.69
13	WP II	2	15	45	3	4	41.8	190.7	57.99	6.46
14	WP II	2	30	10	1	8	149.3	83.2	18.86	8.44
15	WP II	2	45	30	2	12	132.9	99.6	29.96	4.44
16	WP II	3	15	10	2	12	77.7	154.8	65.5	6.76
17	WP II	3	30	30	3	4	89.2	143.3	10.07	6.12
18	WP II	3	45	45	1	8	231.5	1	45.72	7.95

Table 4. X-ray elastic constants.

	Sample I	Sample II
Elastic constants T^{-1} Pa ($1/2S_2$)	6.98	16.84

Equation (4) can be utilized to estimate the shear residual stress in further studies:

$$a^- = \frac{1}{2}\left(Ie_{\varphi\psi+} - Ie_{\varphi\psi-}\right) = \frac{1}{2}S_2 \sin(2\psi)(\tau_\varphi) \quad (4)$$

The sample calibration for the stress test is represented below.

Sample calculation of σ for trial 2 (WP I): The machined specimen was sectioned to 25 × 25 mm using a wire-cut EDM machine. To limit modifications of the machined surface properties by heating in the wire EDM operation, we ensured that the cutting edge was far away from the calibration surface zone. The etching of re-solidified metal on the machined zone resulted in reduced computation errors. Residual stress was estimated in the aluminum matrix phase of the machined specimen. The measurement was accomplished by selecting the isolated peak diffracted at the highest value of 2θ from the plane. Figure 1 shows the X-ray spectra for trial 2. From the obtained X-ray spectra, the peak selected for residual stress determination was at approximately 137.23°.

Table 5 represents the various parameters for trial 2 used to measure residual stress at different 2θ ψ-tilts (positive and negative).

The normal residual stress was analyzed by comparing the linear fit regression equation obtained from the plot of a+ vs. sin2ψ with Equation (3), i.e., the equation obtained was:

$$a^+ = \left(5.21 * \sin^2\psi = 0.291\right) * 10^{-4} 5.21 \times 10^{-4} = \frac{1}{2}S_2\left(\sigma_\varphi\right)$$

where $1/2S_2 = 6.98\ T^{-1}$ Pa, $\varphi = 0$

As such, the normal residual stress (σ_0) for trial 2 was 74.6 MPa.

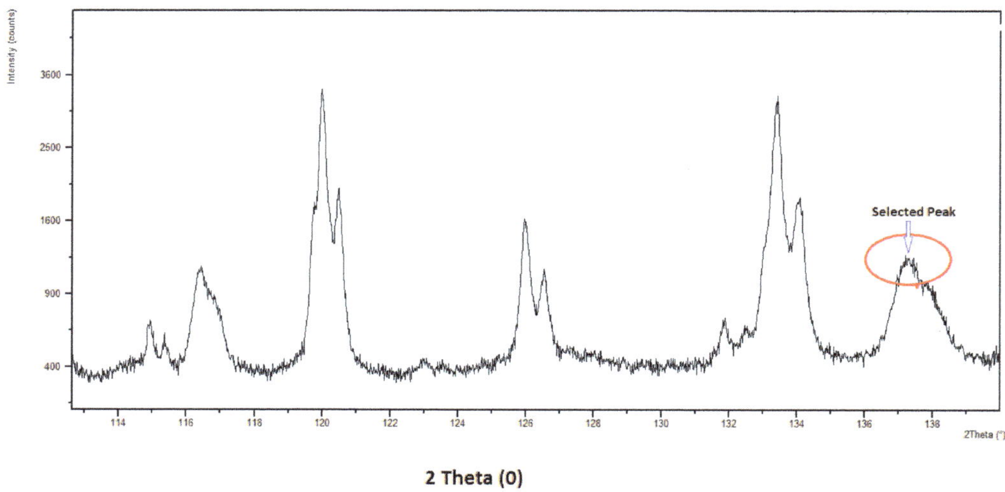

Figure 1. Selected X-ray spectra peaks for residual stress calibration.

Table 5. X-ray spectra of lattice strain intrial 2.

	ψ	$\mathrm{Sin}^2\ \psi$	$d_{\varphi\psi}$	$\psi\varphi$	a+	a−
	0	0	0.827269	0		
Positive ψ	12.92	0.05	0.827286	0.00002050	5.08×10^{-5}	-3.02×10^{-5}
	18.44	0.1	0.827311	0.00005060	-1.70×10^{-5}	6.76×10^{-5}
	22.79	0.15	0.827472	0.00024500	1.97×10^{-4}	4.82×10^{-5}
	26.57	0.2	0.827425	0.00018900	2.48×10^{-4}	-5.90×10^{-5}
	30.00	0.25	0.827450	0.00021900	1.84×10^{-4}	3.52×10^{-5}
	33.21	0.3	0.827499	0.00027800	9.01×10^{-5}	1.88×10^{-4}
	36.27	0.35	0.827564	0.00035700	2.10×10^{-4}	1.47×10^{-4}
	39.23	0.4	0.827434	0.00020000	1.11×10^{-4}	8.91×10^{-5}
	42.13	0.45	0.827673	0.00048800	2.94×10^{-4}	1.94×10^{-4}
	45	0.5	0.827543	0.00033100	3.58×10^{-4}	-2.73×10^{-5}
Negative ψ	12.92	0.05	0.827336	0.00008099		
	18.44	0.1	0.827199	−0.00008462		
	22.79	0.15	0.827392	0.00014868		
	26.57	0.2	0.827523	0.00030703		
	30.00	0.25	0.827392	0.00014868		
	33.21	0.3	0.827188	−0.00009791		
	36.27	0.35	0.827321	0.00006286		
	39.23	0.4	0.827287	0.00002176		
	42.13	0.45	0.827351	0.00009912		
	45	0.5	0.827588	0.00038561		

$$a^{+} = \tfrac{1}{2}\left(Ie_{\varphi\psi+} + Ie_{\varphi\psi-}\right)\ (fornormalstress)$$
$$a^{-} = \tfrac{1}{2}\left(Ie_{\varphi\psi+} - Ie_{\varphi\psi-}\right)\ (forshearstress)$$
$$where\varphi = 0$$

2.4. *Analysis of Variance for MER, SR, Residual Stress*

The *MER*, SR, and σ results were analyzed using analysis of variance (ANOVA). A summary of the ANOVA used for *MER*, SR, and σ is presented in Table 6. The significant parameters were chosen by comparing the F-values with F-critical at a confidence level of 95%. The higher the F-value, the greater the effect of the parameter on the responses.

Table 6. Analysis of variance for *MER*, SR, and σ.

Factors	dof	Sum of Squares			Variance			F-Value		
		MER	SR	σ	*MER*	SR	σ	*MER*	SR	σ
W/Pc	1	1977.26	50.0333	4204.4	1977.26	50.0333	4204.4	10.50 *	80.99 *	9.13 *
Electrode	2	62.50	1.9590	3284.0	31.25	0.9795	1641.98	0.17	1.59	3.57 *
Pulse-off	2	435.50	0.9768	14262.5	217.75	0.4884	7131.27	1.16	0.79	15.49 *
Pulse-on	2	1448.53	4.6520	110.5	724.26	2.3260	55.26	3.85 *	3.77 *	0.12
Dielectric medium	2	10.67	22.5037	6526.1	5.33	11.2519	3263.04	0.03	18.21 *	7.09 *
Current	2	1494.86	10.6571	4725.5	747.43	5.3286	2362.75	3.97 *	8.63 *	5.13 *
Error	6	1129.73	3.7064	2762.6	188.29	0.6177	460.44			
Total	17	6559.04	94.4886	35875.7						

* Significant factor.

MER: The ANOVA results show that the current and pulse-on contributed significantly to changes in *MER*. Additionally, the variations in the workpiece material had significant effects on the *MER*. On the contrary, the dielectric medium, pulse-off time, and electrode material had no significant effect. It was observed that with increases in pulse-on time and current, the *MER* increases significantly, since increases in the current and pulse-on time increase the spark energy duration; thus, with increased heat input, the temperature increases, resulting in the workpiece's higher melting or evaporation rate;

SR: The surface roughness of the machined surface was significantly influenced by factors such as the powder concentration, current, and pulse-on time. Furthermore, the two materials showed quite different SR values. The roughness increased with increases in current and pulse-on time, whereas the powder in the dielectric medium improved the surface finish. An increase in current or pulse-on time increases the spark energy, which drives the formation of bigger and deeper craters, leading to a rough machined surface. The addition of powder consistently improved the finish of the machined surfaces; the spark becomes more uniform with increased frequency and widens the spark gap. This reduced the magnitude of the impact forces, resulting in small and shallow craters and lowering the surface roughness [24,25];

Residual Stress: The ANOVA results show that the pulse off time, powder mixing in the dielectric medium, and current significantly affected the σ. Additionally, the selected MMCs showed different residual stress values for similar parameter settings. It can be seen from the results that the pulse-on time affected the *MER* and SR but had no effect on σ. On the other hand, the pulse-off time hada significant impact on the development of residual stress due to the re-solidification time duration. The presence of suspended particles in the dielectric medium facilitates the easy formation of plasma channels between the electrode and the workpiece, resulting in lower SR and residual stress. The conductivity of suspended particles plays the major role in determining the SR but has no impact on the development of σ.

3. Analytical Hierarchical Process

Optimizing the responses independently results in vastly different parametric combinations of the machining process parameters. For example, if *MER* was optimized separately, this would cause some parameters to increase *MER* (the higher *MER* is, the better the function); however, these parameters may not result in decreased SR. The opposite would be true if SR was optimized individually. The identical condition involves residual stresses optimization. To obtain a result that is close to the target, the responses must be optimized together according to their priority. The analytical hierarchy process (AHP) offers one technique that suggested the best combination of parameters to reach or nearly reach the target. The AHP is simply structured and widely used for multiple-goal decision-making techniques and is classified as a decision-making tool for use under conditions of certainty, i.e., the data are obtained deterministically and the tool is designed for situations in which ideas, feelings, and emotions are quantified into a numerical scale [26]. The main steps used in the implementation of the AHP are as follows:

- Define the objective and evaluation criteria and develop the hierarchical structure, with an objective at the top level, the criteria and sub-criteria at the intermediate level, and the available alternatives at the lowest level;
- Form a pair-wise comparison matrix for each level with respect to the higher level and determine the relative importance of the different alternatives with respect to its immediately superior sub-criteria. The comparison is made on a 9-point "fundamental scale of Saaty", as represented in Table 7.
- Compute the relative weights for the pair-wise comparison matrices using eigenvector methods;
- Judge the scope of inconsistency by using the largest eigenvector. The judgment of the accepted degree of consistency can be checked by means of the consistency ratio (CR) of the consistency index (CI) with the appropriate value of the random index (RI) from Table 8.
- Repeat the above steps for all levels in the hierarchy, with the overall relative value evaluated by the linear addition function.

Table 7. Saaty's fundamental scale.

Scale Value	Explanation
1	Equally preferred
3	Slightly more preferred
5	Strongly preferred
7	Very strongly preferred
9	Extremely preferred
2, 4, 6, 8	Used to reflect compromise between scale values

Table 8. Random consistency index.

k	1	2	3	4	5	6	7	8	9	10	11	12	13
RI	0.00	0.00	0.58	0.90	1.12	1.24	1.32	1.41	1.45	1.49	1.51	1.48	1.48

The steps are summarized in Figure 2.

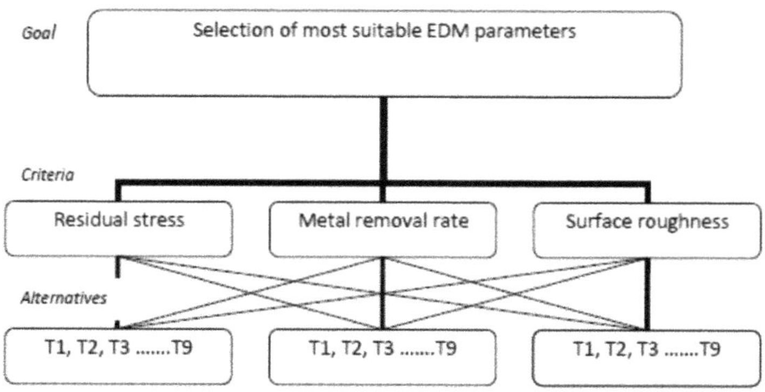

Figure 2. Hierarchy layout of the analytical hierarchy process.

The MMCs used in the present study are used for very high-end applications in aerospace and mining, while the residual stresses developed during the EDM process affect the service life of the product. Residual stress was assigned the maximum weight, followed by the material removal rate and surface roughness. To attain the desired objective, the residual stress results were slightly modified (cost-to-benefit conversion, which can be achieved by using the -ve sign), as follows:

$$\sigma_n = (\sigma_{max} - \sigma_0) + 1 \quad (5)$$

where σ_{max} is the maximum value of the residual stress in the corresponding trial set for each workpiece, σ_0 is the residual stress measured with the X-ray diffraction method, and σ_n is a modified residual stress value (refer to Table 3), which was calculated from Equation (5).

Using the criteria for assigning weights to the residual stress, *MER*, and SR, a (3 × 1) weight column matrix, as shown in Table 9, was established for pair-wise comparison.

Table 9. AHP pair-wise comparison of weighting criteria.

	σ	*MER*	SR	Priority Vector
σ	1	2	5	0.581552
MER	1/2	1	3	0.308996
SR	1/5	1/3	1	0.109452
	λ_{max} = 3.00369	CI = 0.0018473, RI = 0.58, CR = 0.003		

Subsequently a pair-wise comparison of the experimental trials (alternatives) was developed for, σ_n, *MER*, and SR for each workpiece, with the results shown in Tables 10–15. The synthesized matrix to obtain priority vector of σ for WP I is shown in Table 16. It was also ensured during pair-wise comparisons of alternatives that if the values attained during comparison were greater than the maximum limit of Saaty's fundamental scale, the highest value of the scale (9) was selected to avoid inconsistency.

Table 10. Pair-wise comparison of σ values against their alternatives for WP I.

.	T1	T2	T3	T4	T5	T6	T7	T8	T9	Priority Vector
T1	1	1	1	1	1	3	1	1	9	0.130829
T2	1	1	1	1/2	1	3	1	1	9	0.12105
T3	1	1	1	1/2	1	2	1	1	9	0.115577
T4	1	2	2	1	1	5	1	2	9	0.180525
T5	1	1	1	1	1	3	1	1	9	0.130829
T6	1/3	1/3	1/2	1/5	1/3	1	1/4	1/3	9	0.050517
T7	1	1	2	1	1	4	1	1	9	0.136302
T8	1	1	1	1/2	1	3	1	1	9	0.12105
T9	1/9	1/9	1/9	1/9	1/9	1/9	1/9	1/9	1	0.0133204
		λ_{max} = 9.23034				CI = 0.0287927, RI = 1.45, CR = 0.0198				

Table 11. Pair-wise comparison of *MER* values with respect to their alternatives for WP I.

	T1	T2	T3	T4	T5	T6	T7	T8	T9	Priority Vector
T1	1	1/5	1/9	1/9	1/7	1	1/8	1/4	1/4	0.02080
T2	5	1	1/2	1/2	1	5	1/2	1	1	0.09982
T3	9	2	1	1	1	9	1	3	3	0.19547
T4	9	2	1	1	1	8	1	2	2	0.17520
T5	7	1	1	1	1	6	1	2	2	0.15448
T6	1	1/5	1/9	1/8	1/6	1	1/7	1/3	1/3	0.02328
T7	8	2	1	1	1	7	1	2	2	0.17034
T8	4	1	1/3	1/2	1/2	3	1/2	1	1	0.08303
T9	4	1	1/3	1/2	1/2	3	1/2	1	1	0.08303
		λ_{max} = 9.02762				CI = 0.00640, RI = 1.45, CR = 0.0023				

Table 12. Pair-wise comparison of SR values with respect to their alternatives for WP I.

	T1	T2	T3	T4	T5	T6	T7	T8	T9	Priority Vector
T1	1	1	1/2	1	1	1	1/2	1	1/2	0.0834519
T2	1	1	1/3	1	1/2	1	1/2	1	1/2	0.0728733
T3	2	3	1	3	1	2	1	3	1	0.176214
T4	1	1	1/3	1	1/2	1	1/2	1	1/2	0.0728733
T5	1	2	1	2	1	1	1	2	1	0.133865
T6	1	1	1/2	1	1	1	1/2	1	1/2	0.0834519
T7	2	2	1	2	1	1	1	2	1	0.152199
T8	1	1	1/3	1	1/2	1/2	1	1	1/2	0.0728733
T9	2	2	1	2	1	1	1/2	2	1	0.152199
	λ_{max} = 9.10338				CI = 0.0129221, RI = 1.45, CR = 0.008911					

Table 13. Pair-wise comparison of σ values with respect to their alternatives for WP II.

	T10	T11	T12	T13	T14	T15	T16	T17	T18	Priority Vector
T10	1	1	1	1	2	2	1	1	9	0.138612
T11	1	1	1	1/2	2	1	1	1	9	0.119842
T12	1	1	1	1	2	2	1	1	9	0.138612
T13	1	2	1	1	2	2	1	1	9	0.165782
T14	1/2	1/2	1/2	1/2	1	1	1/2	1/2	9	0.0758749
T15	1/2	1	1/2	1/2	1	1	1/2	1/2	9	0.089421
T16	1	1	1	1	2	2	1	1	9	0.129583
T17	1	1	1	1	2	2	1	1	9	0.128872
T18	1/9	1/9	1/9	1/9	1/9	1/9	1/9	1/9	1	0.0134007
	λ_{max} = 9.1803				CI = 0.0225376, RI = 1.45, CR = 0.0155					

Table 14. Pair-wise comparison of *MER* values with respect to their alternatives for WP II.

	T10	T11	T12	T13	T14	T15	T16	T17	T18	Priority Vector
T10	1	1/3	2	1/3	1	1	1/3	2	1/2	0.068095
T11	3	1	6	1	3	2	1	6	1	0.185247
T12	1/2	1/2	1	1/5	1/2	1/3	1/6	1	1/4	0.0330161
T13	3	3	3	1	3	2	1	6	1	0.181600
T14	1	1	1	1	1	1/2	1/3	2	1/2	0.0629946
T15	1	1	1	1	1	1	1/2	3	1/2	0.0923417
T16	3	3	3	3	3	3	1	6	1	0.1852470
T17	1/2	1/2	1/2	1/2	1/2	1/2	1/2	1	1/5	0.0314636
T18	2	2	2	2	2	2	2	2	1	0.159995
	λ_{max} = 9.05235				CI = 0.00654334, RI = 1.45, CR = 0.00451					

Table 15. Pair-wise comparison of SR values with respect to their alternatives for WP II.

	T10	T11	T12	T13	T14	T15	T16	T17	T18	Priority Vector
T10	1	1/2	1	1	1	2	1	1	1	0.107181
T11	2	1	2	2	1	2	2	2	1	0.171361
T12	1	1/2	1	1	1/2	1	1	1	1/2	0.0858196
T13	1	1/2	1	1	1	1	1	1	1	0.0990902
T14	1	1	2	1	1	2	1	1	1	0.126548
T15	1/2	1/2	1	1	1/2	1	1/2	1	1/2	0.0741207
T16	1	1/2	1	1	1	2	1	1	1	0.107181
T17	1	1/2	1	1	1	1	1	1	1	0.0990902
T18	1	1	2	1	1	2	2	1	1	0.12661
	λ_{max} = 9.16155				CI = 0.0201939, RI = 1.45, CR = 0.0139					

Table 16. Synthesized matrix of σ for WP I.

	T1	T2	T3	T4	T5	T6	T7	T8	T9
T1	0.13434	0.10405	0.10405	0.17208	0.13433	0.12442	0.13585	0.11843	0.12329
T2	0.13434	0.10405	0.10405	0.08604	0.13433	0.12442	0.13585	0.11843	0.12329
T3	0.13434	0.10405	0.10405	0.08604	0.13433	0.08295	0.13585	0.11843	0.12329
T4	0.13434	0.20809	0.20809	0.17208	0.13433	0.20737	0.13585	0.23685	0.12329
T5	0.13434	0.10405	0.10405	0.17208	0.13433	0.12442	0.13585	0.11843	0.12329
T6	0.04478	0.05202	0.05202	0.03442	0.04478	0.04147	0.03396	0.03948	0.12329
T7	0.13434	0.20809	0.20809	0.17208	0.13433	0.16590	0.13585	0.11843	0.12329
T8	0.13434	0.10405	0.10405	0.08604	0.13433	0.12442	0.13585	0.11843	0.12329
T9	0.01493	0.01156	0.01156	0.01912	0.01493	0.00461	0.01509	0.01316	0.01370

The pair-wise comparison (9 × 9 matrix) of the alternatives was completed by comparing and rounding off the response ratio obtained in experimental trials. For example, if trial 1 gives a value of 16.5 and trial 2 gives 4.5, then the response ratio is the ratio of the values of the two trials (trial1/trial2), which is 3.66 (rounding off = 4). The same procedure was adopted for all the response parameters to assign the weights. In the work [27] adopted percentage change in the state of tool wear while machining medium carbon steel workpiece. The change in percentage was used to assign the weight in pair-wise comparison matrix.

To illustrate this calculation, the pair-wise matrix for residual stress (σ) (WP I) is considered, as shown in Table 16.

Step 1: The matrix was normalized by dividing each element of the matrix by its column total. For example, for the T(1,1) element, a value of 0.13434 was obtained by dividing 1 by the column total of 7.444 (1 + 1 + 1 + 1 + 1 + 1/3 + 1 + 1 + 1/9). The same procedure was adopted for each element of the matrix, with the results given in Table 16.

Step 2: The estimation of the priority vector was done by taking the row average, i.e., (0.13434 + 0.10405 + 0.10405 + 0.17208 + 0.13433 + 0.12442 + 0.13585 + 0.11843 + 0.12329) and dividing it by 9.

Similarly, the synthesized pair-wise comparison matrix, priority vectors, and validation of the constructed matrix were performed for each response parameter. The overall weight was calculated by multiplying the alternative available priority vectors for each sample with the criteria weight, as given in Table 17.

Table 17. Overall weight matrix of WP I for the priority vector.

Trials	σ (0.581552)	*MER* (0.308996)	SR (0.109452)	Overall Priority Vector	Ideal Weight Vector
T1	0.130829	0.02080	0.0834519	0.091645	0.548453
T2	0.12105	0.09982	0.0728733	0.109219	0.653623
T3	0.115577	0.19547	0.176214	0.146899	0.879124
T4	**0.180525**	**0.17520**	**0.0728733**	**0.167097**	**1.000000 ***
T5	0.130829	0.15448	0.133865	0.138471	0.828682
T6	0.050517	0.02328	0.0834519	0.045707	0.273534
T7	0.136302	0.17034	0.152199	0.148560	0.889066 **
T8	0.12105	0.08303	0.0728733	0.103185	0.617514
T9	0.0133204	0.08303	0.152199	0.049217	0.294540

* 1st rank, ** 2nd rank.

The overall priority for each EDM parameter setting was calculated as demonstrated below:

Overall weight of T1 (WP I): Overall Weight = 0.581552 (0.130829) + 0.308996 (0.02080) + 0.109452 0.0834519) = 0.091645.

Similarly, the overall weight was calculated for each trial conducted for the selected workpieces.

The remaining calculations were completed by combining the assigned criteria weight with the alternative priority weight to obtain the overall priority results (Tables 17 and 18), as per the hierarchical steps given in Figure 1.

Table 18. Overall weight matrix of WPII for the priority vector.

Trials	σ (0.581552)	*MER* (0.308996)	SR (0.109452)	Overall Priority Vector	Ideal Weight Vector
T10	0.138612	0.068095	0.107181	0.113791	0.73091
T11	0.119842	0.185247	0.171361	0.146440	0.940701
T12	0.138612	0.0330161	0.0858196	0.100555	0.645949
T13	**0.165782**	**0.181600**	**0.0990902**	**0.155672**	**1.000000** *
T14	0.0758749	0.0629946	0.126548	0.077638	0.498728
T15	0.089421	0.0923417	0.0741207	0.088929	0.571260
T16	0.129583	0.1852470	0.107181	0.149991	0.963508 **
T17	0.128872	0.0314636	0.0990902	0.095874	0.615871
T18	0.0134007	0.159995	0.12661	0.071110	0.456796

* 1st rank, ** 2nd rank.

The ideal weight vector was obtained by dividing the priority vector with the largest priority weight element in the matrix. The advantage of using an idealized weight vector is that the ranking of trials does not change due to the influence of a newly introduced non-optimal identical alternative [27].

From the calculated overall priority, the trials were ranked for each type of MMC. The maximum overall weight or composite performance score for sample I was obtained for T4 as given in Table 17 (See also Table 3), which was conducted with a graphite electrode; dielectric medium mixed with Cu powder; pulse-off and pulse-on times of 15 μs and 45 μs, respectively; and current of 8 amps. Similarly, the maximum overall weight or composite performance score for sample II as obtained in Table 18 corresponds to trial number 13 (Table 3) of the original L18 array. This trial was also completed with a graphite electrode; dielectric medium mixed with graphite powder; pulse-off and pulse-on times of 15 μs and 45 μs, respectively; and a current setting of 4 amps. It was observed that the presence of powder in the dielectric medium expanded the area of the spark zone between the electrodes, thereby minimizing the impact of thermal shocks on the machined surface and diminishing the induced residual stresses; thus, the solution that globally optimizes residual stresses, *MER*, and SR for the two types of MMCs (WP I and WP II) used in the experiment was obtained (Table 19).

Table 19. Summarized process parameters for the target responses.

Parameter	WP I	WP II
Tool Electrode	Graphite	Graphite
Dielectric medium	PMEDM (Cu)	PMEDM (Gr)
Pulse-off time	15 μs	15 μs
Pulse-on time	45 μs	45 μs
Current	8 Amp	4 Amp

4. Conclusions

In the present study, three output response parameters, namely σ, *MER*, and SR, were optimized using a manageable AHP technique. Due to conflicting parameter settings for different output responses in the EDM process, identifying process parameters is a complex decision-making process. A manageable AHP approach was used in the present study to obtain a more reliable global composite performance score for various trial conditions in powder-mixed electric discharge machining (PMEDM) of MMCs. The process conditions that affected the three responses, namely σ, *MER*, and SR, were identified and optimized using AHP for two different types of MMCs. The current and pulse-on time significantly affected *MER*, while the addition of the powder, current, and pulse-on time influenced

SR. Despite this, the pulse-off duration had no significant effect on *MER* or SR. Still, the pulse-off duration had the most considerable influence on residual stresses, followed by the dielectric medium, current, and type of tool electrode.

The three responses were optimized together according to the predetermined goal using AHP. The optimal process conditions for the selected materials were identified. The overall process for both workpieces revealed that machining the workpiece with a graphite tool electrode and higher pulse-on time setting coupled with lowest pulse-off time in the presence of a suspended particle dielectric medium (PMEDM) contributed to minimizing the residual stress with the desired *MER*. Due to the denser ceramic-reinforced particles in WP I compared to WP II, the target results were achieved at a higher current level (i.e., 8A) than the current required for WP II. The optimal settings for achieving the specified target results involved the graphite tool electrode coupled with pulse-on and -off times of 45 µs and 15 µs, respectively, for both workpieces. The methodology used to obtain optimum EDM process parameters can be extended by prioritizing different responses (i.e., *MER*, SR) according to the end-use application of the product. Overall, the use of AHP will open the horizon for EDM practitioners to determinevarious process parameters, improving their ability to achieve their desired targets.

Author Contributions: Conceptualization, S.S.S., P.S.B., E.S.S. and T.R.A.; methodology, S.S.S., K.R.M. and T.R.A.; validation, S.S.S. and T.R.A.; formal analysis, S.S.S., T.RA. and E.S.S.; investigation, S.S.S., E.S.S. and T.R.A.; resources, S.S.S., E.S.S., T.R.A., and K.R.M.; data curation, E.S.S., K.R.M. and V.V.S.; software, V.V.S.; writing—original draft preparation, S.S.S., P.S.B., E.S.S. and T.R.A.; writing—review and editing, S.S.S., K.R.M., T.R.A. and V.V.S.; visualization, S.S.S., E.S.S. and T.R.A.; supervision, S.S.S. and T.R.A.; project administration, S.S.S., K.R.M. and T.R.A. funding acquisition, T.R.A., E.S.S. and K.R.M. All authors have read and agreed to the published version of the manuscript.

Funding: Financial support was provided by the Russian Ministry of Science and Higher Education (project FSNM-2020-0028).

Conflicts of Interest: The authors declare no conflict of interest. The funders had no role in the design of the study; in the collection, analyses, or interpretation of data; in the writing of the manuscript, or in the decision to publish the results.

References

1. Callister, W.D. Composite. In *Material Science and Engineering Material*, 6th ed.; Wiley: New York, NY, USA, 2004; pp. 527–559.
2. Bains, P.S.; Sidhu, S.; Payal, H.S. Fabrication and Machining of Metal Matrix Composites: A Review. *Mater. Manuf. Process.* **2016**, *31*, 553–573. [CrossRef]
3. Ablyaz, T.; Bains, P.; Sidhu, S.; Muratov, K.; Shlykov, E. Impact of Magnetic Field Environment on the EDM Performance of Al-SiC Metal Matrix Composite. *Micromachines* **2021**, *12*, 469. [CrossRef]
4. Ahamed, A.R.; Asokan, P.; Aravindan, S. EDM of hybrid Al–SiCp–B4Cp and Al–SiCp–Glassp MMCs. *Int. J. Adv. Manuf. Technol.* **2009**, *44*, 520–528. [CrossRef]
5. Sidhu, S.S.; Kumar, S.; Batish, A. Metal Matrix Composites for Thermal Management: A Review. *Crit. Rev. Solid State Mater. Sci.* **2015**, *41*, 132–157. [CrossRef]
6. Dufour, B.; McNulty, M.; Miller, S. Microwave multi-chip module utilizing aluminum silicon carbide with in-situ cast components and high density interconnect technology. *Int. J. Microcircuits Electron. Packag.* **1997**, *20*, 303–308.
7. Novich, B.E.; Adams, R.W.; Occhionero, M.A. Low-Cost MCM-D Cavity Substrates for Packaging High Density Si and GaAs Devices. In Proceedings of the 10th European Microelectronics Conference, ISHM, Copenhagen, Denmark, 14–17 May 1995; pp. 475–481.
8. Sidhu, S.S.; Batish, A.; Kumar, S. Fabrication and electrical discharge machining of metal–matrix composites: A review. *J. Reinf. Plast. Compos.* **2013**, *32*, 1310–1320. [CrossRef]
9. Sidhu, S.; Batish, A.; Kumar, S. Study of Surface Properties in Particulate-Reinforced Metal Matrix Composites (MMCs) Using Powder-Mixed Electrical Discharge Machining (EDM). *Mater. Manuf. Process.* **2014**, *29*, 46–52. [CrossRef]
10. Bains, P.S.; Singh, S.; Sidhu, S.S.; Kaur, S.; Ablyaz, T.R. Investigation of Surface Properties of Al–SiC Composites in Hybrid Electrical Discharge Machining. In *Structural Materials*; Springer: Cham, Switzerland, 2018; pp. 181–196.
11. Rafaqat, M.; Mufti, N.A.; Ahmed, N.; AlAhmari, A.M.; Hussain, A. EDM of D2 Steel: Performance Comparison of EDM Die Sinking Electrode Designs. *Appl. Sci.* **2020**, *10*, 7411. [CrossRef]
12. Khanra, A.; Sarkar, B.; Bhattacharya, B.; Pathak, L.; Godkhindi, M. Performance of ZrB2–Cu composite as an EDM electrode. *J. Mater. Process. Technol.* **2007**, *183*, 122–126. [CrossRef]

13. Withers, P.; Bhadeshia, H.K.D.H. Residual stress. Part 1—Measurement techniques. *Mater. Sci. Technol.* **2001**, *17*, 355–365. [CrossRef]
14. Welzel, U.; Ligot, J.; Lamparter, P.; Vermeulen, A.C.; Mittemeijer, E.J. Stress analysis of polycrystalline thin films and surface regions by X-ray diffraction. *J. Appl. Crystallogr.* **2005**, *38*, 1–29. [CrossRef]
15. Cuullity, B.D. *Elements of X-ray Diffraction*, 2nd ed.; Addison-Wesley: New York, NY, USA, 1978.
16. Sidhu, S.; Batish, A.; Kumar, S. Neural network–based modeling to predict residual stresses during electric discharge machining of Al/SiC metal matrix composites. *Proc. Inst. Mech. Eng. Part B J. Eng. Manuf.* **2013**, *227*, 1679–1692. [CrossRef]
17. Singh, P.; Raghukandan, K.; Rathinasabapathi, M.; Pai, B. Electric discharge machining of Al–10%SiCP as-cast metal matrix composites. *J. Mater. Process. Technol.* **2004**, *155-156*, 1653–1657. [CrossRef]
18. Tzeng, Y.-F.; Chen, F.-C. Multi-objective optimisation of high-speed electrical discharge machining process using a Taguchi fuzzy-based approach. *Mater. Des.* **2007**, *28*, 1159–1168. [CrossRef]
19. Sidhu, S.S.; Batish, A.; Kumar, S. EDM of Metal Matrix Composite for Parameter Design Using Lexicographic Goal Programming. *Mater. Manuf. Process.* **2013**, *28*, 495–500. [CrossRef]
20. Saaty, T.L. A scaling method for priorities in hierarchical structures. *J. Math. Psychol.* **1977**, *15*, 234–281. [CrossRef]
21. Al-Harbi, K.M.-S. Application of the AHP in project management. *Int. J. Proj. Manag.* **2001**, *19*, 19–27. [CrossRef]
22. Whitaker, R. Validation examples of the Analytic Hierarchy Process and Analytic Network Process. *Math. Comput. Model.* **2007**, *46*, 840–859. [CrossRef]
23. Banerjee, S.; Sutradhar, G.; Sahoo, P. Design of experiment analysis of elevated temperature wear of Mg-WC nano-composites. *Rep. Mech. Eng.* **2021**, *2*, 202–211. [CrossRef]
24. Singh, G.; Ablyaz, T.R.; Shlykov, E.S.; Muratov, K.R.; Bhui, A.S.; Sidhu, S.S. Enhancing Corrosion and Wear Resistance of Ti6Al4V Alloy Using CNTs Mixed Electro-Discharge Process. *Micromachines* **2020**, *11*, 850. [CrossRef]
25. Haque, R.; Sekh, M.; Kibria, G.; Haidar, S. Improvement of Surface Quality of Ti-6Al-4VAlloy by Powder Mixed Electrical Discharge Machining Using Copper Powder. *Facta Univ. Ser. Mech. Eng.* **2021**. [CrossRef]
26. Taha, H.A. Operation research an introduction. In *Decision Analysis and Games*, 7th ed.; Pearson Education: New York, NY, USA, 2002; pp. 503–511.
27. Das, S.; Chattopadhyay, A. Application of the analytic hierarchy process for estimating the state of tool wear. *Int. J. Mach. Tools Manuf.* **2003**, *43*, 1–6. [CrossRef]

micromachines

MDPI

Article

Impact of Magnetic Field Environment on the EDM Performance of Al-SiC Metal Matrix Composite

Timur Rizovich Ablyaz [1,*], Preetkanwal Singh Bains [2], Sarabjeet Singh Sidhu [3], Karim Ravilevich Muratov [1] and Evgeny Sergeevich Shlykov [1]

1 Mechanical Engineering Faculty, Perm National Research Polytechnic University, 614000 Perm, Russia; karimur_80@mail.ru (K.R.M.); kruspert@mail.ru (E.S.S.)
2 Mechanical Engineering Department, IKG Punjab Technical University, Kapurthala 144603, India; preetbains84@gmail.com
3 Mechanical Engineering Department, Beant College of Engineering and Technology, Gurdaspur 143521, India; sarabjeetsidhu@yahoo.com
* Correspondence: lowrider11-13-11@mail.ru

Abstract: In the present work, a hybrid magnetic field assisted powder mixed electrical discharge machining had been carried out on the Aluminum-Silicon Carbide (Al-SiC) metal matrix composite. The aim of the study was to obtain higher surface finish, and enhanced material removal rate. The dielectric mediums employed were plain EDM oil, SiCp mixed and graphite powder mixed EDM oil for flushing through the tube electrode. The magnetic field intensity, discharge current, T-on/off duration and type of dielectric were the control variables used for present investigation. From the results, it was observed that the machining variables for instance, discharge current, T-on/off duration and type of dielectric conditions remarkably affected the material removal rate, micro-hardness and surface roughness of the machined composite material. The MRR augmented considerably with an increase in the magnetic field intensity along with peak current. Subsequently, the composite with lesser vol.% of SiC particulates witnessed sharp rise in MRR in maximum magnetic field environment (0.66T). In addition, quality of the machined surface improved significantly in graphite powder mixed dielectric flushing condition with intermediate external magnetic field environment. Besides, an enhancement of micro-hardness was quantified as compared to base material due to the transfer of the material (SiCp) during powder mixed ED machining.

Keywords: electrical discharge machining; metal matrix composite; Taguchi; micro-hardness; surface roughness; material removal rate

Citation: Ablyaz, T.R.; Bains, P.S.; Sidhu, S.S.; Muratov, K.R.; Shlykov, E.S. Impact of Magnetic Field Environment on the EDM Performance of Al-SiC Metal Matrix Composite. *Micromachines* **2021**, *12*, 469. https://doi.org/10.3390/mi12050469

Academic Editor: Irene Fassi

Received: 5 April 2021
Accepted: 19 April 2021
Published: 21 April 2021

1. Introduction

Metal Matrix Composites (MMCs) are the lightweight materials possessing enormous exceptional mechanical and physical properties such as high specific strength, stiffness and wear resistance [1–3]. SiC particles reinforced aluminum matrix (Al-SiC) is one of the categories of MMCs finding huge applications in the aerospace and automotive sectors and various other industries. The machining of difficult-to-cut materials such as ceramics, super alloys and composites has been a major challenge for manufacturing industries in today's era. New and advanced machining techniques as "non-conventional machining methods" have been the breakthrough to curb the engineering challenges posed by the rapid growth in the development of such materials. These days, these non-conventional machining processes are more commonly employed in the manufacturing for machining such hard-to-process materials class. One such machining method- EDM, is being widely used for machining all kinds of metallic materials to the desired size and shape accurately regardless of their mechanical properties [4]. It is a thermoelectric technique wherein, material removal and desired surface roughness from the workpiece is achieved by the spark generated repeatedly between tool and workpiece submerged in dielectric. Indeed, the selection of suitable machining parameters is of utmost important to achieve optimal machining responses [5,6]. For instance, as reported [7] surface roughness depended capillary imbibitions phenomenon is very important parameters in many applications of petroleum engineering, textile industries or fuel cell etc. Thus, these desired roughness can easily controlled with the suitable tuning of EDM spark energy parameters. Furthermore, aerospace related prototypes and products, automobile industry, electronics, medical and surgical components are the few prominent fields where applications of this manufacturing technology can be witnessed [8]. Additionally, desired surface modification or topology of numerous parts in medical, aeronautical and aerospace industry has been achieved using the optimized EDM process [9].

However, the reliability and safety of conventional ED machined components is being questioned repeatedly. The surface cracking of the machined components owing to the development of recast layer (also known as white layer) and heat affected zone underneath it has been the prominent issue. The hybrid machining (mixing of two or more conventional machining processes) has proved to be a key future technology to solve this issue. Owing to this, it's never been an uphill task to achieve desired surface integrity as complex shapes and precision with the advent of hybrid EDM. The hybrid EDM is a mechanism to enhance the machining characteristics of a conventional EDM by combining various individual processes together to achieve desired surface integrity and stability of process and to overcome the limitations of the individual constituents [10,11]. From the research studies [12], it is witnessed that the magnetic field incorporation with conventional EDM technique significantly improves the thermoelectric properties, enhancing its machining capability. In addition, various researchers have experimented with EDM to improve for its low MRR, and enhance the surface properties of workpiece by mixing various types of powders in dielectric fluid (known as powder mixed EDM), which has proved to be a major breakthrough [13–15]. Likewise, an increased MRR and micro-hardness of machined surface along with reduced roughness has been witnessed in powder mixed hybrid EDM by researchers [16]. The research community has also been successful in achieving mirror-finish surfaces with a powder-mixed dielectric [17,18].

A novel hybrid EDM methodology using the external magnetic field has showcases its process effectiveness by enhancing the effective debris expulsion from the melt pool. Some researchers and academicians have put forward the impact of magnetic field on MRR and Surface Roughness (SR) [19] experimentally in the spark machining process. Moving a step forward, Heinz et al. [20] fully investigated the possibility to employ external magnetic field to generate a 'Lorentz Force' (LF) affecting the melt pool as a resultant of mutually perpendicular electric and magnetic fields. Recently, Rouniyar and Shandilya [21] experimented with an identical hybrid EDM process assisted by external magnetic field and highlighted that a resultant Force (RF) was witnessed. They demonstrated that the RF was due to the interplay of Magnetic Field (MF) and Electric Field (EF), which as a result increased the density of electrons and enhanced MRR.

This elaborative literature study encouraged the authors to dive deep into this magnetic and electric field interference to explore and uncover its process capabilities. The review of past work recognized the noteworthy effects of magnetic field on the EDM process parameters. Therefore, we aimed the present study towards the investigation of the effects of Magnetic field on the output parameters of Al-SiC composite when coupled with EDM. The quest for more efficient machining techniques necessitated the requirement of hybridization of state-of-the-art technologies with exceptional efficiency and stability. The main motive behind this study was to analyze the performance of magnetic field environment on powder mixed assisted EDM operation on machining characteristics for instance, MRR, MH, and SR of Al-SiC metal matrix composite. The prominent process parameters like peak current, duration of T-on/off, and magnetic field intensity were selected to analyze their effects on process performance.

2. Experimentation

In this study, aluminum matrix-based composite (Table 1) variants reinforced with silicon carbide (SiCp, electronic grade manufactured by CPS Technologies, Norton, MA, USA) were used as workpiece. The distribution of SiC (yellow arrows) in alumnium matrix is presented in Figure 1.

Table 1. MMCs properties (Source: CPS Tech, Norton, MA, USA).

Property	W/P-1	W/P-2	W/P-3
Aluminum Alloy (%)	63	45	37
SiC (%)	37	55	63
Thermal conductivity (W/mk)	170	190	190
Density (g/cc)	2.89	2.96	3.01
Specific Heat (J/gK) at 25 °C	0.808	0.786	0.741
Young's Modulus (GPa)	167	167	188

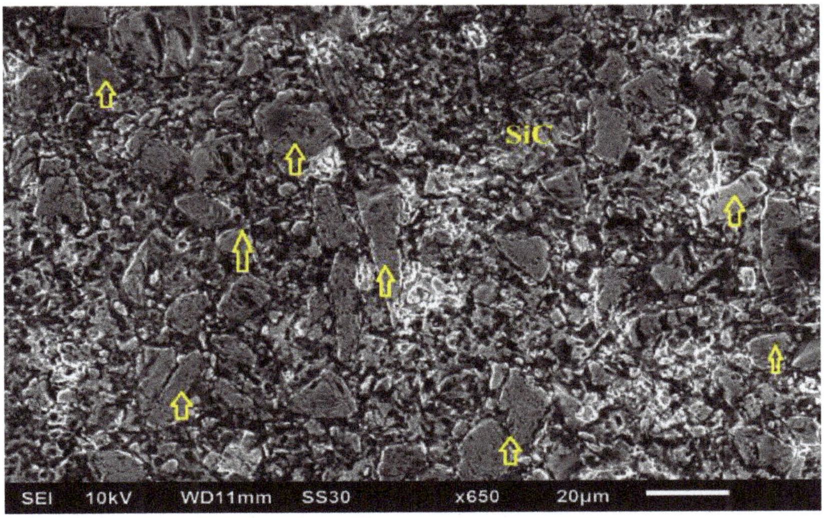

Figure 1. SEM (Make: JSM-6610LV Joel, Tokyo, Japan) of un-machined surface of Al-SiC workpiece at X650 (SiC particulates are highlighted with yellow color arrows).

The electrolytic copper tube electrode (Figure 2) having 15 mm outer and 4 mm inner diameter was used to allow the powder mixed dielectric to flow through the tool. The experimentation was carried out employing the ZNC EDM (Make: OSCARMAX, Taichung City, Taiwan) (Figure 3a). As the past studies witnessed the application of magnetic field enhanced the plasma ionization in EDM spark zone and controlled its expansion, same principle was implemented in present study. Two sets of permanent magnets (0.33 T each) were fixed in such a manner so as to achieve desired magnetic field effect into the machining zone, as shown in Figure 3b. Commercial grade EDM oil (specific gravity = 0.763, freezing point = 94 °C) was used as the dielectric medium. Figure 3c presents the arrangement of dielectric flow through the hollow electrode. Figure 3d shows the machined zone on the Al-SiC workpiece. The merits of powder mixed EDM have been witnessed from literature. Hence, the SiC (220 mesh) abrasive particles and graphite particulates (400 mesh) were used in 30 g/L for circulating through tool while machining. All the experimentation trials were conducted with magnetic field intensity, current, duration of pulse, volume percentage of SiC and type of dielectric medium (Table 2) as prominent parameters. The material removal rate, average surface roughness (Ra), and micro-hardness were evaluated as the response outcomes in this study. The surface roughness value was recorded with surface roughness tester (Surftest SJ-400, Mitutoyo America Corporation, Aurora, IL, USA) at three distinct positions of the machined surface and mean was considered.

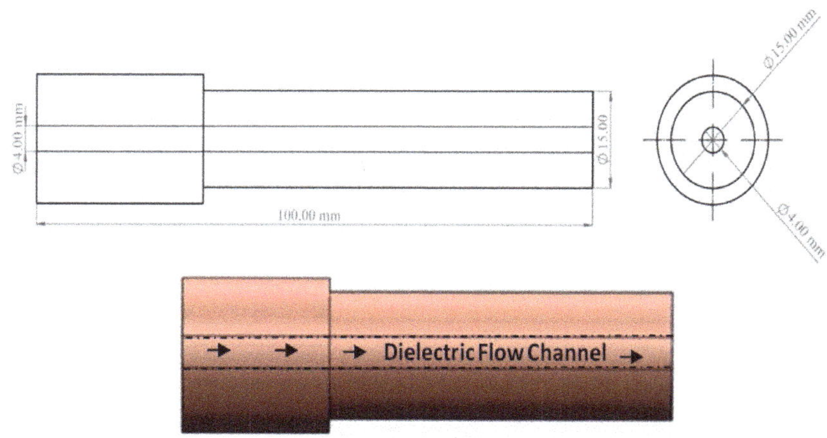

Figure 2. Copper electrode with inner hole for dielectric flushing.

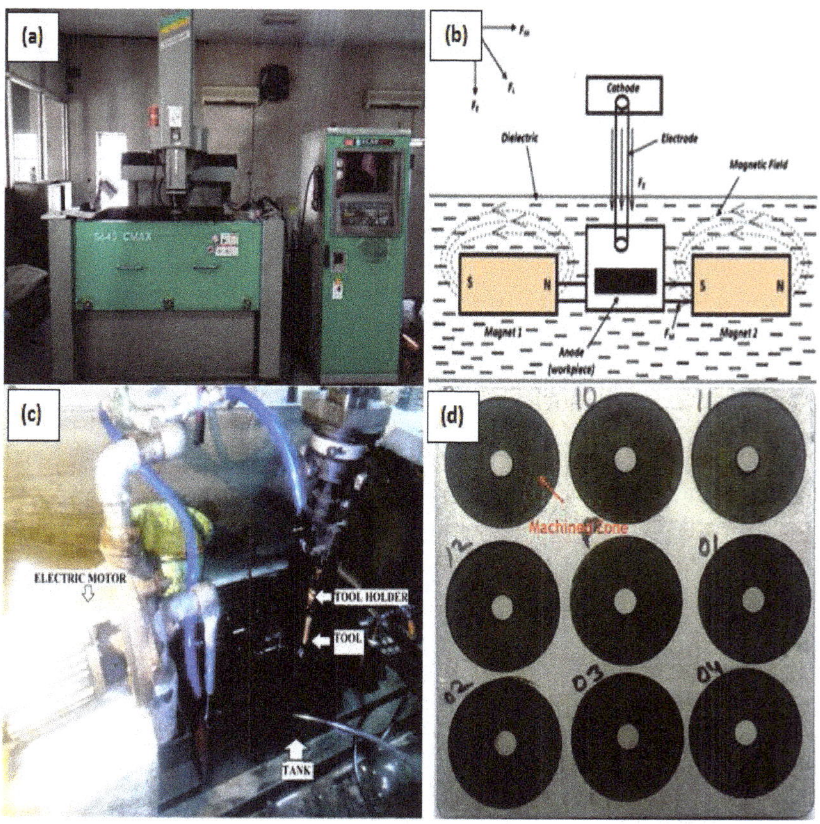

Figure 3. (**a**) EDM used for experimentation (**b**) Schematic diagram of MFAPEDM set-up and (**c**) Powder mixed dielectric set-up (**d**) machined Al-SiC workpiece.

Table 2. Input parameters and their levels.

Variables/Notations	Level		
	1	2	3
Current (A)/I	4	10	16
Pulse-on (μs)/T-on	30	45	90
Pulse-off (μs)/T-off	30	45	90
Magnetic field intensity (T)/B	0	0.33	0.66
Dielectric medium	Plain dielectric	SiC mixed (220 mesh)	Graphite mixed (400 mesh)
Workpiece (W/P)	Al-37% SiC (W/P-1)	Al-55% SiC (W/P-2)	Al-63% SiC (W/P-3)

To achieve accuracy in output responses, two replications were performed at random order to find out the mean. The Taguchi's experimental design matrix [22] assisted the authors to identify and scrutinize the prominent parameters for this study. The orthogonal array of Taguchi's experimental design reduces number of experimental trials to measure the effect of parameters included in the study. Valid conclusions during ED machining of Al-SiC MMCs were drawn and the factor assignment was done using Minitab-17 software.

The selection of Taguchi's orthogonal array depends on the number of factors (i.e., process parameters, herein 6 factors) and interactions of interest (herein, 1 interaction) and the number of levels of process parameters (i.e., 3 levels, refer Table 2). The total degree of freedom (fa) for each factor is the number of levels (la) minus one i.e., $f_a = l_a - 1$ (i.e., 2 for each factor) and the degree of freedom for interaction is the product of interacting factors degrees of freedom i.e., $f_{aXb} = (f_a)(f_b)$(Herein $2 * 2$). The minimum required degree of freedom in the experimental design is the sum of the entire factors and the interaction's degree of freedom. Thus, the degree of freedom selected array (f_{OA}) has must satisfy the inequality $f_{OA} \geq (f_{aXb} + f_a)$. Thus L27 orthogonal array was selected for the present study.

For the critical discussion on MRR, MH and SR of the powder mixed electrical discharge machining process; the results of the investigations are represented graphically. The various outcomes were analyzed through ANOVA to test the significance of model adopted. Prior to experimentation, the workpiece was designated as negative and tool with positive polarity. A precision electronic balance (Citizen, CY220, Mumbai, India) was used to measure workpiece weights before and after the machining. The investigation consisted of 27 distinct trials that helped to investigate the material removed, micro-hardness and surface quality of machined Al-SiC metal matrix composite using various input variables as tabulated in Table 3. This corresponding table constituted the input variables opted as a part of standard L27 orthogonal array control log and recorded values of responses for individual machined surface. The observed values, with and without magnetic field were recorded in designated columns as MRR, MH and SR output.

Table 3. Experimental Factors and Results.

	Process Parameters						Responses		
Trials	Magnetic Field (T)	Current (A)	T-on (μs)	T-off (μs)	Flushing Type	W/P	MRR (mg/min)	MH (HV)	SR (μm)
1	0	4	30	30	1	1	28.815	209.3	0.21
2	0	4	45	45	3	2	21.320	306.6	0.19
3	0	4	90	90	2	3	19.739	353.9	0.39
4	0	10	30	45	3	3	16.012	309.0	0.21
5	0	10	45	90	2	1	24.584	289.8	0.67
6	0	10	90	30	1	2	18.542	389.0	0.79
7	0	16	30	90	2	2	36.255	225.0	0.18
8	0	16	45	30	1	3	28.400	360.9	0.39
9	0	16	90	45	3	1	31.128	262.9	0.12
10	0.33	4	30	45	2	2	23.620	202.6	0.22

Table 3. *Cont.*

	Process Parameters						Responses		
Trials	Magnetic Field (T)	Current (A)	T-on (μs)	T-off (μs)	Flushing Type	W/P	MRR (mg/min)	MH (HV)	SR (μm)
11	0.33	4	45	90	1	3	15.683	241.8	0.19
12	0.33	4	90	30	3	1	28.631	245.0	0.22
13	0.33	10	30	90	1	1	30.523	134.2	0.13
14	0.33	10	45	30	3	2	20.675	290.7	0.21
15	0.33	10	90	45	2	3	29.850	302.8	0.31
16	0.33	16	30	30	3	3	29.518	316.2	0.20
17	0.33	16	45	45	2	1	43.520	262.9	0.54
18	0.33	16	90	90	1	2	41.663	202.6	0.32
19	0.66	4	30	90	3	3	31.231	241.8	0.22
20	0.66	4	45	30	2	1	53.524	245.0	0.92
21	0.66	4	90	45	1	2	30.575	234.2	0.14
22	0.66	10	30	30	2	2	49.850	290.7	1.95
23	0.66	10	45	45	1	3	39.411	302.8	0.30
24	0.66	10	90	90	3	1	53.620	216.2	0.36
25	0.66	16	30	45	1	1	51.653	202.9	1.31
26	0.66	16	45	90	3	2	38.661	262.6	0.55
27	0.66	16	90	30	2	3	29.573	341.8	0.80

3. Results and Discussion

3.1. *Influence on the Material Removal Rate (MRR)*

The experimental design matrix and results obtained, thereafter, are demonstrated in the Table 3. It is clear from Table 4 and Figure 4 that MRR is a function of current, duration of T-on/off and externally applied magnetic field. It has been exhibited that the magnetic field intensity was the most significant input parameter that affected the MRR steeply, followed by peak current and type of workpiece. Duration of pulse and dielectric medium, on the contrary, did not turn to be the significant parameters to influence MRR. At high discharge current, the enhanced MRR can be attributed to an improved volume of erosion in the maximum magnetic field intensity due to the high spark energy. The workpiece variant 1 (37 %vol. fraction of SiC) with comparatively lesser vol.% of SiC particulates resulted in enhanced MRR. This is attributed to the increased conductivity of workpiece owing to decreased vol.% of SiC particulates resulting in abrupt material removal. Furthermore, Figure 5 represented the surface morphology of work pieces of Al-37% SiC (Figure 5a) and Al-63% SiC (Figure 5b) after electrical discharge treatment. Figure 5a morphology witnesses the ablation as a mechanism of material removal of machined surface, whereas the Figure 5b indicates the melting and re-solidification of molten workpiece. The dense vol.% of SiC in the matrix exhibits shielding effect while machining and diminishes the MRR.

Table 4. ANalysis of VAriance (ANOVA) for Material Removal Rate (MRR).

Scheme	DF	Seq SS	Adj SS	Adj MS	F-Value	*p*-Value
Magnetic Field (T)	2	1411.29	1411.29	705.646	12.95	0.001 **
Current (A)	2	336.98	336.98	168.488	3.09	0.077 *
T-on (μs)	2	12.71	12.71	6.357	0.12	0.891
T-off (μs)	2	1.61	1.61	0.806	0.01	0.985
Flushing Type	2	89.80	89.80	44.898	0.82	0.459
W/P	2	640.96	640.96	320.480	5.88	0.014 *
Residual Error	14	762.99	762.99	54.500		
Total	26	3256.35				

** Most significant, * Significant

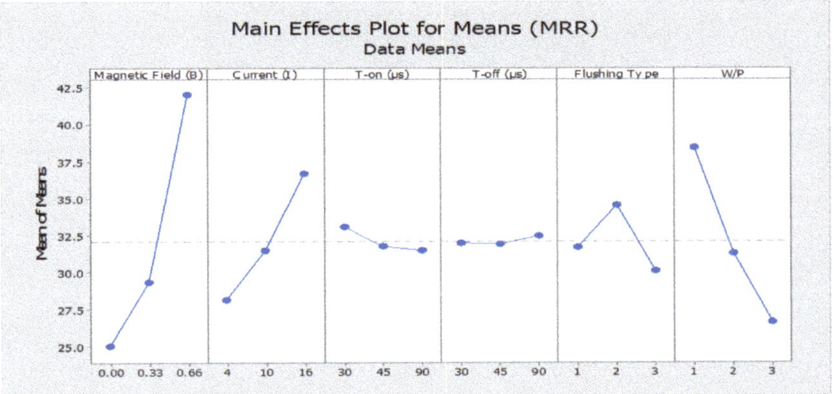

Figure 4. Main effect plot for material removal rate (mg/min).

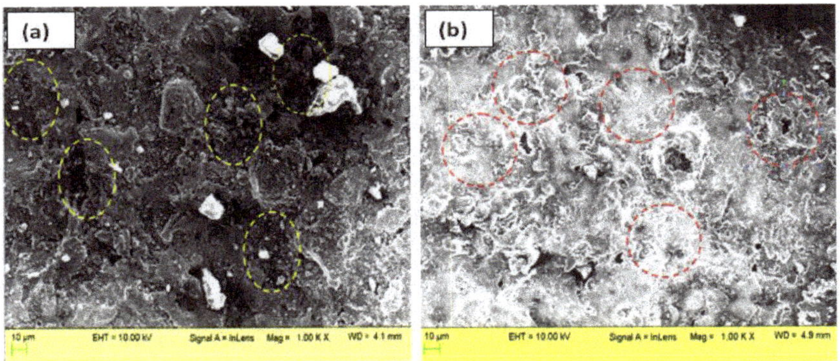

Figure 5. Scanning electron microscope (SEM) analysis of workpiece after machining (**a**) Al-37% SiC shows ablation mechanism of material removal; (**b**) Al-63% SiC shows melting as the dominating material removal mechanism.

Moreover, the maximum material removal was witnessed in Trial 24, wherein 118% increase in MRR was observed in graphite powder mixed dielectric flushing in extreme magnetic field intensity compared to similar parametric conditions in Trial 5. The probable reason for increase in MRR could be the effective molten debris flushing on the machined surface due to the generation of Lorentz forces by the interplay of magnetic as well as electrical fields which enhanced the plasma pressure. Apart from this, injecting the powder mixed dielectric through the hole in the tube electrode enhanced its pressure, augmented the debris flushing. The interaction between different variables was not evident to be significant while powder mixed magnetic field assisted EDM.

3.2. *Influence on the Micro-Hardness (MH)*

The effect of various machining parameters and the surface conditions of the machined workpiece were analyzed by measurement of the indentation hardness. The micro-hardness measurements were carried out at the different locations of the machined surface as well as parent material, cross-sectionally. The machining parameters like magnetic field intensity, type of workpiece and pulse-on/off duration posed considerable effect on MH values, as shown in Table 5, except the discharge current and type of flushing which were the in-significant factors affecting the surface hardness of ED machined MMCs. It is clear from the graph (Figure 6) that machining of W/P-3 without magnetic field effect and in high-end pulse-on time resulted in the enhanced micro-hardness of the machined surface at minimum pause time. This is due to the reason that the higher concentration of SiC particulates resulted in oxides formation at high frequency (i.e., minimum pulse-off time) of sparks generated while machining.

Table 5. ANalysisof VAriance (ANOVA) for Micro-Hardness (MH).

Source	DF	Seq SS	Adj SS	Adj MS	F-Value	*p*-Value
Magnetic Field (T)	2	15,287	15,287	7643.6	10.91	0.001 *
Current (A)	2	3426	3426	1713.0	2.44	0.123
T-on (μs)	2	13,332	13,332	6665.9	9.51	0.002 *
T-off (μs)	2	15,191	15,191	7595.3	10.84	0.001 *
Flushing Type	2	3338	3338	1669.2	2.38	0.129
W/P	2	27,458	27,458	13,729.2	19.59	0.000 **
Residual Error	14	9811	9811	700.8		
Total	26	87,844				

** Most significant, * Significant.

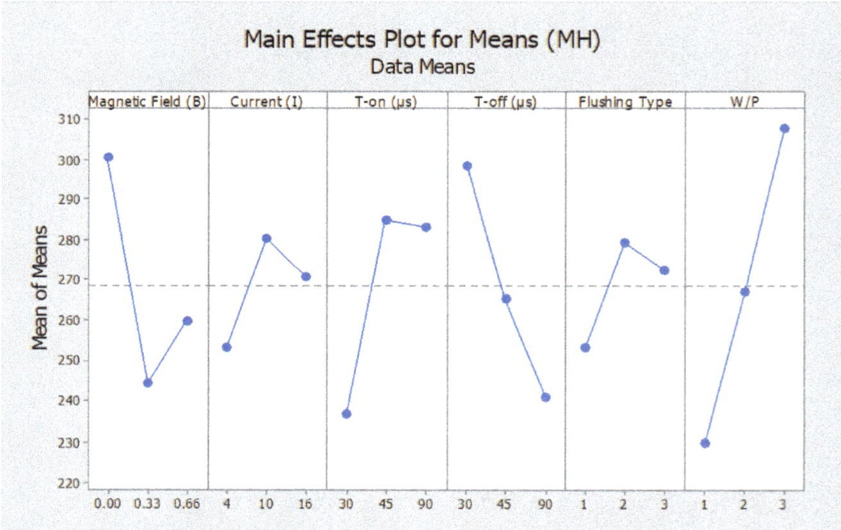

Figure 6. Main Effects plot for means for Microhardness (HV).

The machined surface was further analyzed using X-ray diffraction method on PANalytical's X'PertPro MPD (Netherlands). The X-ray spectra of the workpiece was characterized using Cu-Kα radiations (λ = 1.5406 A^0) with the generator setting of 40 mA and 45 kV. The X-ray diffractograms of the machined surfaces (Trial 8 and Trial 16) are shown in Figure 7. The peak pattern shows the formation of oxides such as Al_2SiO_5, SiO_2, Si_3W_5 on the surface of machined MMCs, thus resulted in the enhanced hardness of machined surface [23]. On the contrary, the reduced surface hardness could be advantageous, wherein post-EDM additional processing such as grinding, grit blasting, etching etc., could be avoided [24]. Moreover, without magnetic field interference and in high current, more material deposition occurred that led to the enhancement in the micro-hardness (613.6%) as recorded (360.9 HV in Trial 8 as compared to 50.57 HV of parent workpiece) and 19.18% high as observed in Trial 23 (302.8 HV). There was no confinement of plasma and therefore more molten material deposited on the machined surface in the absence of magnetic field which was in line with previous studies related to magnetic field application. However, the interplay between various parameters and magnetic field were not significant enough affecting the output of the experimentation.

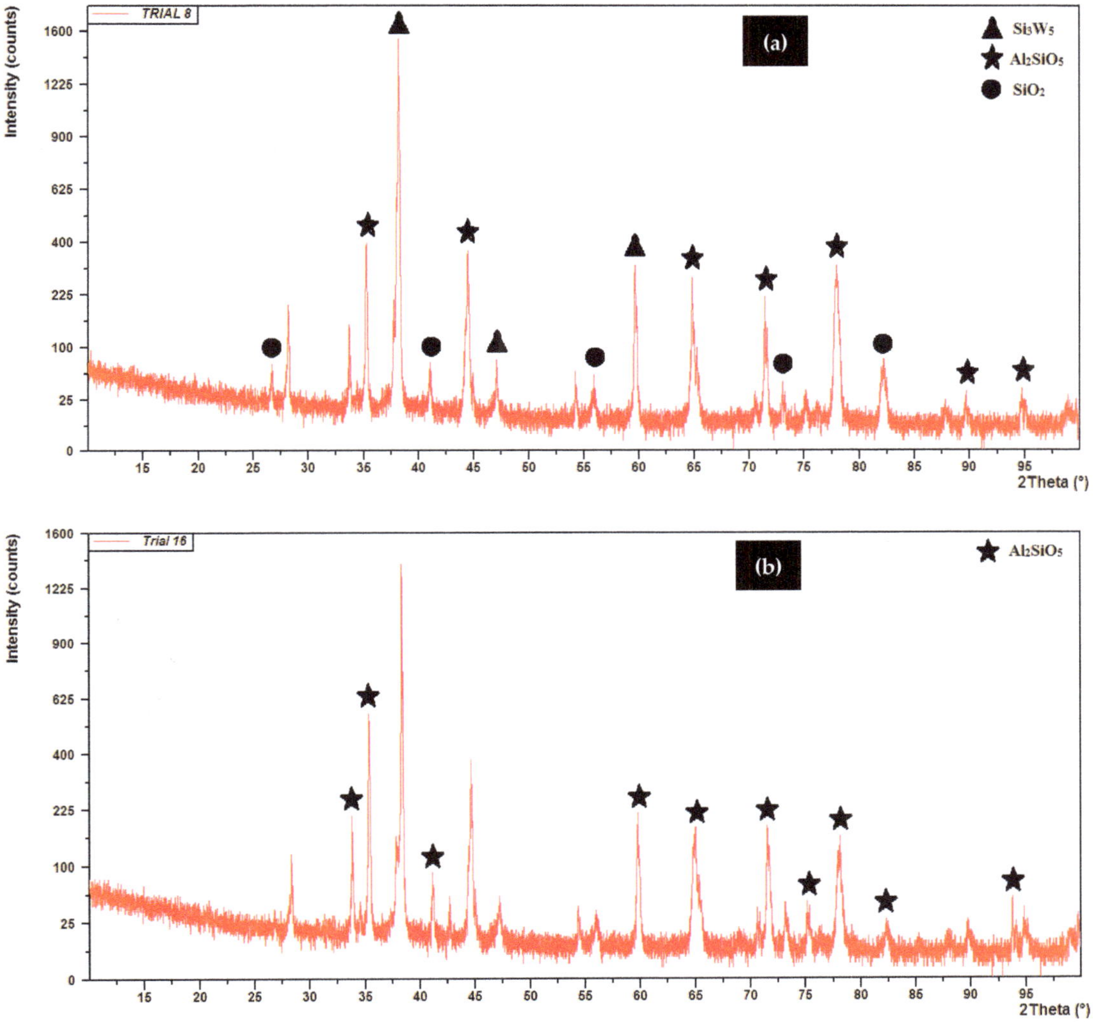

Figure 7. X-ray diffraction of (**a**) Trial 8 and (**b**) Trial 16.

3.3. Influence on the Surface Roughness (SR)

The surface morphology of machined surface portrays a true signature of the tool-workpiece interaction while machining and material removal mechanism. The surface roughness was significantly influenced by magnetic field intensity and type of dielectric medium. The relations of SR with various process parameters are shown in Table 6 and Figure 8. The average surface roughness value for graphite powder mixed dielectric was witnessed as the lowest value. Alternatively, a rougher surface is achieved when machining is executed in 0.66 T magnetic field influence in SiC powder mixed dielectric medium. This can be related to the fact that at high magnetic field intensity abrupt material removal took place that turned the surface rough. Comparatively, a smoother machined surface is achieved when EDM is carried out in magnetic field intensity of 0.33 T. In this scenario,

Lorentz forces developed while machining for little less duration comparatively with graphite kept the machining zone smoother and decreased the roughness.

Table 6. ANalysisof VAriance (ANOVA) for Surface Roughness (SR).

Source	DF	Seq SS	Adj SS	Adj MS	F-Value	p-Value
Magnetic Field (T)	2	1.10890	1.10890	0.55445	8.30	0.008 *
Current (A)	2	0.30250	0.30250	0.15125	2.26	0.154
T-on (μs)	2	0.07783	0.07783	0.03891	0.58	0.576
T-off (μs)	2	0.47459	0.47459	0.23729	3.55	0.068 *
Flushing Type	2	0.76963	0.76963	0.38481	5.76	0.022 *
W/P	2	0.16805	0.16805	0.08403	1.26	0.326
B x T-on	4	0.92913	0.92913	0.23228	3.48	0.050 *
Residual Error	10	0.66784	0.66784	0.06678		
Total	26	4.49845				

* Significant.

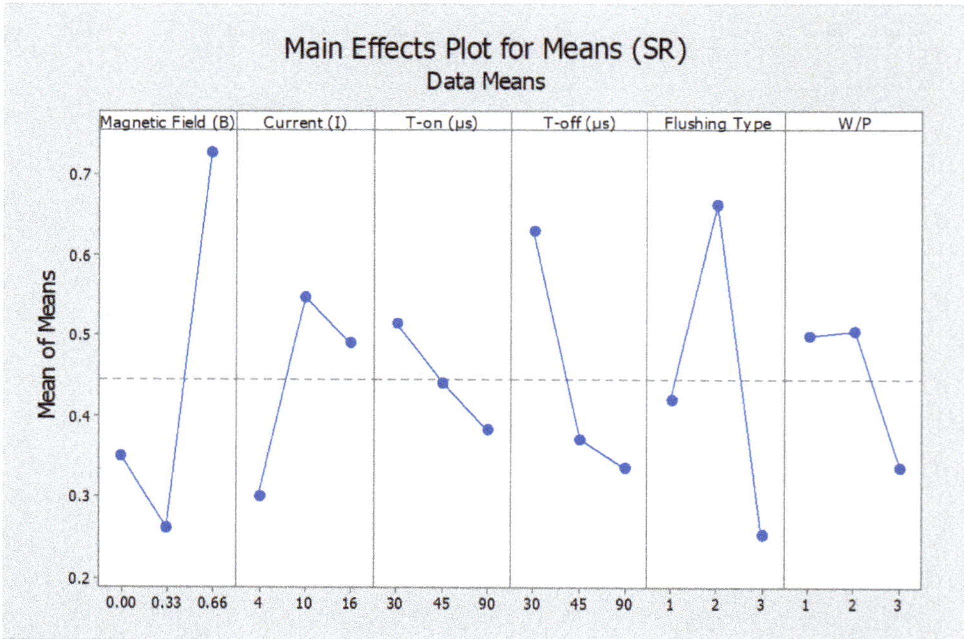

Figure 8. Main effects plot for surface roughness (μm).

The interaction between various process parameters was statistically analyzed at 95% confidence level (i.e., $p \leq 0.05$). It was observed that the interaction between magnetic field intensity (B) and T-on significantly affected the surface roughness. From Figure 9, it is depicted that the surface roughness enhanced at high magnetic field intensity and low T-on value. From the SEM analysis (Figure 10a), it can be observed that non-uniform surface with significant undulations is visible prominently during machining in SiC mixed dielectric medium in comparison to graphite mixed dielectric (Figure 10b). The high thermal conductivity of graphite particulates is responsible for distributing and dissipating uniform heat to the workpiece surfaces, thereby limiting the size of the craters thus produced. Owing to the externally applied magnetic field in powder mixed EDM process, the graphite powder particles under the effect of Lorentz forces crowded together to bridge the machining void in the dielectric fluid. Multiple plasma discharges are developed, as a result, from a unit pulse input avoiding deeper craters, resulting in improved surface quality [25]. Moreover, the generation of Lorentz forces with interference of external magnetic field into the EDM current density assisted in channelized path of current carriers from the tool to workpiece.

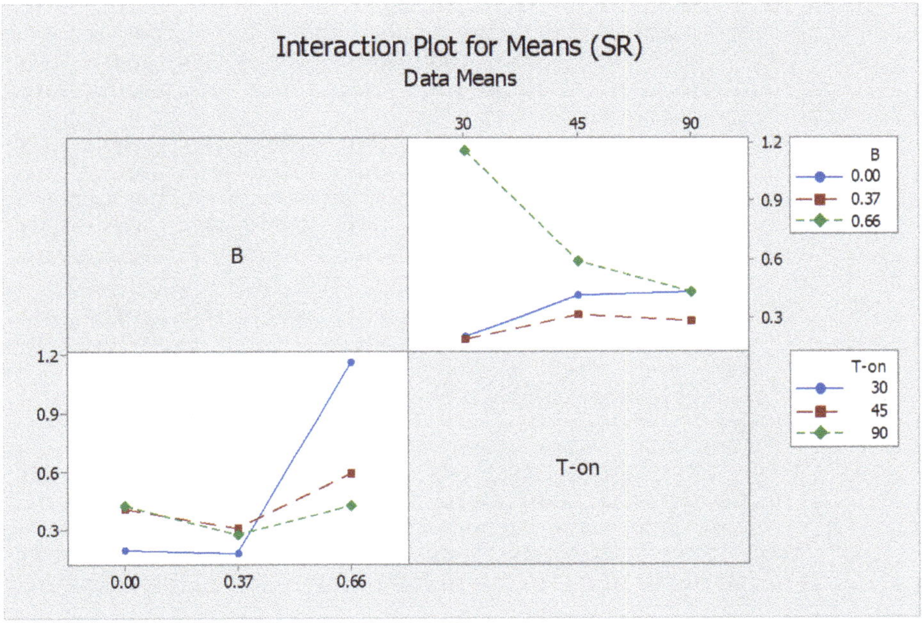

Figure 9. Interaction Plot between Magnetic field intensity (B) and T-on.

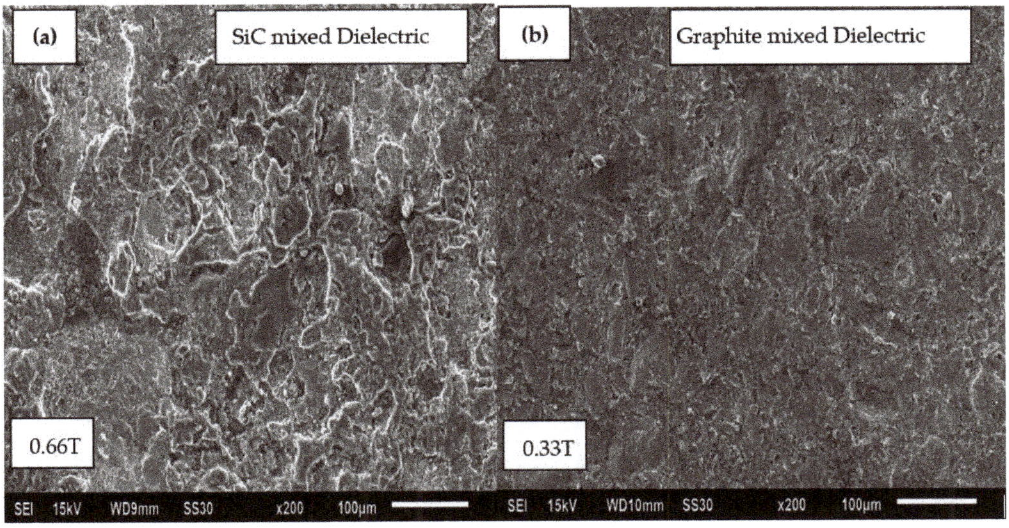

Figure 10. SEM analysis for comparative surface texture of various machined samples. (**a**) SiC mixed dielectric at 0.66T (**b**) Graphite mixed dielectric 0.33T.

4. Conclusions

The study was carried out on EDM of aluminum composite with plain and powder mixed flushing conditions. The following observations are enlisted.

- The MRR is significantly affected by the machining parameters such as magnetic field environment, peak current and SiC% content of workpiece.
- The removal rate increased significantly with the incorporation of magnetic field intensity along with peak current.
- It is also evident that the decreased vol.% of SiC particulates led to a sharp rise in MRR. A 118% increase in MRR under the influence of magnetic field was observed in plain dielectric flushing when compared to identical parametric conditions in trials without magnetic field.
- An enhancement (613.6%) in the micro-hardness was witnessed due to the transfer of materials and formation of new phases while ED machining.
- The surface finish of machined MMCs was greatly affected by magnetic field intensity as well as type of dielectric. The surface finish improved steeply in graphite powder mixed dielectric flushing conditions at intermediate (0.33 T) magnetic field.

Future Scope: The literature studies depicted that aluminum matrix composites have been the major choice in the field of EDM of MMCs, but no significant work is available for the composite with other matrix phase such cobalt, steel etc. In addition to this, the experimental research can be extended and analyzed using various other levels of magnetic field of permanent magnets as well as rotating electromagnets.

Author Contributions: Conceptualization, P.S.B. and S.S.S.; methodology, P.S.B., and S.S.S.; software, P.S.B., S.S.S. and T.R.A.; validation, P.S.B., S.S.S. and T.R.A.; formal analysis, K.R.M., E.S.S.; investigation, P.S.B., S.S.S. and T.R.A.; validation, P.S.B., S.S.S.; resources, K.R.M., E.S.S.; data curation, K.R.M., E.S.S.; writing—original draft preparation, P.S.B. and T.R.A.; writing—review and editing, P.S.B., S.S.S. and T.R.A.; visualization, T.R.A.; supervision, S.S.S.; project administration, T.R.A.; funding acquisition, T.R.A. and K.R.M. All authors have read and agreed to the published version of the manuscript.

Funding: This work was supported by Russian Science Foundation, grant number 20-79-00048.

Institutional Review Board Statement: Not applicable.

Informed Consent Statement: Not applicable.

Data Availability Statement: Not applicable.

Conflicts of Interest: The authors declare no conflict of interest. The funders had no role in the design of the study; in the collection, analyses, or interpretation of data; in the writing of the manuscript, or in the decision to publish the results.

References

1. Bains, P.S.; Sidhu, S.S.; Payal, H.S. Fabrication and Machining of Metal Matrix Composites: A Review. *Mater. Manuf. Process.* **2016**, *31*, 553–573. [CrossRef]
2. Chatterjee, P.; Mondal, S.; Boral, S.; Banerjee, A.; Chakraborty, S. A novel hybrid method for non-traditional machining process selection using factor relationship and multi-attributive border approximation method. *Facta Univ. Series: Mech. Eng.* **2017**, *15*, 439–456. [CrossRef]
3. Das, S.; Acharya, U.; Rao, S.S.; Paul, S.; Roy, B.S. Assessment of the surface characteristics of aerospace grade AA6092/17.5 SiCp-T6 composite processed through EDM. *CIRP J. Manuf. Sci. Technol.* **2021**, *33*, 123–132. [CrossRef]
4. Singh, G.; Sidhu, S.S.; Bains, P.S.; Bhui, A.S. Surface evaluation of ED machined 316L stainless steel in TiO_2 nano-powder mixed dielectric medium. *Mater. Today Proc.* **2019**, *18*, 1297–1303. [CrossRef]
5. Bhui, A.S.; Singh, G.; Sidhu, S.S.; Bains, P.S. Experimental investigation of optimal ed machining parameters for Ti-6Al-4V biomaterial. *Facta Univ. Ser. Mech. Eng.* **2018**, *16*, 337–345. [CrossRef]
6. Singh, G.; Bhui, A.S.; Singh, L. On the assessment of material removal rate and surface characteristics of Ti-6Al-4V processed by WEDM. *Eng. Res. Express* **2019**, *1*, 015038. [CrossRef]
7. Xiao, B.; Huang, Q.; Chen, H.; Chen, X.; Long, G. A fractal model for capillary flow through a single tortuous capillary with roughened surfaces in fibrous porous media. *Fractals* **2021**, *29*, 2150017. [CrossRef]

8. Sidhu, S.S.; Kumar, S.; Batish, A. Electric Discharge Machining of 10 vol% Al_2O_3/Al Metal Matrix Composite - An Experimental Study. *Mater. Sci. Forum* **2013**, *751*, 9–19. [CrossRef]
9. Singh, G.; Sidhu, S.S.; Bains, P.S.; Bhui, A.S. Improving microhardness and wear resistance of 316L by TiO_2 powder mixed electro-discharge treatment. *Mater. Res. Express* **2019**, *6*, 086501. [CrossRef]
10. Zhu, Z.; Guo, D.; Xu, J.; Lin, J.; Lei, J.; Xu, B.; Wu, X.; Wang, X. Processing Characteristics of Micro Electrical Discharge Machining for Surface Modification of TiNi Shape Memory Alloys Using a TiC Powder Dielectric. *Micromachines* **2020**, *11*, 1018. [CrossRef]
11. Zhang, Y.; Wang, C.; Wang, Y.; Ni, Q.; Ji, L. Geometric Accuracy Improvement by Using Electrochemical Reaming with a Helical Tube Electrode as Post-Processing for EDM. *Materials* **2019**, *12*, 3564. [CrossRef]
12. Yan, B.-H.; Chang, G.-W.; Chang, J.-H.; Hsu, R.-T. Improving Electrical Discharge Machined Surfaces Using Magnetic Abrasive Finishing. *Mach. Sci. Technol.* **2004**, *8*, 103–118. [CrossRef]
13. Shaikh, M.B.N.; Arif, S.; Siddiqui, M.A. Fabrication and characterization of aluminium hybrid composites reinforced with fly ash and silicon carbide through powder metallurgy. *Mater. Res. Express* **2018**, *5*, 046506. [CrossRef]
14. Singh, G.; Sidhu, S.S.; Bains, P.S.; Singh, M.; Bhui, A.S. On surface Modification of Ti Alloy by Electro Discharge Coating Using Hydroxyapatite Powder Mixed Dielectric with Graphite Tool. *J. Bio-Tribo-Corrosion* **2020**, *6*, 1–11. [CrossRef]
15. Arif, S.; Alam, M.T.; Ansari, A.H.; Shaikh, M.B.N.; Siddiqui, M.A. Analysis of tribological behaviour of zirconia reinforced Al-SiC hybrid composites using statistical and artificial neural network technique. *Mater. Res. Express* **2018**, *5*, 056506. [CrossRef]
16. Singh, G.; Lamichhane, Y.; Bhui, A.S.; Sidhu, S.S.; Bains, P.S.; Mukhiya, P. Surface morphology and microhardness behavior of 316l in Hap-PMEDM. *Facta Univ. Series: Mech. Eng.* **2019**, *17*, 445–454. [CrossRef]
17. Wong, Y.; Lim, L.; Rahuman, I.; Tee, W. Near-mirror-finish phenomenon in EDM using powder-mixed dielectric. *J. Mater. Process. Technol.* **1998**, *79*, 30–40. [CrossRef]
18. Bains, P.S.; Sidhu, S.S.; Payal, H.S.; Kaur, S. Magnetic Field Influence on Surface Modifications in Powder Mixed EDM. *Silicon* **2019**, *11*, 415–423. [CrossRef]
19. Rattan, N.; Mulik, R.S. Experimental Investigations and Multi-response Optimization of Silicon Dioxide (Quartz) Machining in Magnetic Field Assisted TW-ECSM Process. *Silicon* **2017**, *9*, 663–673. [CrossRef]
20. Heinz, K.; Kapoor, S.G.; DeVor, R.E.; Surla, V. An investigation of magnetic-field-assisted material removal in micro-EDM for non-magnetic materials. *J. Manuf. Sci. Eng.* **2011**, *133*, 021002. [CrossRef]
21. Rouniyar, A.K.; Shandilya, P. Fabrication and experimental investigation of magnetic field assisted powder mixed electrical discharge machining on machining of aluminum 6061 alloy. *Proc. Inst. Mech. Eng. Part B: J. Eng. Manuf.* **2019**, *233*, 2283–2291. [CrossRef]
22. Montgomery, D.C. *Design and Analysis of Experiments*; John Wiley & Sons: Hoboken, NJ, USA, 2017.
23. Sidhu, S.S.; Batish, A.; Kumar, S. Study of Surface Properties in Particulate-Reinforced Metal Matrix Composites (MMCs) Using Powder-Mixed Electrical Discharge Machining (EDM). *Mater. Manuf. Process.* **2014**, *29*, 46–52. [CrossRef]
24. Holmberg, J.; Wretland, A.; Berglund, J. Grit Blasting for Removal of Recast Layer from EDM Process on Inconel 718 Shaft: An Evaluation of Surface Integrity. *J. Mater. Eng. Perform.* **2016**, *25*, 5540–5550. [CrossRef]
25. Hu, F.Q.; Cao, F.Y.; Song, B.Y.; Hou, P.J.; Zhang, Y.; Chen, K.; Wei, J.Q. Surface properties of SiCp/Al composite by pow-der-mixed EDM. *Procedia CIRP* **2013**, *6*, 101–106. [CrossRef]

micromachines

Article

Micro Electrical Discharge Machining of Ultrafine Particle Type Tungsten Carbide Using Dielectrics Mixed with Various Powders

Sai Dutta Gattu and Jiwang Yan *

Faculty of Science and Technology, Keio University, Yokohama 223-8522, Japan; gattu.46@keio.jp
* Correspondence: yan@mech.keio.ac.jp; Tel.: +81-45-566-1445

Abstract: Electrical discharge machining (EDM) is widely used to machine hard materials, such as tungsten carbide; however, the machining rate and surface quality are low. In this research, the effects of mixing electrically conductive carbon nanofiber (CnF), semiconductive silicon (Si) powder, and insulative alumina powder (Al_2O_3) at different concentrations in a dielectric fluid were studied by observing single discharge craters and hole machining performance in the EDM of ultrafine particle type tungsten carbide. Craters obtained using carbon nanofiber and alumina were much smaller than in oil-only conditions. In contrast, The results show that adding CnF significantly improved the material removal rate under all conditions. Si and Al_2O_3 powders only improved the machining performance at a high discharge energy of 110 V. Furthermore, improvement in surface roughness was observed prominently at high voltages for all the powders. Among the three powders, alumina was found to improve the surface roughness the most.

Keywords: micro-EDM; tungsten carbide; carbon nanofiber; silicon; alumina

Citation: Gattu, S.D.; Yan, J. Micro Electrical Discharge Machining of Ultrafine Particle Type Tungsten Carbide Using Dielectrics Mixed with Various Powders. *Micromachines* **2022**, *13*, 998. https://doi.org/10.3390/mi13070998

Academic Editors: Irene Fassi and Francesco Modica

Received: 29 May 2022
Accepted: 23 June 2022
Published: 25 June 2022

1. Introduction

Tungsten carbide (WC) and its composites (WC-Co) possess excellent physical properties, such as high melting and boiling points, wear, and corrosion resistance. As a result, it is useful in various industrial applications, notably in die/mold making, cutting, and surgical tools. Consequently, various studies have focused on machining tungsten carbide economically with high precision. Conventionally, tungsten carbide is machined using turning and grinding [1–3] using PCD and CBN tools. However, the tool wear is high, increasing costs [4]. In recent years, there has been a growing trend of using non-conventional methods, such as electric discharge machining (EDM) [5], electrochemical machining (ECM) [6], and laser machining [7], to process tungsten carbide products.

EDM is a thermo-electric process that removes material by thermal erosion due to repeated high-energy electrical sparks. Since there is no physical contact between the workpiece and electrode in this process, the machinability depends only on the tool and the workpiece's electrical and thermal properties. Thus, this process can be used to machine any given conductive material regardless of its hardness. In addition, this method can also be used to directly transfer complex 3D shapes with high precision onto the workpiece without force-induced deformation.

Consequently, various studies have been performed on the EDM of tungsten carbide. Studies by Hourmand et al. [8] and Assarzadeh et al. [9] have used analytical models, such as response surface methodology and ANOVA analysis, to determine the optimum parameter values of peak current, spark time on and off, and gap voltage. Research on the effect of EDM on hardness and sub-surface damage on tungsten carbide has also been performed [10]. Furthermore, Jahan et al. [11] have studied the effect of different electrodes on EDM efficiency by comparing the material removal and electrode wear rate. However, it has been found that the EDM of WC-Co has low machining efficiency, with a high tool wear rate and surface defects. Recently, using tool rotation or vibration [12,13], servo movement and flushing strategies [14,15], and adding the powder to the EDM of WC-Co, have been studied to improve these shortcomings.

The dielectric separating the electrodes serves many functions during the EDM process. In addition to insulation, it helps transport debris away and acts as a coolant during the discharge phenomenon [16]. It has been studied that adding the powder to the dielectric during the EDM process can alter the machining process by dispersing the discharges and thus, improving machining efficiency [17]. Various studies have shown that powder addition improved material removal rate, electrode wear rate, and surface roughness. Furthermore, a few studies also focused on optimizing the powder-mixed EDM (PMEDM) of WC-Co using analytical models [18,19]. Jahan et al. [20] have also developed a mathematical model to study powder behavior during EDM. Their model underlines that the presence of any additive within the dielectric results in an increase in the discharge gap.

The most commonly used powders in EDM are aluminum (Al) and graphite (Gr). Their high electrical conductivity, low density, availability of particles on a nanometer scale, and low hazardous nature allow them to be widely used. However, in recent years, the effects of many other particles, such as silicon (Si), graphene, molybdenum sulfide (MoS_2), and alumina (Al_2O_3) have also been studied. Pecas et al. [20] have shown that under a few conditions (powder concentration of 2 g/L for 1 mm^2 electrode), silicon powder addition can generate a mirror surface finish on AISI H13 steel. Liew et al. [21] have used carbon nanofiber-assisted EDM of reaction-bonded silicon carbide and have significantly improved material removal and electrode wear rates. Sahu et al. [22] and Jahan et al. [23] have used nano-alumina powder in the EDM of Inconel and tungsten carbide, respectively. They have shown an increase in the material removal rate and a decrease in the surface roughness for specific parametric values.

Many studies have been performed on the EDM of tungsten carbide. However, up to date, the effects of using these newly attempted types of powders on the EDM of WC-Co, especially the ultrafine particle type WC-Co, which is a more promising material in the industry, are still unknown. In particular, the effect of powders in the micro-EDM process, where the discharge gap is much smaller than in conventional EDM is unclear. To evaluate the impact of characteristics of different powders, we studied the single discharge craters and their progression during EDM drilling. The finding of this study reveals the process mechanisms and parameters to improve the machinability of ultrafine particle-type tungsten carbide using micro-EDM.

2. Process Mechanisms of Powder-Mixed EDM

During EDM, an essential concern is the expulsion of debris outside the discharge gap. The debris generated during machining is generally removed by two mechanisms: bubble expansion and dielectric flow [12,16]. When the debris is not properly flushed out, it will result in (a) secondary discharges, where the debris is machined instead of the workpiece, or (b) short-circuiting. This results in frequent tool backtracking, increasing machining time, and increased costs. If the discharge gap widens, the dielectric can easily flow in, promoting the flushing of debris and improving process stability.

When a particle is introduced in an electric field (E), as shown in Figure 1a, the polarization of the molecule occurs. This polarization results in an electric field aberration (E_1) within the vicinity of the particle, as shown in Figure 1b. During EDM, when the electrodes are brought closer to each other, this aberration results in field intensification near the particle, creating a more accessible pathway for electrons to flow, thus, promoting dielectric breakdown. Moreover, when several particles get closer to each other, as shown in Figure 1c, the field is intensified further (E_2), creating 'bridges' for ions to flow, promoting even faster breakdown. As a result, the interelectrode gap required to form sparks is larger. Several studies [21,23,24] have reported that these promote process stability, thus improving machining efficiency.

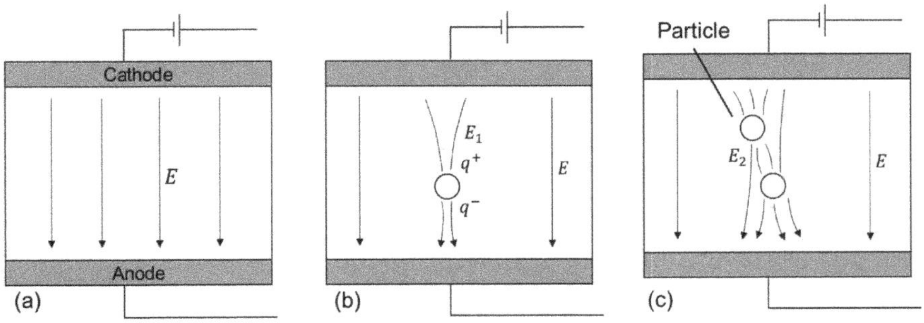

Figure 1. Schematic of (**a**) electric field in between two electrodes and (**b**) distortion due to presence of a single particle, (**c**) bridge formation with multiple particles.

Since the presence of particles alters the discharge characteristic, it is vital to study the effect of the properties of different powders. It has been reported that particles generally exhibit four different behaviors within the discharge gap: reciprocation, bridging, adherence to the electrode, and agglomeration [25]. Polarized molecules reciprocate between electrodes when an electric field is applied due to electromotive force. The physical and electrical properties of molecules determine the extent of this movement. At the same time, bridging occurs when particles align themselves parallel to the electric field, forming chain-like structures (Figure 1c). Reciprocation and bridging promote dielectric breakdown.

In contrast, the adhesion of particles on the workpiece surface inhibits the machining by preventing energy transfer. Moreover, it has been observed that nanoparticles tend to agglomerate due to high surface energy or Van der Walls forces. The particle agglomeration will decrease polarizability, inhibiting discharge channel formation and increasing the possibility of short-circuiting.

Carbon nanofibers are incredibly light and highly conductive materials. It has been reported that due to these properties, the nanofibers form bridges quickly, improving the machining output [21]. On the other hand, alumina is a dense material with very low electric conductivity. Due to these properties, the bridging of particles will not occur, and the sedimentation of the particles on the bottom surface will be prominent, inhibiting machining. Silicon is a semiconductor with a density and electrical conductivity between alumina and carbon nanofibers. Semiconductors exhibit a unique property of decreasing resistivity with an increase in temperature. Thus, depending on the condition within the working gap, silicon particles may exhibit all kinds of behaviors mentioned above.

Based on the process mechanism and powder behavior during EDM, the differences in the material removal, tool wear, and surface roughness during PMEDM of WC-Co are elucidated.

3. Experimental Methods

3.1. EDM Setup

In this study, a micro-EDM machine, Panasonic MG-ED72, was used. The device has a resolution of 0.1 μm in the x, y, and z axes. The machining is performed using a resistor-capacitor (RC) circuit, which is suitable for micromachining [26]. The control parameters are voltage ranging from ±50 V to ±110 V and capacitance of 3300, 1100, 200, and 100 pF. The discharge gap, pulse duration, and peak current are not controllable. A spindle is attached to the micro-EDM machine, rotating at 3000 rpm. An acrylic tank was fabricated to contain the dielectric. Discharge progress was monitored using an oscilloscope. The photograph of the machining setup is shown in Figure 2. During experimentation, no external stirring or circulation of the dielectric was applied.

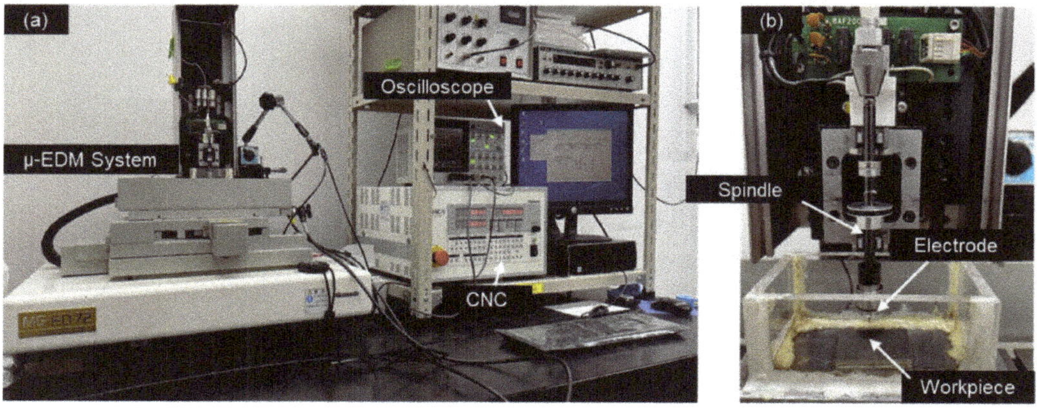

Figure 2. Photograph of (**a**) μ-EDM system, (**b**) electrode and workpiece arrangement.

3.2. Electrode and Workpiece

Ultra-fine grade cemented Tungsten carbide (WC-Co) KM-10, produced by Toyo Tool, Japan, was used as the workpiece material. The average grain size was around 0.7 μm, and the cobalt binder concentration was about 6%. The mechanical properties of the workpiece are shown in Table 1. The surface of the workpiece was mirror-polished to 0.01 μmSa surface roughness using a diamond powder slurry. A copper (Cu) electrode of a diameter of 1 mm from Nilaco Corporation, Japan, was used as a tool material. The electrodes were deburred and dressed using reverse-polarity EDM before each machining test. The scanning electron micrograph (SEM) of the surface of the workpiece is shown in Figure 3a.

Table 1. Properties of tungsten carbide.

Property	Value
Material type	Ultrafine particle type (avg. size 0.69 μm)
Binder type and content ratio	Co (6%)
Density [g/cm^3]	14
Melting point [K]	3140
Hardness (HRA)	91.5
Thermal conductivity [W/mK]	110

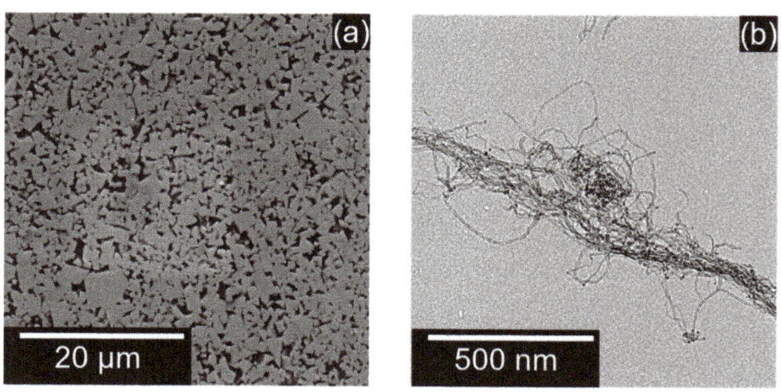

Figure 3. *Cont.*

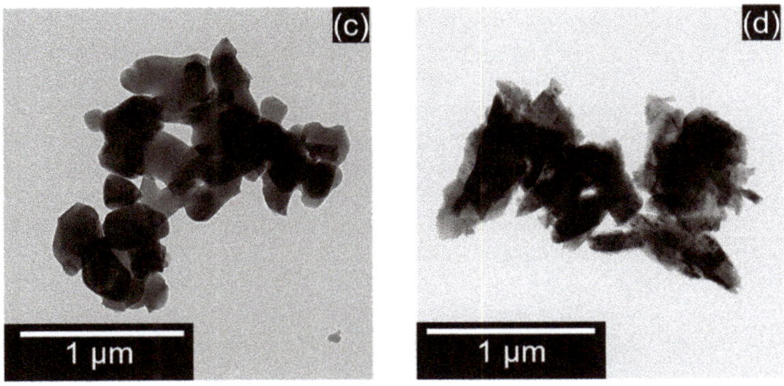

Figure 3. SEM of (**a**) tungsten carbide, TEM of (**b**) carbon nanofiber, (**c**) alumina, (**d**) silicon particles.

3.3. Powder Materials

Three different powders of varying shapes, densities and electrical properties were used in this experiment. Commercially available polishing grade alumina powder, Baikalox 1CR, manufactured by Baikowski International Co. Ltd., was used as the insulator powder. Semiconductive silicon powder, manufactured by Kyocera, Japan, was used. Lastly, carbon nanofibers of diameter 10–20 nm and height 0.1–10 μm, produced by Mitsubishi Chemicals, Japan, were used as a conductive powder in this experiment. The transmission electron micrographs [TEM] of carbon nanofiber, alumina, and silicon are shown in Figure 3b–d, respectively. A laser particle analyzer from Horiba semiconductors, LA-960, was used to measure the particle size. Figure 4a,b show the size distribution of the silicon and alumina powders, respectively. The thermal and electric properties of the powder are listed in Table 2.

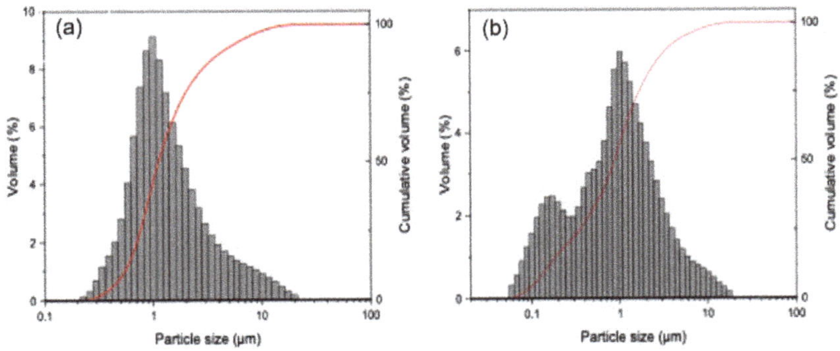

Figure 4. Particle size distribution of (**a**) silicon and (**b**) alumina used in this study.

Table 2. Properties of different powders.

Property	Carbon	Silicon	Alumina
Density [g/cm^3]	1.3–2	2.329	3.965
Electrical Conductivity [μΩ/cm]	50	1×10^5	1×10^{-9}
Thermal Conductivity [W/mK]	400	150	30
Melting Point [K]	3823	1687	2345

3.4. Machining Conditions

Two different EDM tests were performed to determine the effect of powder addition on EDM: fundamental characteristics and hole machining tests. The micro-EDM machine used in this study uses an in-built contact detection program. An edge is detected if a threshold number of discharges occurs within a resistance-capacitance (RC) cycle. This method was used to obtain isolated single craters. Next, machining tests on the evaluation of PMEDM were performed by drilling holes of a depth [h] of 50 μm under different conditions, as shown in Table 3. The discharge energies used in this experiment are used for rough-phase micromachining. Full factorial experiments were performed with two repetitions for each condition. A hydrocarbon-based dielectric, Casty Lube EDS was used. As seen in Figure 3b–d, powders tend to agglomerate. Hence, the dielectric and powder mixture was subjected to an ultrasonic bath for 15 min before each machining to obtain a homogenous mixture.

Table 3. Machining conditions.

Property	Value
Tool	Copper (Ø 1 mm)
Workpiece	WC-Co
Polarity	Tool (-ve)
Voltage [V]	90, 100, 110
Capacitance [pF]	3300
Feed rate [μm/s]	0.5
Powders	CnF, Al_2O_3. Si
Concentration [g/L]	0, 0.25, 0.5, 0.75, 1.0

3.5. Evaluation Conditions

The machining output of EDM with various powders was evaluated with three parameters: material removal rate, electrode wear rate, and surface roughness. Material removal rate (*MRR*) is defined as the ratio of the volume of the material removed (Vol_m) to machining time (M_t) Equation (1). A laser microscope from Olympus, Japan, was used to map the three-dimensional topography of the machined hole. The analysis software TalyMap, Amtek Corp. was used to determine the mean depth plane [*d*], and the diameter [*D*] at (z = 1/2 *d*) of the hole machined.

$$MRR = \frac{\pi D^2 d}{4\ M_t} \tag{1}$$

Electrode wear rate is defined as the ratio of the volume of electrode worn Vol_e to machining time (M_t). Electrode wear is an essential factor in determining the economic feasibility of the machining process. Since the micro-EDM machine used in this study does not use any wear compensation mechanism, the difference between set depth [*h*] and actual machined depth [*d*] is the electrode wear. Since machining time varies based on machining conditions, studying the electrode's relative wear to that of the workpiece is considered in this research. The equation to determine relative electrode wear rate (*REWR*) is shown in Equation (2). The machined region's average surface roughness (Sa) was measured over an area of 256 × 256 μm at intervals of 0.25 μm. The ISO 25178 standard of measurement with no filters, measured over three locations on each machined surface was performed.

$$REWR = \frac{{D_e}^2[h-d]}{D^2 d} \tag{2}$$

4. Results and Discussion

4.1. Single Discharge Experiments

Figure 5a–c shows the single discharge craters obtained using EDM oil without powder addition for 90, 100, and 110 V, respectively. Unlike the craters in the EDM of other materials, such as silicon carbide [12,21,27], which show clear circular boundaries, craters in WC-Co were irregular. Furthermore, enlarged images of the crater surface, as shown in Figure 5d,e, indicate that the machining has not completely progressed to the crater's edge since the polishing marks from the original sample are retained after EDM (A1 and B1). These marks indicated that at these discharge energies, the heat generated is not sufficient to melt the material at the edges; instead, the grains are pushed outwards. Additionally, as seen in A2–B2, redeposited materials are observed. Since tungsten carbide is a highly dense material, it is difficult to eject the molten materials entirely from the discharge gap, which redeposits onto the surface. The formation of micro-cracks was also observed, as seen in B3. Crack formation is due to the preferential removal of cobalt binder, which has a lower melting point than tungsten carbide, forming voids, which develop into cracks.

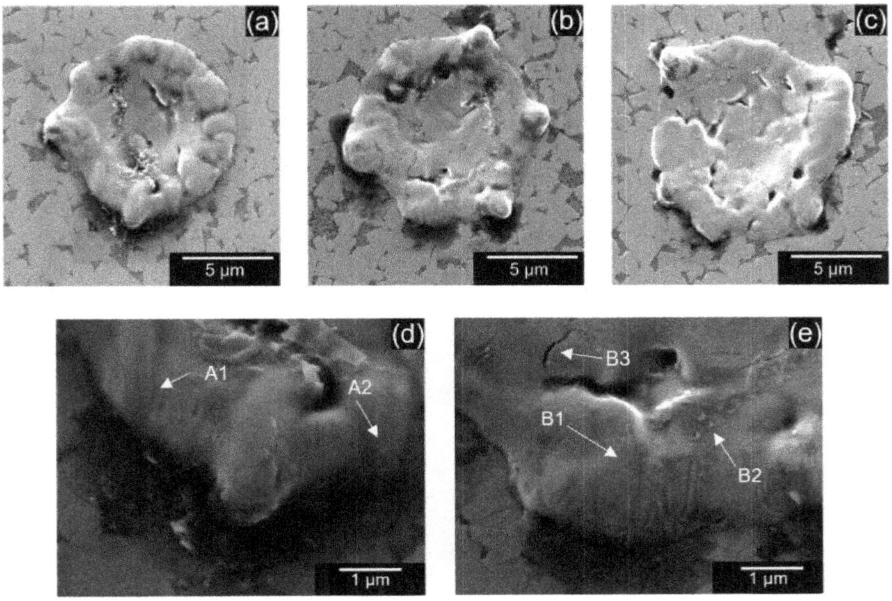

Figure 5. Single discharge craters for (**a**) 90 V, (**b**) 100 V, (**c**) 110 V using EDM oil without powder addition. Crater edge at (**d**) 90 V, (**e**) 100 V.

Figure 6a–i show the SEM images of single discharge craters observed at various PMEDM conditions. The average diameter of these craters is summarized in Figure 7. In the case of carbon nanofiber and alumina mixed EDM, the average crater diameter reduced, whereas it increased in the case of silicon. In addition, the crater diameter increased with voltage for all conditions.

According to previous research, the discharge gap increased drastically when the carbon nanofibers were added to the dielectric [21]. Furthermore, as explained earlier, the breakdown strength is significantly reduced due to its high conductivity. The plasma, which is formed, has lower energy; as it travels for a longer distance, it weakens further. This results in a narrow crater. In the case of alumina-mixed EDM, the increase in the discharge gap is not significant due to its low electrical conductivity. Additionally, alumina particles absorb plasma energy due to their high thermal conductivity. Since lower energy is transmitted to the workpiece, lower material is removed. However, in the case of silicon, contrary results were observed. Silicon exhibits thermal and electrical properties in between carbon nanofiber and alumina. Although the discharge gap is increased compared to the no powder condition, it is not as large as carbon nanofibers. As a result, an expanded plasma channel is formed, which results in a wider crater. Thus, it can be seen that the properties of powder influence the discharge process significantly.

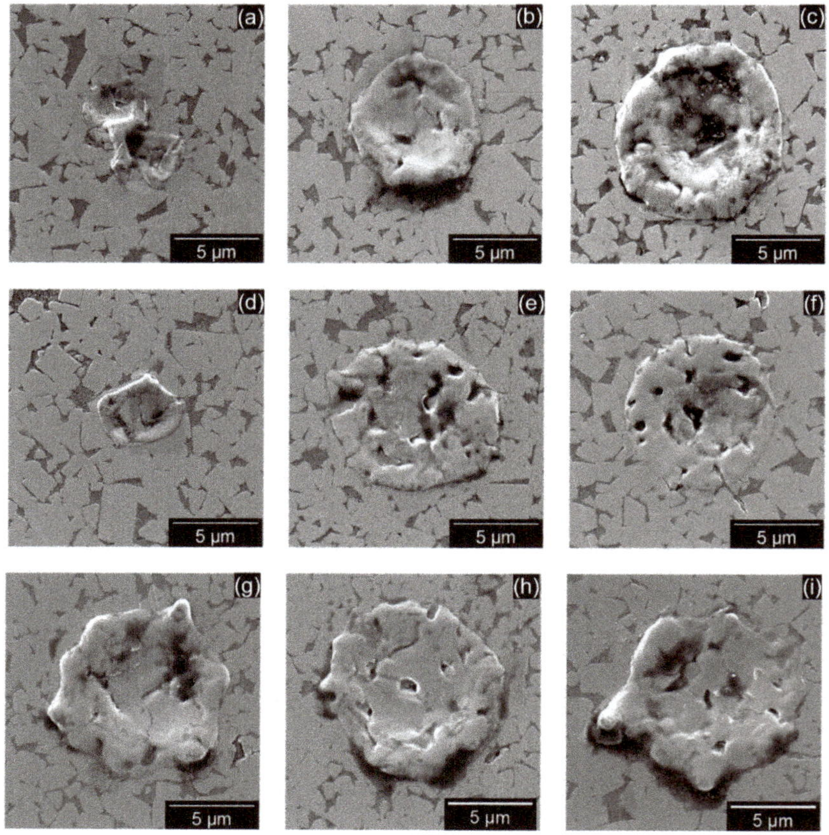

Figure 6. Single discharge craters for (**a**) 90 V, (**b**) 100 V, (**c**) 110 V using EDM oil with carbon nanofiber powder, (**d**) 90 V, (**e**) 100 V, (**f**) 110 V with alumina powder and (**g**) 90 V, (**h**) 100 V, (**i**) 110 V with silicon powder.

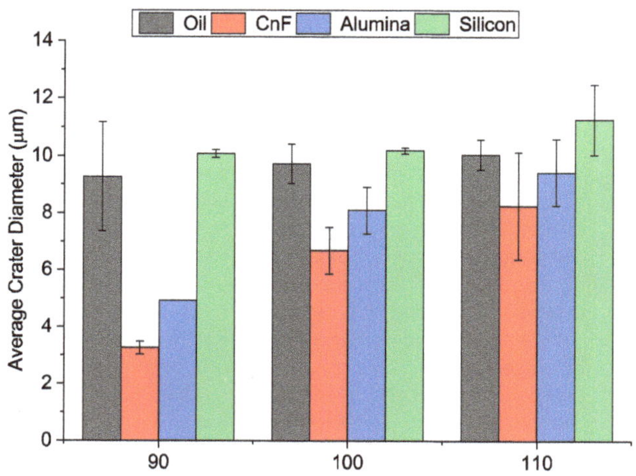

Figure 7. Average single discharge crater diameter under different conditions.

4.2. Hole Machining by Carbon Nanofiber-Mixed EDM

The machined profiles and the material removal rate with oil and carbon nanofiber-mixed EDM are summarized in Figures 8 and 9, respectively. In the only EDM oil condition, the maximum MRR obtained was 0.0026 mm^3/min at 110 V condition, which decreased with the decrease in voltage. With an increase in voltage (V), the maximum discharge energy per spark (W) increases, and as a result, the material removed per discharge also increases, Equation (3). This can be seen evidently in Figure 7, where the crater diameter increased with an increase in voltage.

$$W = \frac{1}{2}CV^2 \quad (3)$$

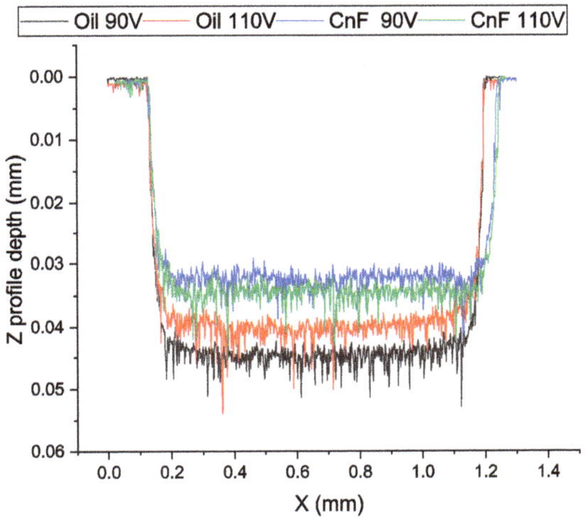

Figure 8. Profiles of machined holes at 90 and 110 V using oil and 0.5 g/L CnF-mixed EDM.

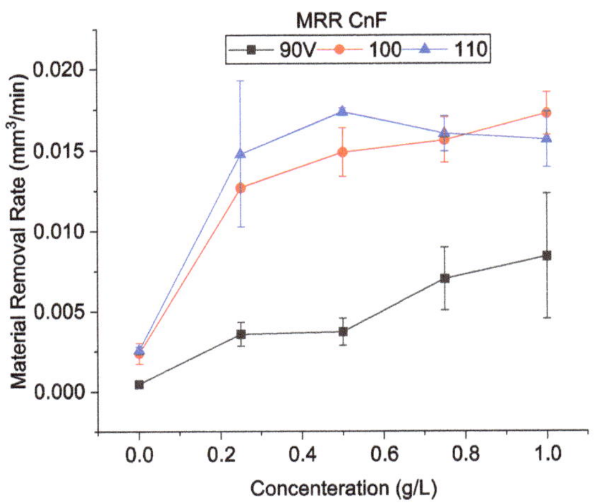

Figure 9. Material removal rate using CnF-mixed EDM.

It can be seen from the machined profiles in Figure 8 that the machined holes were shallower and wider with the addition of the carbon nanofiber. The increase in machined hole diameter, called overcut, is due to an increase in the discharge gap due to the addition of nanofibers.

It was observed that even though the machined depth with the oil-only condition was large, the MRR increased for all concentrations of CnF and voltages. At a higher discharge energy of 110 V, the maximum MRR observed was 0.017 mm^3/min at a concentration of 0.5 g/L, over 6.5 times that of the oil-only condition. This tremendous improvement in MMR can be attributed to the ease of discharge initiation and increased discharge gap width. It was also observed that at 110 V, the machining speed was equal to the set feed speed (0.5 μm/s) at all concentrations. This indicates no ineffective discharge period in this condition, where the tool is forced to retract to prevent short-circuiting, could be observed. This can be monitored by observing the waveforms on the oscilloscope, as seen in Figure 10a. It was observed that during EDM with oil, there was a lot of arcing which had the waveform M1 in contrast to normal EDM (M2). It was also observed that if arcing persisted, it led to the formation of complex waveforms [28] as seen in M3, followed by short-circuiting and tool retraction. In contrast, as seen in Figure 10b, the addition of carbon nanofiber increased the number of sparks, and complex waveforms were not observed.

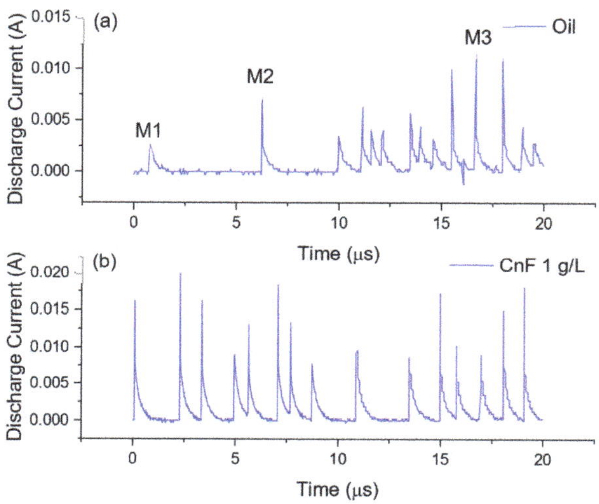

Figure 10. Current waveforms during EDM using (**a**) oil and (**b**) carbon nanofiber mixed EDM (1 g/L) at 110 V.

It can be seen from Figure 10 that the MRR increases with an increase in concentration and then decreases after a peak at 0.5 g/L. In EDM with oil without process alteration, it has been reported that the spark tends to accumulate in one region [12,29]. As debris concentration increases, spark formation in the vicinity is likely, as shown in Figure 1b. Whereas, in powder-EDM, sparks are not localized since particles are dispersed throughout the surface. In CnF-mixed EDM, initially, as the particle concentration increases, there is an increase in the number of discharge pathways dispersing EDM, increasing process stability, and thus increasing MRR. However, as the particles' concentration increased, many nanofibers started to either accumulate or adhere to the electrodes, inhibiting the discharge process.

Similarly, an improved machining efficiency was observed when the voltage was reduced to 90 V. A machining rate of 0.008 mm^3/min was achieved, over 18 times that of only oil. Generally, the discharge gap decreases considerably at low voltages, increasing the possibility of debris accumulation within the working gap. This increases the frequency of ineffective discharges, such as short-circuiting due to debris accumulation and secondary discharges on the machined debris. Thus, flushing away debris becomes a prime concern. During CnF-mixed EDM, the machined speed increased from 0.18 to 0.53 times the set machining speed when the concentration increased from 0.25 to 1.0 g/L. There was a significant no discharge period. However, during the tool backtracking period, the fresh dielectric with powder flows in, removing the debris. This period is essential at low voltage, where the discharge gap is lower, as it resets the dielectric. The proportion of powder entering the gap is higher at higher concentrations so that machining can be restarted quickly. Although the machining efficiency increased, it can be considered that at a feed rate of 0.5 μm/s and 90 V using carbon nanofibers, uninterrupted machining cannot be achieved entirely.

An improvement of over 7.7 times in the MRR without powder addition was observed at 100 V. Similar to 110 V, the machining was almost continuous, the machining speed increasing from 0.68 to 0.92 times the feed speed. Furthermore, the MRR increased with concentration. At 1.0 g/L, the MRR was even higher than 110 V. The brief intermissions during EDM at 100 V allowed new dielectrics to flow in, which is considered to improve the machining rate compared with 110 V.

Figure 11 compares electrode wear for different machining conditions in CnF-mixed EDM. It was observed that, unlike in previous studies, electrode wear rate increased with the addition of CnF to the dielectric, which increases further with the increase in the powder concentration.

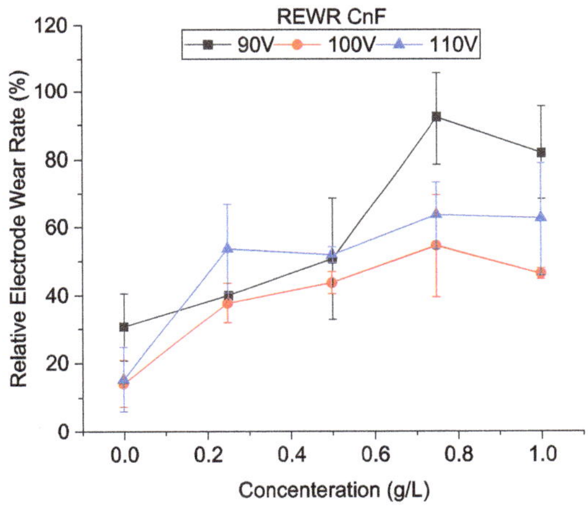

Figure 11. Relative electrode wear rate using CnF-mixed EDM.

During the EDM process, plasma travels from the cathode and bombards the anode, removing material by heat transfer and abrasive action. However, some electrons also travel in the opposite direction removing material from the anode. Since the kinetic energy of ions is higher than electrons, the amount of material removed from the anode is much higher. In addition, the temperature in the discharge gap can exceed 3000 K [30], removing material from both electrodes. Thus, the tool wear depends on the electrodes' electrical and thermal properties. The melting point of copper is much lower than that of tungsten carbide [1631 K], and thus, the melting of copper will be higher at high temperatures. At a high voltage or discharge energy, the fraction of electrons is high, and the temperature in the working gap is enormous. Hence a higher wear rate is observed. The dielectric flow into the discharge gap, which acts as a coolant and reduces the heat transferred to the electrodes is necessary to prevent electrode wear.

As mentioned above, the machining at higher voltages [100 and 110 V] was almost continuous. Since no external flushing was used, if no electrode backtracking occurs, the flow of the new dielectric is limited. The debris removed is only due to bubble explosion and not debris flow, and as a result, the temperatures within the gap will continue to rise. The electrode wear will increase correspondingly with an increasing machining rate, as seen in Figure 11. Since the discharge energy at 100 V is smaller than 110 V, the temperature reached within the gap is expected to be smaller, resulting in a lower tool wear rate.

In contrast, very high electrode wear was observed at 90 V. This result is similar to previous studies, in which the electrode wear rate decreased with the discharge energy and then increased. Due to improper flushing at lower energies, arcing will occur, which results in more material being removed from the electrode than the workpiece. A considerable variation was also observed in the relative electrode wear rate due to this instability, which is more prominent at 90 V. In addition to arcing, with an increase in concentration, the tool wear rate also increased due to continued machining similar to higher energies. A decrease in wear rate was observed at 1 g/L at all voltages. It is believed that at a very high concentration, adherence of particles to the surface of the electrode is increased. This adherence may have prevented the heat from being transferred from the dielectric to the electrode.

Figure 12 shows the SEM of the surface of the machined holes at different voltages using EDM oil. The craters are narrow and shallow at lower voltages (Figure 12a,b). Many surface defects, such as micro-cracks (C1), micro-pores (C2), and redeposited materials (C3) were observed on the surface. The number of redeposited materials increased at higher voltages, as seen in Figure 12c. At high voltages, the gap temperatures were larger, and molten debris could not be cool down quickly, so they redeposit onto the surface. Similar surface defects could also be seen in carbon nanofiber-mixed EDM as seen in Figure 13a–d. However, the redeposited materials were more extensive (D1) and could also be seen at lower voltages Figure 13a,b.

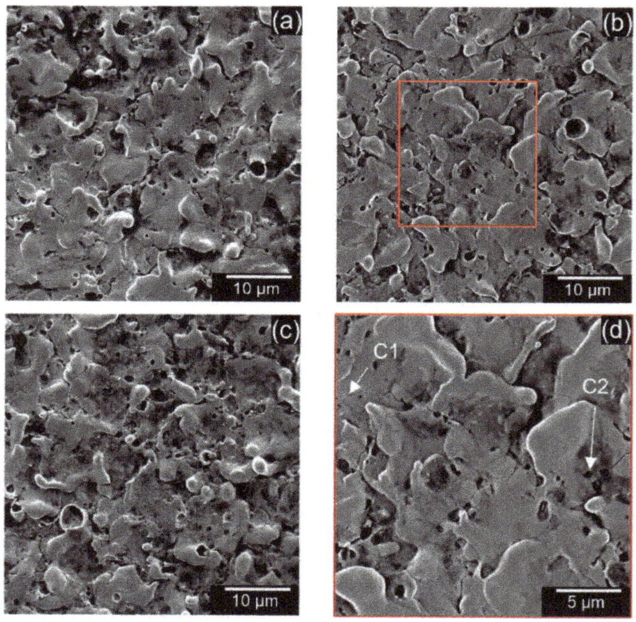

Figure 12. SEM of the machined surface using oil-only at (**a**) 90 V, (**b**) 100 V, (**c**) 110 V, and (**d**) enlarged region of 100 V.

Previous results have indicated that adding powder increases discharge dispersion, creating a uniform surface. In addition, it has also been reported that using powder-mixed EDM produces wide yet shallow craters [31]. These two factors are considered to result in lower surface roughness. Figure 14 shows the result of surface roughness. At 110 V, adding powder decreased the surface roughness for all concentrations, whereas the surface roughness has shown various trends for lower voltages.

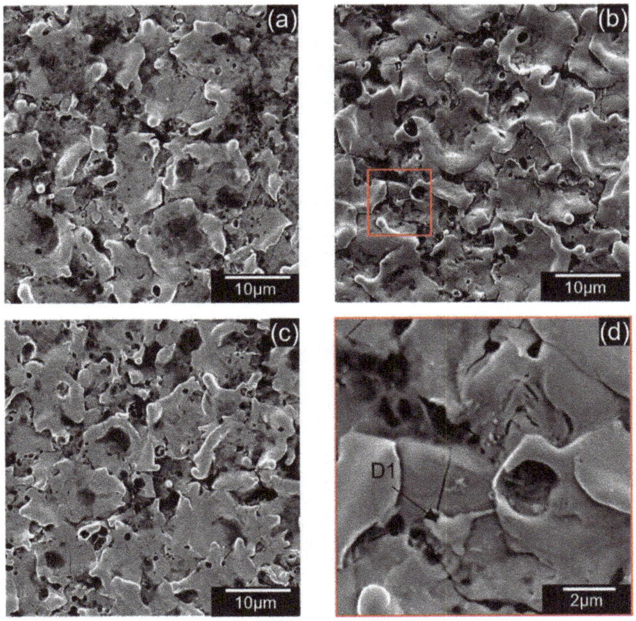

Figure 13. SEM of machined surface using 0.5 g/L carbon nanofibers at (**a**) 90 V (**b**) 100 V, (**c**) 110 V, (**d**) enlarged region of 100 V.

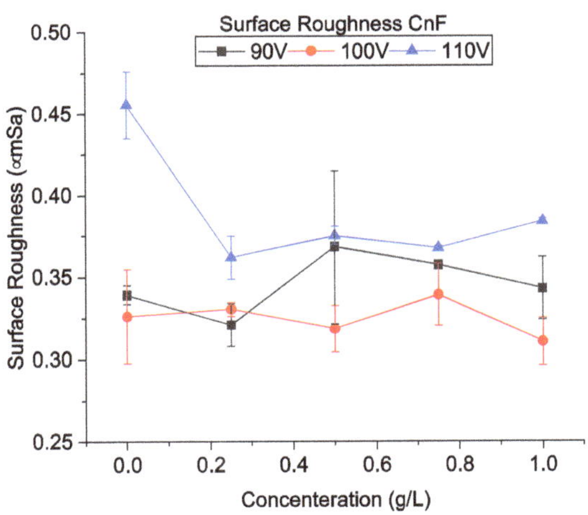

Figure 14. Surface roughness using CnF-mixed EDM.

At higher discharge energy, the material removed per discharge is increased. This results in a deep crater, as seen in Section 4.1. As the discharge energy decreases, the crater size decreases, resulting in a smoother surface finish. This phenomenon is also observed in this study when the voltage is reduced from 110 to 100 V. However, the surface roughness increases when the voltage is further reduced. At the 90 V condition, redeposited materials could have increased surface roughness when compared to only the EDM oil condition, as seen in Figure 13a.

In summary, using carbon nanofiber powder-mixed EDM of tungsten carbide improved the machining rate significantly under all machining conditions. However, an increased tool wear rate was observed. Under certain machining conditions, a fine surface was observed. Among the machining conditions, at 100 V, a high material removal, low tool wear, and a smoother finish can be simultaneously achieved at concentrations of 0.5–0.75 g/L. These results also agree with a previous study [21], which showed improvement in the MRR using carbon nanofiber on reaction-bonded silicon carbide.

4.3. Hole Machining by Alumina-Mixed EDM

Figure 15 shows the MRR obtained at various machining conditions using alumina. At high discharge energy, similar to CnF, there was a higher MRR when compared to the no powder condition. The highest MRR obtained was 0.00971 mm^3/min at 0.5 g/L, about 3.8 times of the oil-only condition. It was also observed that the MRR increases initially and then decreases with an increase in concentration. As mentioned in Section 4.2, it has been proven theoretically and experimentally that particle addition increases the discharge gap regardless of the particle's properties. However, unlike CnF, due to its insulating nature, the alumina particles do not get polarized easily in an electric field due to the absence of free electrons, and dipoles are not created. The formation of 'bridges' is inhibited. However, under strong electric fields, even insulative materials become polarized. This indicates that the alumina particles can participate in spark formation at high voltages. This effect reduces as the voltage is reduced. These results agree with previous studies in which alumina has been shown to improve the MRR in transistor circuits at high peak voltages [22,32]. However, in fine-finishing micro-EDM using RC circuits, such as this study, the improvement in the MRR is not significant. Additionally, it was found that the machined feed speed varied between 2.32 at 0.5 g/L and 3.28 at 0.75 g/L to that of the set feed speed. These results show that although there is an improvement in the MRR at 110 V compared to oil, there was still a significant amount of time in which machining did not occur.

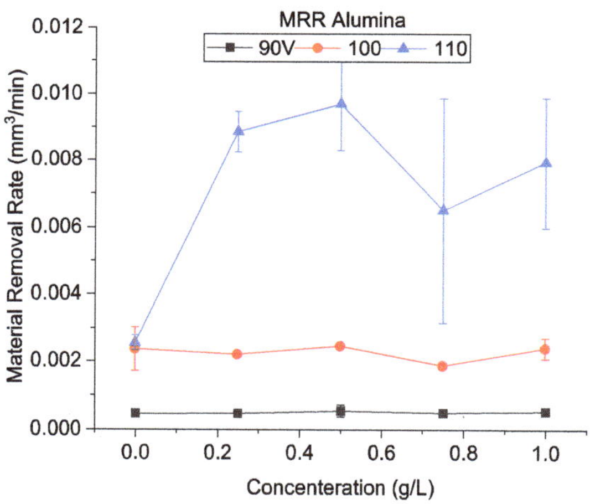

Figure 15. Material removal rate using alumina-mixed EDM.

At the low discharge energy of 90 V, adding alumina increased the MRR at all concentrations. However, the improvement was not substantial. The highest MRR was 0.00055 mm^3/min, 1.19 times the only oil condition. It can be seen from Figure 7 that the average crater size of alumina-mixed EDM was much lower than that of oil-only conditions at all conditions, indicating that the material being removed per spark is very low. Thus, the machining rate improvement is due to the lower no-discharge period. The machined feed speed was also similar to the no powder added condition. It was observed that, due to the higher density of alumina than the dielectric, the powder settling was a significant phenomenon. Since no external stirring or circulation was used, the powder sedimentation cannot be overcome.

Furthermore, as the machining time increased, the fraction of powder suspended in the dielectric decreased. During electrode backtracking, the fresh dielectric contained a limited amount of alumina. As a result, when machining time is extended, the MRR becomes similar to that of oil only. In addition, as mentioned earlier, the alumina addition does not directly improve spark formation. An increase in particle concentration will only inhibit the EDM process. Additionally, similar to that at 90 V, no significant improvement in the MRR was observed at 100 V.

Figure 16 shows the relative electrode wear of alumina mixed-EDM. Similar to CnF, the tool wear rate was generally found to increase with the addition of powder and was more significant at 90 V. Unlike CnF addition, where tool wear is due to high gap temperature due to continuous machining, the tool wear in alumina is expected to be due to frequent arcing and short-circuiting. As a result, a significant variation in the electrode wear results was observed. This is evident as electrode wear increases with a decrease in voltage and concentration for most conditions.

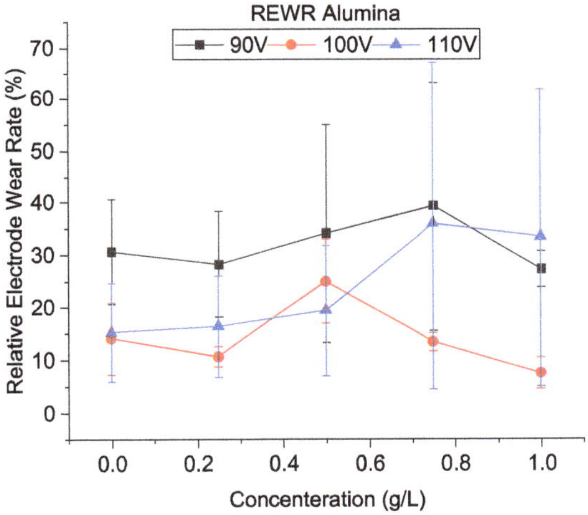

Figure 16. Relative electrode wear using alumina-mixed EDM.

Figure 17a–c shows the SEM of the machined surfaces with alumina mixing. The surface roughness of the machined holes using alumina is shown in Figure 18. The addition of alumina decreased the surface roughness for all concentrations at 110 and 100 V. Small and shallow craters, as seen in Figure 6d–f and uniform discharge distribution, helped to achieve a smoother surface. Moreover, as the voltage decreases to 90 V, frequent arcing and short-circuiting make the surface rough. It has also been observed that the effect of concentration does not significantly alter the surface roughness. As mentioned above, as the machining time increases, the powder concentration suspended within the dielectric becomes lower due to powder sedimentation, resulting in similar surfaces.

Effective machining using alumina can be achieved by using a high voltage of 110 V for rough machining and 100 V for fine finishing. The effect of concentration is not significant. In their study on using nano-alumina powders at finish phase machining [23], the authors indicated that there was no significant increase in the MRR but an improvement in surface roughness. In our study using micro-alumina at rough machining conditions, we found a similar roughness improvement. However, we have found that alumina can improve the material removal process at higher machining energies.

Figure 17. SEM of machined surface using 0.5 g/L alumina at (**a**) 90 V (**b**) 100 V, (**c**) 110 V.

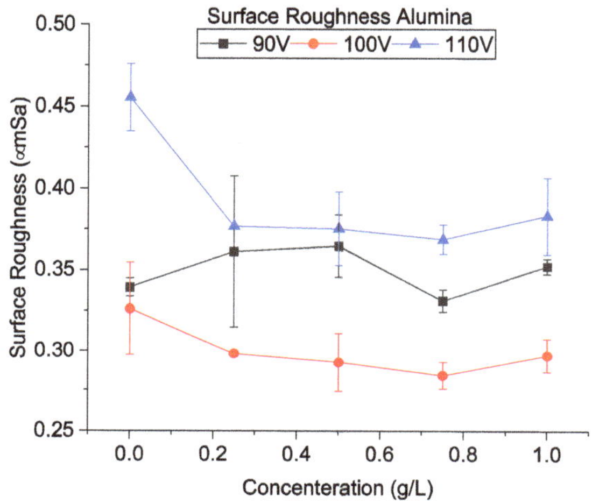

Figure 18. Surface roughness using alumina-mixed EDM.

4.4. Hole Machining by Silicon Nanofiber-Mixed EDM

The material removal rate obtained during the machining using silicon is shown in Figure 19. It was observed that the MRR increased for all conditions in the presence of silicon. The highest MRR of 0.0197 mm^3/min was obtained at 1 g/L at 110 V. It was also observed that the MRR increases with an increase in particle concentration at higher voltages. Like CnF, silicon particles within the discharge gap cause electric field aberrations. However, silicon's aberration is expected to be lower compared to nanofibers with very high electrical conductivity. Furthermore, silicon particles are much smaller than nanofibers that are 0.1–10 μm in length. Due to Van der Walls forces, particle agglomeration is also higher at smaller sizes.

As seen in Figure 7, the crater diameter of silicon, unlike CnF and alumina, is higher than that of oil. Additionally, the crater diameter did not vary significantly with voltage, similar to oil-only conditions. The material removed per discharge is thus higher than that of oil-only conditions. Moreover, the presence of silicon also improves the distribution of discharges. These two factors are expected to increase MRR.

Furthermore, semiconductors show a decrease in resistivity with an increase in temperature. Thus, if the machining is continuous and the working gap's temperature increases, silicon's conductivity would increase drastically. A highly conductive powder would further enhance the discharge breakdown for the following discharges, and the machining would proceed rapidly. The temperature in the working gap is generally higher at high discharge energies. Furthermore, an increase in concentration can enhance this effect. This is evident in the MRR results shown in Figure 19. It was observed that an increase in voltage increased the material removal rate. Furthermore, the increase in concentration improved it further.

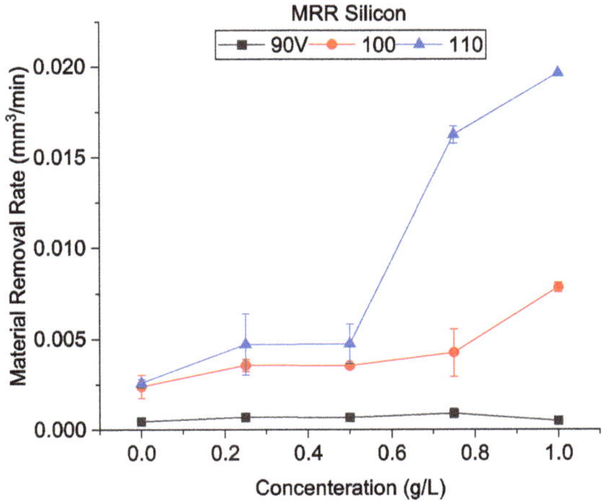

Figure 19. Material removal rate using silicon-mixed EDM.

However, at 90 V, the increase in concentration did not improve the MRR considerably and was found to be reduced at very high concentrations. At low voltage, the gap size is significantly smaller. The size of the agglomerated Si particles becomes similar to that of the gap itself. This inhibits the machining process, which reduces furthermore with an increase in powder concentration.

The relative electrode wear for different machining conditions is shown in Figure 20. Like CnF and alumina, high tool wear was observed at 90 V, which decreased with an increase in voltage. It was due to the increased amount of arcing at low voltages. This instability in machining also resulted in a high fluctuation in electrode wear value at lower voltages.

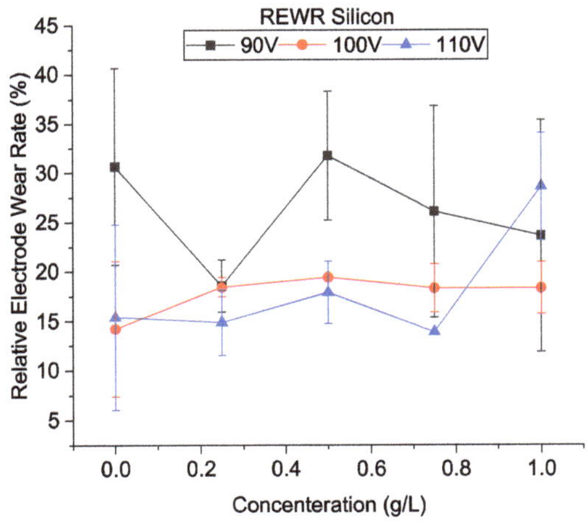

Figure 20. Relative electrode wear rate using silicon-mixed EDM.

Figure 21a–c shows the SEM of the machined surfaces with silicon mixing. Surface roughness is shown in Figure 22. It was observed that the surface roughness decreased for most conditions during PMEDM with Si powder. The roughness value decreases initially with voltage and then increases. Like CnF and Al_2O_3, the initial decrease in surface roughness was due to the reduction in crater size, as seen in Figure 7. However, unlike other powders, the difference in crater size between 90 V and 100 V is insignificant. The roughness increases due to higher proportions of short-circuits and arcing, causing unevenness throughout the surface.

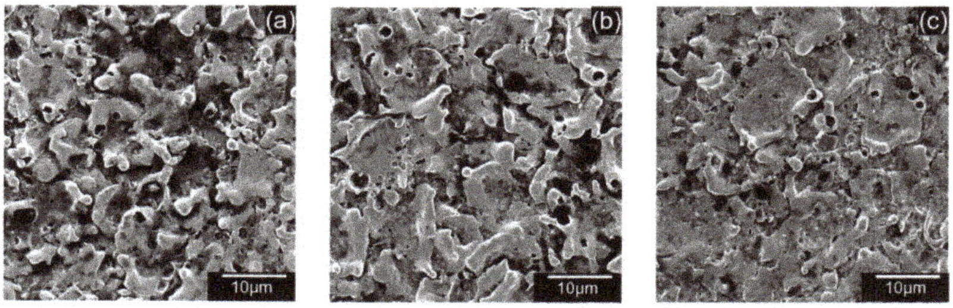

Figure 21. SEM of machined surface using 0.5 g/L silicon at (**a**) 90 V (**b**) 100 V, (**c**) 110 V.

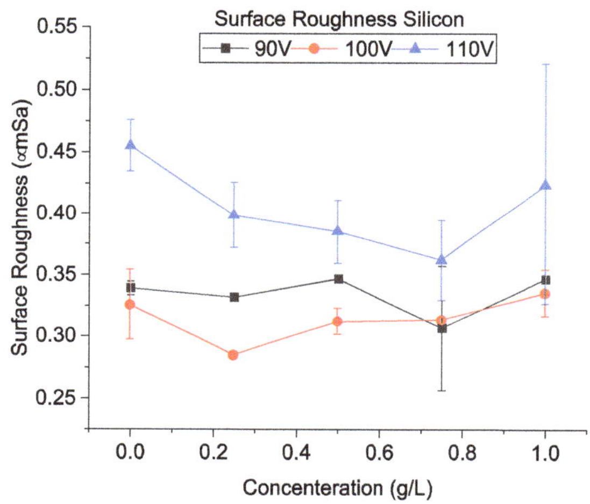

Figure 22. Surface roughness using silicon-mixed EDM.

Furthermore, it was observed that the roughness value decreases with increased concentration. Initially, with increases in concentration, the discharges become uniformly distributed due to a larger number of possible discharge pathways. However, at very high concentrations, many particles start to agglomerate, resulting in discharge concentration in the vicinity of the agglomerated particles and an uneven surface.

Silicon-mixed EDM of tungsten carbide can be an effective alternative to conductive powders. However, they can yield high material removal rates only at high discharge energy and powder concentrations. The electrode wear rate is also lower than CnF for most conditions. In their study, Pecas et al. [20] used silicon powder (10 μm) in the EDM of AISI H13. They have indicated that the addition of silicon improved the roughness values. We have also found that silicon addition can also improve the roughness in micro-machining conditions.

5. Conclusions

Three kinds of powders, conductive carbon nanofibers (CnF), semiconductive silicon (Si), and insulative alumina (Al_2O_3), were mixed in a dielectric fluid during electric discharge machining (EDM) of ultrafine particle type tungsten carbide. Single discharge and hole machining tests were performed, and their effects on material removal rate (MRR), relative electrode wear rate (REWR), and surface roughness were studied. When compared to conventional EDM without powder mixing, the following conclusions were made:

1. CnF and Al_2O_3-mixed EDM results in a crater with a smaller size when compared to oil EDM, which increases with an increase in voltage. Si-mixed EDM, like oil EDM, has large craters, which do not vary significantly with changes in voltage.
2. The MRR increased with the addition of all powders. Because of its electrical properties, CnF improved the MRR tremendously. Such effects were observed in Al_2O_3 and Si only at high voltages. The MRR increased initially, with an increase in concentration, then decreased.
3. A higher relative electrode wear rate (REWR) was also high using powder-mixed EDM. In particular, using CnF, large amounts of electrode wear were observed. Due to a smaller discharge gap at 90 V, an increased amount of process instability was observed and, as a result, higher electrode wear.
4. The surface finish was improved for all powders and concentrations at higher voltages of 100 and 110 V compared to EDM with oil only. However, such improvement was not observed when the voltage was reduced further, owing to much faster machining.
5. It was also found that the increase in powder concentration improved the surface roughness. Due to a much faster machining rate, carbon nanofibers produced a rougher surface compared to alumina and silicon.

At rough machining conditions, the objective is to achieve high machining material removal with repeatability. The results of this study indicate that conductive carbon nanofiber is the best choice for powder-mixed EDM of tungsten carbide since it can machine at a very high rate for all voltages and concentrations. Careful selection of machining parameters is required to obtain desired output using silicon and alumina powders. It is presumable that the addition of suitable powder could aid in improving the micro-processing of materials having similar microstructures to ultrafine particle tungsten carbide, whose machinability is lower.

Author Contributions: Conceptualization, S.D.G. and J.Y.; methodology, S.D.G. and J.Y.; software, S.D.G.; validation, S.D.G.; formal analysis, S.D.G.; investigation, S.D.G.; resources, J.Y.; data curation, S.D.G.; writing—original draft preparation, S.D.G.; writing—review and editing, J.Y.; visualization, S.D.G.; supervision, J.Y.; project administration, J.Y.; funding acquisition, S.D.G. and J.Y. All authors have read and agreed to the published version of the manuscript.

Funding: This research was partially supported by a grant from the Keio Leading-edge Laboratory of Science and Technology (KLL) and the Keio University Doctorate Student Grant-in-Aid Program from the Ushioda Memorial Fund.

Institutional Review Board Statement: Not applicable.

Informed Consent Statement: Not applicable.

Data Availability Statement: Not applicable.

Conflicts of Interest: The authors declare no conflict of interest.

References

1. Nath, C.; Rahman, M.; Neo, K.S. Machinability Study of Tungsten Carbide Using PCD Tools under Ultrasonic Elliptical Vibration Cutting. *Int. J. Mach. Tools Manuf.* **2009**, *49*, 1089–1095. [CrossRef]
2. Nath, C.; Rahman, M.; Neo, K.S. A Study on Ultrasonic Elliptical Vibration Cutting of Tungsten Carbide. *J. Mater. Process. Technol.* **2009**, *209*, 4459–4464. [CrossRef]
3. Teramachi, A.; Tarvainen, S.; Yan, J. Drilling of Cemented Carbide by Ultrasonic Vibration-Assisted Grinding on a Flexible Stage. *J. Jpn. Soc. Abras. Technol.* **2019**, *64*, 320–325.
4. Liu, K.; Li, X.P.; Rahman, M.; Liu, X.D. CBN Tool Wear in Ductile Cutting of Tungsten Carbide. *Wear* **2003**, *255*, 1344–1351. [CrossRef]
5. Jahan, M.P.; Rahman, M.; Wong, Y.S. A Review on the Conventional and Micro-Electrodischarge Machining of Tungsten Carbide. *Int. J. Mach. Tools Manuf.* **2011**, *51*, 837–858. [CrossRef]

6. Schubert, N.; Schneider, M.; Michaelis, A.; Manko, M.; Lohrengel, M.M. Electrochemical Machining of Tungsten Carbide. *J. Solid State Electrochem.* **2018**, *22*, 859–868. [CrossRef]
7. Marimuthu, S.; Dunleavey, J.; Smith, B. Picosecond Laser Machining of Tungsten Carbide. *Int. J. Refract. Met. Hard Mater.* **2020**, *92*, 105338. [CrossRef]
8. Hourmand, M.; Sarhan, A.A.D.; Sayuti, M. Characterizing the Effects of Micro Electrical Discharge Machining Parameters on Material Removal Rate during Micro {EDM} Drilling of Tungsten Carbide ({WC}-Co). *IOP Conf. Ser. Mater. Sci. Eng.* **2017**, *241*, 12005. [CrossRef]
9. Assarzadeh, S.; Ghoreishi, M. Mathematical Modeling and Optimization of the Electro-Discharge Machining (EDM) Parameters on Tungsten Carbide Composite: Combining Response Surface Methodology and Desirability Function Technique. *Sci. Iran.* **2015**, *22*, 539–560.
10. Lee, S.H.; Li, X. Study of the Surface Integrity of the Machined Workpiece in the EDM of Tungsten Carbide. *J. Mater. Process. Technol.* **2003**, *139*, 315–321. [CrossRef]
11. Jahan, M.P.; Wong, Y.S.; Rahman, M. A Study on the Fine-Finish Die-Sinking Micro-EDM of Tungsten Carbide Using Different Electrode Materials. *J. Mater. Process. Technol.* **2009**, *209*, 3956–3967. [CrossRef]
12. Shitara, T.; Fujita, K.; Yan, J. Direct Observation of Discharging Phenomena in Vibration-Assisted Micro-Electrical Discharge Machining. *Int. J. Adv. Manuf. Technol.* **2020**, *108*, 1125–1138. [CrossRef]
13. Abdullah, A.; Shabgard, M.R.; Ivanov, A.; Shervanyi-Tabar, M.T. Effect of Ultrasonic-Assisted EDM on the Surface Integrity of Cemented Tungsten Carbide (WC-Co). *Int. J. Adv. Manuf. Technol.* **2008**, *41*, 268. [CrossRef]
14. Singh, N.K.; Singh, Y.; Sharma, A.; Prasad, R. Experimental Investigation of Flushing Approaches on EDM Machinability during Machining of Titanium Alloy. *Mater. Today Proc.* **2021**, *38*, 139–145. [CrossRef]
15. Wang, Z.; Tong, H.; Li, Y.; Li, C. Dielectric Flushing Optimization of Fast Hole EDM Drilling Based on Debris Status Analysis. *Int. J. Adv. Manuf. Technol.* **2018**, *97*, 2409–2417. [CrossRef]
16. Kunieda, M.; Lauwers, B.; Rajurkar, K.P.; Schumacher, B.M. Advancing EDM through Fundamental Insight into the Process. *CIRP Ann.* **2005**, *54*, 64–87. [CrossRef]
17. Kung, K.Y.; Horng, J.T.; Chiang, K.T. Material Removal Rate and Electrode Wear Ratio Study on the Powder Mixed Electrical Discharge Machining of Cobalt-Bonded Tungsten Carbide. *Int. J. Adv. Manuf. Technol.* **2009**, *40*, 95–104. [CrossRef]
18. Singh, J.; Sharma, R.K. Multi-Objective Optimization of Green Powder-Mixed Electrical Discharge Machining of Tungsten Carbide Alloy. *Proc. Inst. Mech. Eng. Part C J. Mech. Eng. Sci.* **2017**, *232*, 2774–2786. [CrossRef]
19. Singh, J.; Kharub, M.; Sahoo, S.K. Analysis of Powder-Mixed EDM Process Characteristics of Tungsten Carbide Alloy by Using GRA Technique. *CVR J. Sci. Technol.* **2020**, *19*, 102–106.
20. Peças, P.; Henriques, E. Effect of the Powder Concentration and Dielectric Flow in the Surface Morphology in Electrical Discharge Machining with Powder-Mixed Dielectric (PMD-EDM). *Int. J. Adv. Manuf. Technol.* **2008**, *37*, 1120–1132. [CrossRef]
21. Liew, P.J.; Yan, J.; Kuriyagawa, T. Carbon Nanofiber Assisted Micro Electro Discharge Machining of Reaction-Bonded Silicon Carbide. *J. Mater. Process. Technol.* **2013**, *213*, 1076–1087. [CrossRef]
22. Sahu, D.R.; Kumar, A.; Roy, B.K.; Mandal, A. *Parametric Investigation into Alumina Nanopowder Mixed EDM of Inconel 825 Alloy Using RSM BT—Advances in Industrial and Production Engineering*; Shanker, K., Shankar, R., Sindhwani, R., Eds.; Springer: Singapore, 2019; pp. 175–184.
23. Jahan, M.P.; Rahman, M.; Wong, Y.S. Modelling and Experimental Investigation on the Effect of Nanopowder-Mixed Dielectric in Micro-Electrodischarge Machining of Tungsten Carbide. *Proc. Inst. Mech. Eng. Part B J. Eng. Manuf.* **2010**, *224*, 1725–1739. [CrossRef]
24. Chow, H.M.; Yang, L.D.; Lin, C.T.; Chen, Y.F. The Use of SiC Powder in Water as Dielectric for Micro-Slit EDM Machining. *J. Mater. Process. Technol.* **2008**, *195*, 160–170. [CrossRef]
25. Furutania, K.; Saneto, A.; Takezawa, H.; Mohri, N.; Miyake, H. Accretion of Titanium Carbide by Electrical Discharge Machining with Powder Suspended in Working Fluid. *Precis. Eng.* **2001**, *25*, 138–144. [CrossRef]
26. Jahan, M.P.; Wong, Y.S.; Rahman, M. A Study on the Quality Micro-Hole Machining of Tungsten Carbide by Micro-EDM Process Using Transistor and RC-Type Pulse Generator. *J. Mater. Process. Technol.* **2009**, *209*, 1706–1716. [CrossRef]
27. Tan, T.H.; Yan, J. Atomic-Scale Characterization of Subsurface Damage and Structural Changes of Single-Crystal Silicon Carbide Subjected to Electrical Discharge Machining. *Acta Mater.* **2017**, *123*, 362–372. [CrossRef]
28. Liao, Y.S.; Chang, T.Y.; Chuang, T.J. An On-Line Monitoring System for a Micro Electrical Discharge Machining (Micro-EDM) Process. *J. Micromech. Microeng.* **2008**, *18*, 035009. [CrossRef]
29. Kitamura, T.; Kunieda, M.; Abe, K. Observation of Relationship between Bubbles and Discharge Locations in EDM Using Transparent Electrodes. *Precis. Eng.* **2015**, *40*, 26–32. [CrossRef]
30. Natsu, W.; Ojima, S.; Kobayashi, T.; Kunieda, M. Temperature Distribution Measurement in EDM Arc Plasma Using Spectroscopy. *JSME Int. J. Ser. C Mech. Syst. Mach. Elem. Manuf.* **2004**, *47*, 384–390. [CrossRef]
31. Jahan, M.P.; Rahman, M.; Wong, Y.S. Study on the Nano-Powder-Mixed Sinking and Milling Micro-EDM of WC-Co. *Int. J. Adv. Manuf. Technol.* **2011**, *53*, 167–180. [CrossRef]
32. Biing Hwa, Y.; Sung Long, C. Characteristics of SKDll by Complex Process of Electrical Discharge Machining Using Liquid Suspended with Alumina Powder. *J. Jpn. Inst. Met.* **1994**, *58*, 1067–1072. [CrossRef]

micromachines

MDPI

Article

Direct Observation of Discharge Phenomena in Vibration-Assisted Micro EDM of Array Structures

Gero Esser and Jiwang Yan *

Department of Mechanical Engineering, Faculty of Science and Technology, Keio University, Yokohama 223-8522, Japan
* Correspondence: yan@mech.keio.ac.jp; Tel.: +81-45-566-1445

Abstract: The batch mode electrical discharge machining (EDM) method has been developed to improve the throughput and accuracy in fabricating array structures, but the process suffers from insufficient debris removal caused by the complex electrode geometry. Tool vibration has been used to improve flushing conditions, but to date the underlying mechanism of the tool vibration on the micro EDM of array structures remains unclear. This study aimed to investigate the effect of tool vibration on the machining process by direct observation of the discharge phenomena in the discharge gap by using a high-speed camera. Micro EDM experiments using 9 and 25 array electrodes were performed, and the effect of tool vibration on the discharge uniformity and tool wear was evaluated. It was found that tool vibration improved the uniformity of the discharge distribution, increased the machining efficiency, and suppressed the tool wear. The discharges occurred in periodic intervals, and the intensity increased with the amplitude of tool vibration. The results of this study indicate that the vibration parameters determine the discharge period duration and intensity to achieve optimum stability and efficiency of the machining process.

Keywords: micro electrical discharge machining; vibration assisted electrical discharge machining (EDM); high-speed camera; array electrode; structured surface

Citation: Esser, G.; Yan, J. Direct Observation of Discharge Phenomena in Vibration-Assisted Micro EDM of Array Structures. *Micromachines* **2022**, *13*, 1286. https://doi.org/10.3390/mi13081286

Academic Editors: Irene Fassi and Francesco Modica

Received: 9 July 2022
Accepted: 8 August 2022
Published: 10 August 2022

1. Introduction

Electrical Discharge Machining (EDM) is a non-traditional machining method with outstanding performance for precision machining of hard and brittle materials [1]. The material removal is based on the thermal effect of electrical discharges in a narrow gap between the tool and workpiece electrode. The gap is filled with a dielectric fluid to enable controlled discharges and the transportation of removed material [2]. Accumulation of removed material as debris particles in the discharge gap is a significant problem in EDM. A higher debris concentration leads to a decreased dielectric breakdown strength, causing unstable machining and short-circuiting [3]. As a result, the machining time increases, and the machining quality can be reduced [4,5]. Uniform debris distribution and timely removal of particles from the gap are thought to be the main factors contributing to a stable machining process [6].

Micro array fabrication is a specific application of micro EDM that has gained increasing attention in recent years. Li et al. [7] used wire Electrical Discharge Grinding (WEDG) to fabricate micro electrodes for micro EDM drilling of arrays with 2 by 128 holes for inkjet printer nozzles. Li et al. [8] fabricated micro hole arrays for the micro embossing of lens arrays by micro EDM drilling. As the accuracy of the tool electrode determines the machining quality of the array, several studies have been carried out to improve the fabrication process and wear performance of micro electrodes for micro EDM drilling of arrays [9,10]. Debnath and Patowari [11] proposed wire EDM to fabricate arrays of rectangular micro pillars as a heat exchanger for small-scale electronic devices. However, the machining accuracy of the arrays was low, with a pillar thickness error of about 25 μm.

The serial process for micro EDM drilling single holes of the array can be parallelized by using array electrode tools. This batch mode machining method is a promising alternative to overcome the challenges of limited throughput and severe wear in the micro EDM drilling of arrays [12,13]. Yi et al. [13] demonstrated the machining of steel shadow masks by batch mode micro EDM with arrays of 4 by 4 rectangular electrodes to fabricate organic thin-film transistors. Zeng et al. [14] studied the performance of ultrasonic vibration-assisted micro EDM machining with arrays of 5 by 5 electrodes. Other applications of micro arrays include filters, fuel nozzles, semiconductor and electronics components, and biotechnology devices [9,12]. The fabrication process of array tools with complex geometry was addressed, and electrode fabrication by reverse EDM, wire EDM, lithography, electroplating, and molding (Lithographie, Galvanoformung, and Abformung (LIGA)), and micro milling was reported [15–18]. Chen [16] designed a multidirectional wire EDM machine tool and fabricated arrays of electrodes with 21 μm width and 700 μm length. Wang et al. [18] developed an integrated machine tool for on-machine micro milling of 3 by 3 rectangular array electrodes and micro EDM of cavities for improved machining accuracy. Sun et al. [19] fabricated array electrodes by reverse EDM and demonstrated that the effect of tool wear could be utilized to fabricate hole arrays with rectangular inlets and circular outlets.

One of the main issues in batch mode manufacturing of arrays is the difficult debris removal due to the complex shaped workpiece–electrode gap. Process instability and nonuniformity caused by debris accumulation is a major problem for array EDM. A variation of the hole diameter and depth across the array has been a common problem in several studies [12,14,20]. To minimize debris accumulation in the discharge gap, methods have been developed to improve debris flushing conditions. Tool rotation is often used for micro EDM drilling, but it is limited to tool geometries with cylindrical shapes and cannot be applied to array tools in batch mode machining [21]. As an alternative, vibration of the tool or workpiece was proposed to improve micro EDM machining of array structures and other non-cylindrical geometries [4]. Unlike other flushing methods, the vibration assistance is effective even when the gap is extremely small, which makes the tool vibration method suitable for micro EDM [22]. Several studies have found that tool vibration can increase the machining stability and reduce the machining time [23,24]. Tong et al. [6] observed an improved accuracy by workpiece vibration in the machining of micro gears. It has been demonstrated that vibration also affects tool wear and surface roughness [23–25]. However, the fundamental mechanism of vibration-assisted EDM and the effect of vibration frequency and amplitude on the machining process for micro EDM of array structures are not yet fully understood. In addition, optimization of the vibration conditions is necessary for industrial applications.

The extremely rapid character of the discharge phenomena in the small gap complicates the investigation of micro EDM mechanisms [2]. Many different physical/chemical processes such as electrical discharges, transfer of heat and material, and chemical reactions interact in the gap. This complexity intricates analytical and numerical modeling [26]. Kitamura et al. [2] developed a direct observation method of the discharge gap that has been used for the clarification of gap phenomena in EDM. The discharge gap of a rectangular electrode was studied. It was found that most of the gap was filled by bubbles, but discharges were mostly observed in the dielectric liquid [2]. In [27], the discharge locations in wire EDM were observed, and the authors suggested that a larger distance of consecutive discharges contributed to a better stability of the machining process. Yue et al. [28] demonstrated that the phenomena in the discharge plasma could be made visible by using bandpass filters or laser illumination. An influence of the gap flow field on the material removal was concluded from the observations. The discharge gap in vibration-assisted micro EDM using cylindrical electrodes was observed by Shitara et al. [24]. The position of discharge spots was evaluated, and the distribution of discharge locations was found to be more uniform with the vibration of the electrode. A so-called interrupted discharge phenomenon was observed where discharges only occurred during a short period of the electrode movement.

The schematic in Figure 1 illustrates the flow field in the gap in micro EDM of single and array electrodes with/without tool vibration. A higher flow resistance results from the array electrode geometry with a long, curved gap. This is associated with a more difficult machining without tool vibration. In general, the tool vibration generates a symmetrical fluid motion in the center region of the electrode end towards the side gap. In the gap of the array electrode, the flow is symmetrical in the center but one-directional towards the side areas. The effects of the difference in local flow direction and flow resistance on the non-uniformity of machined surface accuracy are unknown.

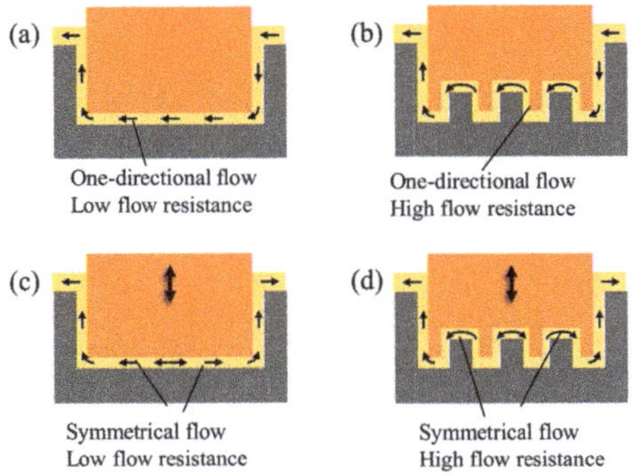

Figure 1. Effect of vibration and difference of the flow field (**a**) single electrode without vibration, (**b**) array electrode without vibration, (**c**) single electrode with vibration, (**d**) array electrode with vibration.

To date, most studies on the gap observation of EDM, especially vibration-assisted EDM, used simple tool geometries such as single rectangular or circular electrodes. Though the application of tool vibration shows its full potential in micro EDM of complex geometries, such as multiple-cavity arrays, the mechanisms underlying this specific situation have not been clarified.

In this study, the direct observation of the discharge gap is applied to micro EDM with array electrodes. The study aims to investigate the fundamental effects of tool vibration assistance by evaluating the distribution and time-dependent change of the discharges. The machining performance improvement mechanism will be examined, and the potential of vibration assistance for micro EDM of complex geometries will be demonstrated. The findings of the discharge phenomena and mechanisms will be useful for the development of optimal vibration conditions for high-uniformity micro fabrication of large-area array structures.

2. Materials and Methods

Micro EDM experiments were performed, and the discharges were recorded by direct gap observation. Image processing was applied to evaluate the discharge locations.

An actuator was required to generate the vibration motion of the tool electrode. A piezoelectric actuator was selected due to the fast response time and high acceleration. The design and verification process described in [29] was followed. The vibration unit illustrated in Figure 2 was designed using a piezoelectric transducer (AE0505D16F, Tokin Co., Sendai, Japan). The frequency and amplitude of the driving voltage were controlled to set the desired displacement. The axial and radial tool motion was analyzed by a laser displacement sensor (LK-H025, Keyence, Osaka, Japan), and low radial movement was confirmed. The vibration modes in Table 1 were selected for the experiments. Throughout this paper, we use the term amplitude for the peak-to-valley amplitude of the vibration motion.

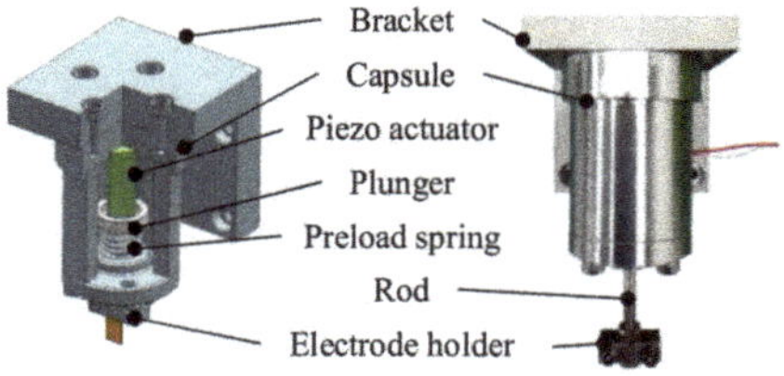

Figure 2. Cross-section and overview of the components of the vibration unit.

Table 1. Vibration parameters selected for the experiments.

Frequency (Hz)	Axial Amplitude (μm)	Radial Amplitude (x/y) (μm/μm)
50	3	0.12/0.11
	5	0.24/0.2
400	3	0.58/0.58
	5	1.05/1.05
1000	3	0.94/0.81
	5	1.1/0.63

A resistor-capacitance (RC) type EDM machine (MG-ED72, Panasonic, Kadoma, Japan) was used to conduct the experiments. As seen in Figure 3, the discharge energy was controlled by setting the voltage and capacitance of the EDM circuit. The experimental parameters of the machine are detailed in Table 2.

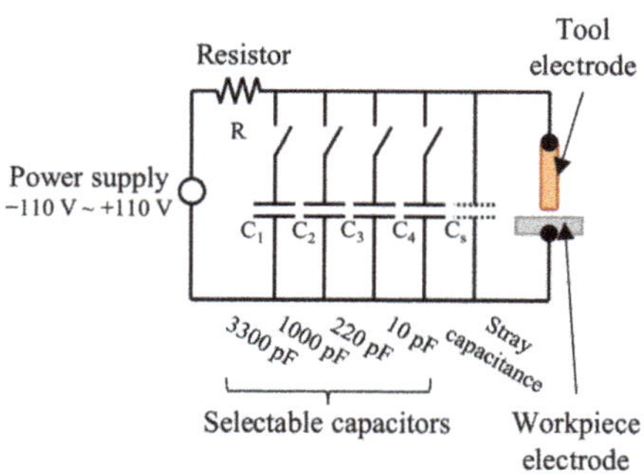

Figure 3. Schematic of the resistor–capacitor type EDM circuit.

Table 2. Experimental parameters of micro EDM experiments.

Parameter	Value
Voltage (V)	100
Capacitance (pF)	220
Feed rate (μm/s)	5
Machining depth (μm)	80
Workpiece electrode	Anode: 4H-SiC (+)
Tool electrode	Cathode: Cu (−)
Working fluid	EDM oil (CASTY-LUBE EDS)

Micro EDM experiments with arrays of 9 and 25 electrodes were performed, and the effect of tool vibration on the machining process was investigated. Figure 4a illustrates the tool array of 9 rectangular electrodes with a size of 100 μm by 100 μm and a height of 200 μm. The electrodes were fabricated from a copper rod of 2 mm diameter by micro milling as it is economical and less time-consuming than other manufacturing methods while achieving sufficient accuracy for the observation experiments. The top surface of the array was hand polished to remove burrs from micro milling and to increase the surface reflectivity of the electrodes to ease the high-speed camera observation. Before the experiments, the tool and workpiece were cleaned in an ultrasonic bath with EDM oil. Tool arrays of 25 cylindrical electrodes are illustrated in Figure 4c. The arrays were fabricated from copper material by reverse EDM to achieve higher accuracy and smaller spacing of the electrodes. An electrode diameter of 80 μm, spacing of 200 μm, and electrode length of 200 μm was achieved.

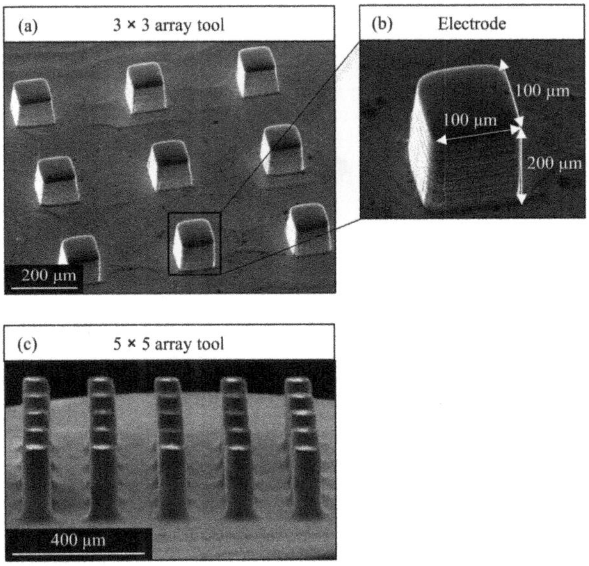

Figure 4. SEM image of the tool array electrodes used in this study. (**a**) Tool array of 9 rectangular electrodes (**b**) Detailed view of an electrode of the array (**c**) Tool array of 25 cylindrical electrodes.

A workpiece material with sufficient electrical conductivity for EDM processing and transparency to enable observation of the gap phenomena was required. Semiconductor materials such as SiC and Ga_2O_3 have been used in previous studies [30]. Despite lower machinability [30], SiC was selected in this work due to its lower cost and better availability. Rectangular pieces of the 4H-SiC wafer (10 mm × 10 mm × 365.7 μm) were used as workpiece electrodes. 4H-SiC is a polar material due to the bilayer stacking sequence of Silicon and Carbon [31,32]. The machining was performed on the Carbon face of the SiC material.

The observation was conducted by a high-speed camera (FASTCAM Mini AX50, Photron Ltd., Tokyo, Japan). The settings of the high-speed camera are shown in Table 3. The experiment setup is illustrated in Figure 5 and was based on previous studies on direct gap observation [2,24,30]. As seen from the schematic in Figure 6, the vibration unit was connected to a signal generator and amplifier and mounted to the z-axis of the micro EDM machine. The tool electrode was attached to the vibration unit. The SiC workpiece was installed in an acrylic tank and submerged in the dielectric fluid. A mirror was placed on an inclined stand beneath the tank. The high-speed camera and high luminance lights were set up in front of the micro EDM machine. The discharge gap was observed through the transparent SiC workpiece from below via the mirror.

Table 3. High-speed camera settings.

Parameter	Value
Framerate (fps)	20,000
Recording time (s)	8.7345
Resolution	256 × 256 pixel

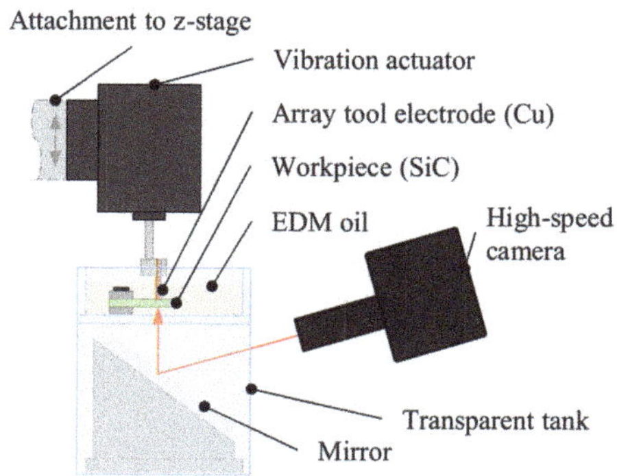

Figure 5. Schematic of the setup for the direct observation of the discharge gap.

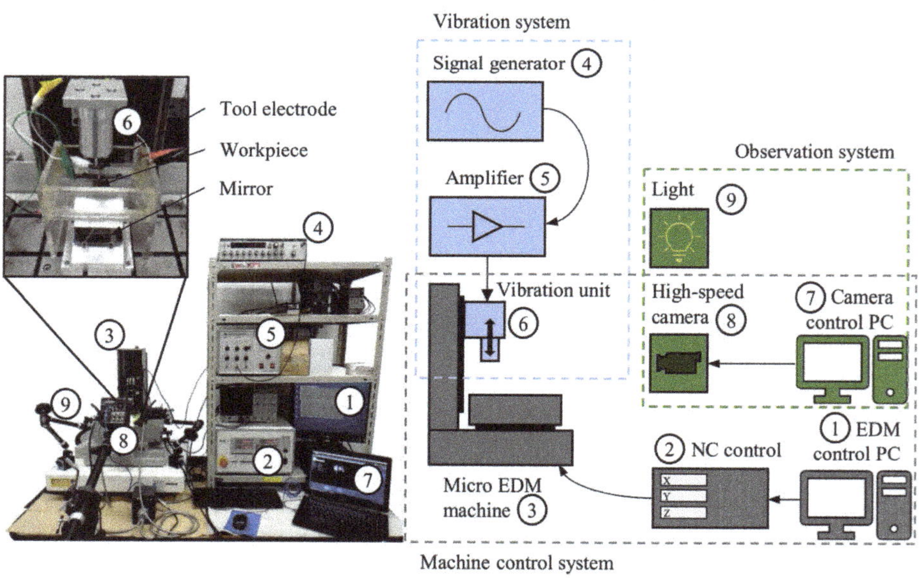

Figure 6. Block diagram representation and photo of the experiment setup.

Die sinking EDM was performed, and the machining progress was tracked. The gap observation settings were optimized to show only the discharges and minimize the interference of bubbles, debris, and other effects. The video recording was started after the electrode reached a machining depth of 50 µm. Machining showed significant differences at this depth, and suitable evaluation results were obtained due to low interference of reflections on the recorded videos.

After the experiments, the gap videos and the machined cavities were analyzed. An image-processing script was developed in MATLAB to evaluate the discharges in the recorded videos. The machined surface of the SiC samples was scanned by a non-contact laser profilometer (NH-3SPs, Mitaka Kohki Co., Tokyo, Japan) and inspected by SEM (Inspect F50, FEI Co., Hillsboro, OR, USA).

3. Results and Discussion

3.1. Discharge Distribution States

In contrast to machining without vibration, where discharges occurred randomly, tool vibration led to a periodic discharge. In regular intervals, the discharges were stopped or only observed sporadically. In the experiments, three different discharge states were observed through the videos. Generally, multiple discharge spots were distributed over several electrodes of the array. Figure 7a shows a single frame recorded during this regular machining state. Occasionally discharges occurred concentrated on a single electrode in a high number. This discharge concentration is visible as a bright area in Figure 7b. The discharge concentration usually occurred at the end of the regular discharge period. Intense discharge continued for an extended time in the vibration cycle, even after regular discharge on the other electrodes was stopped. The discharge concentration phenomenon often appeared repeatedly over multiple vibration cycles. Infrequently, single bright spots, as shown in Figure 7c, that remained for several consecutive frames were observed. In most cases, this phenomenon followed a discharge concentration. The bright spot was assumed to be caused by arcing or short-circuiting when debris particles formed a bridge between the tool and the workpiece. The bright spot phenomenon often observed after the discharge concentration in the videos provides additional support for the findings of Shitara et al. [24] that discharge concentration might be a precursor to short-circuiting.

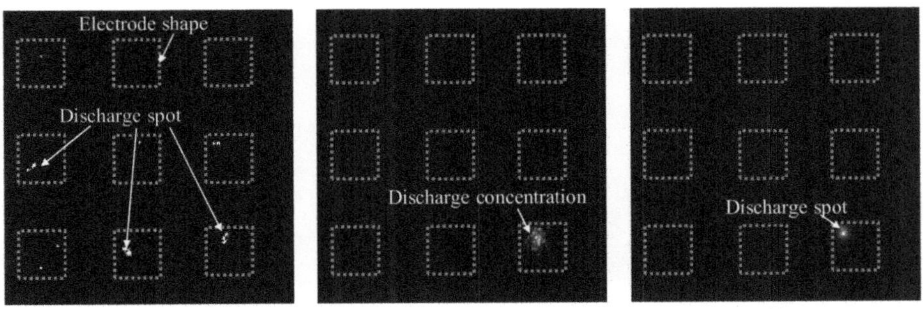

(a) (b) (c)

Figure 7. Different discharge states observed at 50 Hz and 5 μm vibration with the EDM parameters specified in Table 2: (**a**) unform distribution, (**b**) concentration at a single electrode, (**c**) single discharge spot.

3.2. Effect of Vibration on Discharge Distribution

The video was evaluated by thresholding the grayscale video frames in MATLAB. The location and number of the bright discharge spots in each frame were detected. The number of discharges with tool vibration is illustrated in Figure 8. The highest discharge rate was recorded for vibration at 50 Hz and 3 μm, but irregularities were seen twice. All other vibration modes showed good linearity. The discharge rates at 50 Hz, 400 Hz, and 1000 Hz with an amplitude of 5 μm were higher than the discharge rates at 400 Hz and 1000 Hz with 3 μm amplitude.

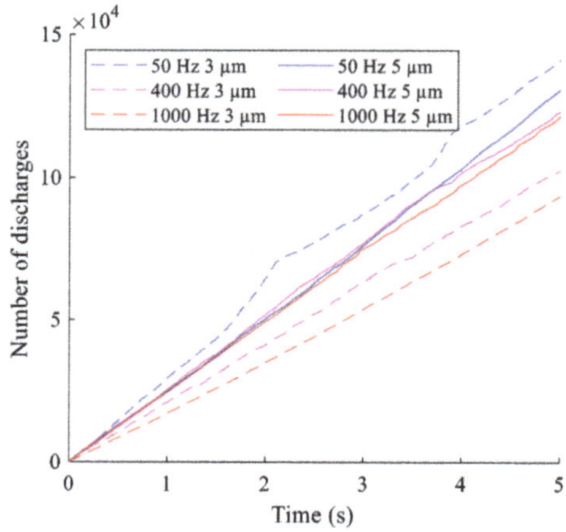

Figure 8. Number of discharges over the recording time.

The frames were partitioned in a grid of 50 by 50 elements, and the number of discharge spots in each element was counted. Without tool vibration, only a small number of discharges were observed, concentrated on a few electrodes of the array. This resulted in the highly non-uniform distribution, as shown in Figure 9.

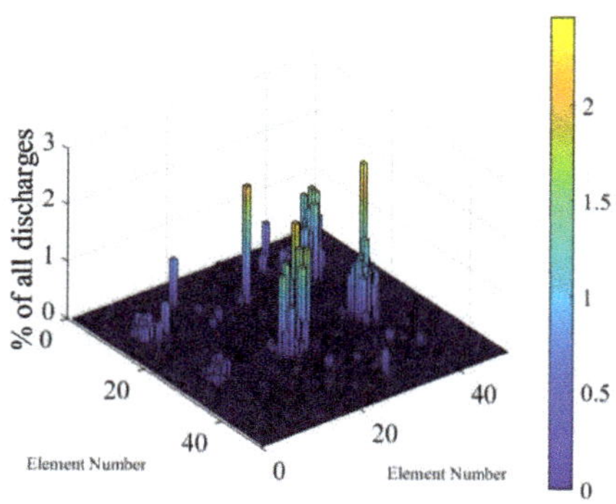

Figure 9. Distribution of discharges without tool vibration and the EDM parameters specified in Table 2.

Such non-uniform distribution was also observed during vibration-assisted EDM at 50 Hz and 3 μm amplitude. As illustrated in Figure 10a, a much higher number of discharges was observed on the top-right electrode, while the number of discharges on the center electrode was slightly increased. The gap observation video showed that this was a result of discharge concentration. The discharge concentration was observed twice on the top-right and once on the center electrode during the recorded timespan.

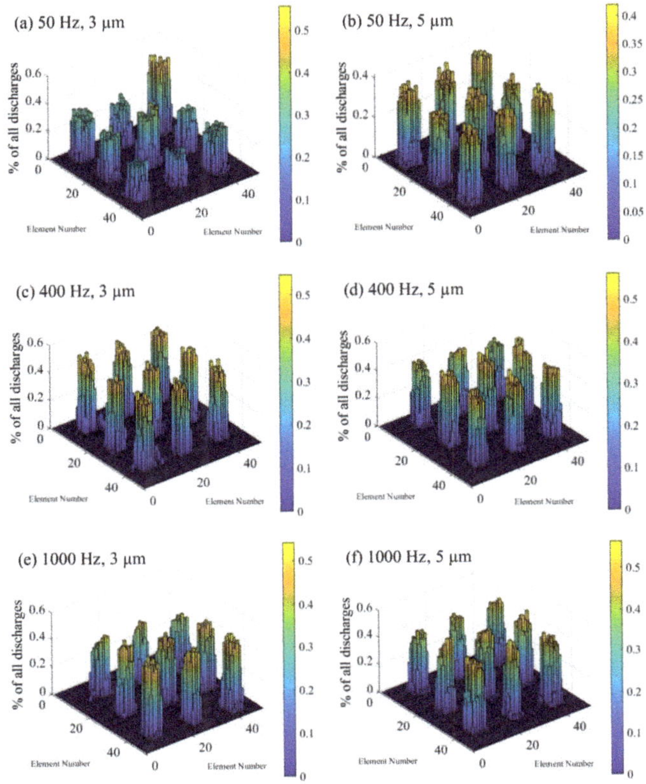

Figure 10. Distribution of discharges with tool vibration and the EDM parameters specified in Table 2. The vibration frequencies and amplitudes are: (**a**) 50 Hz, 3 μm (**b**) 50 Hz, 5 μm (**c**) 400 Hz, 3 μm (**d**) 400 Hz, 5 μm (**e**) 1000 Hz, 3 μm (**f**) 1000 Hz, 5 μm.

The distribution of discharge spots for the vibration mode of 50 Hz and 5 μm amplitude is given in Figure 10b. The uniformity of the discharge distribution was significantly increased. This indicates higher stability of the machining process. The recorded video showed that discharge concentration occurred once on the bottom-right electrode. However, the concentration continued only for a few vibration cycles and did not influence the discharge distribution. Subsequently, the bright spot phenomenon was observed in the location where the discharge concentration occurred.

The discharge distributions at 400 Hz and 1000 Hz with both 3 μm and 5 μm amplitude are illustrated in Figure 10c–f. Machining at these vibration parameters showed good uniformity. A correlation of the discharge uniformity to the frequency or amplitude of the vibration was not identified for these parameters. Compared to the 50 Hz and 5 μm vibration result, the distribution was slightly less uniform. The reasons for this can be found in the higher variation of the discharge numbers in the elements on individual electrode positions. An uneven electrode surface may have resulted in regions with more discharges. Elevations could have been caused by the adhesion of debris or deposited carbon from the EDM oil. A gradual decrease of discharges across the array was observed in some cases, such as at 400 Hz and 5 μm. A higher concentration of debris may have increased the discharge number in one electrode area. It is also possible that some discharges were not visible due to a brightness gradient of the lighting in the gap.

3.3. *Effect of Vibration on the Discharge Period*

The duration and intensity of the periods with/without discharge were analyzed. The number of discharge spots detected per frame at an arbitrary position is displayed in Figure 11. With increasing frequency, the beginning and end of the discharge periods became more apparent as the number of stray discharges decreased. The duration and number of discharges of the discharge periods was evaluated, and the average discharge intensity over the recorded video is illustrated in Table 4. There was a significant correlation between the vibration amplitude and the intensity of the discharge rate. At all observed vibration frequencies, a higher rate of discharge spots during the discharge period was found when the vibration amplitude was increased from 3 μm to 5 μm.

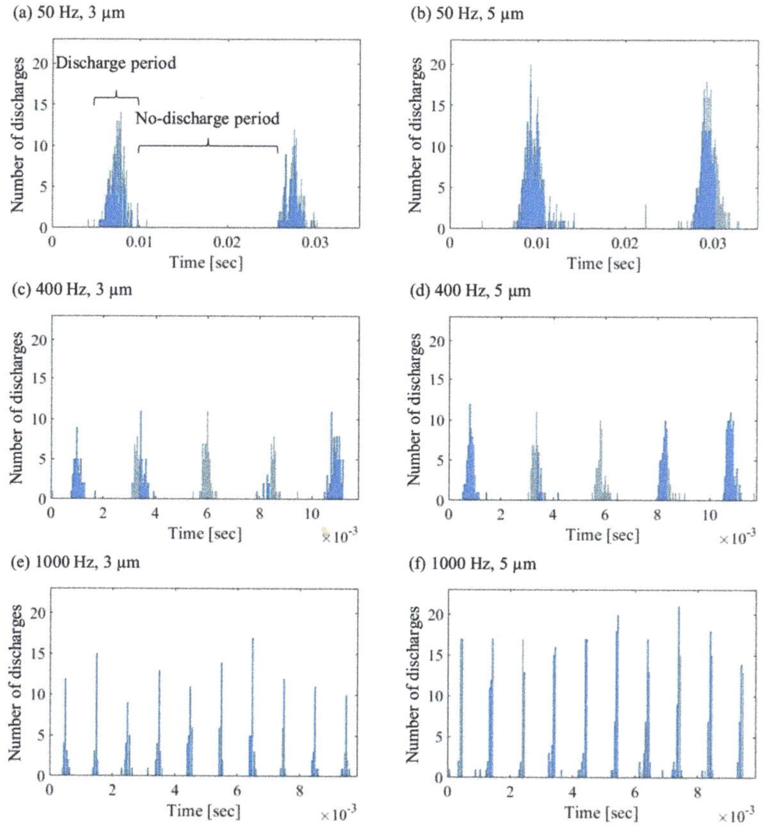

Figure 11. Discharge periods with the parameters specified in Table 2. The vibration frequencies and amplitudes are: (**a**) 50 Hz, 3 μm (**b**) 50 Hz, 5 μm (**c**) 400 Hz, 3 μm (**d**) 400 Hz, 5 μm (**e**) 1000 Hz, 3 μm (**f**) 1000 Hz, 5 μm.

Table 4. Average intensity of discharge during the discharge periods.

Frequency (Hz)	Number of Discharges Per ms Amplitude 3 μm	Amplitude 5 μm
50	100	123
400	63	80
1000	76	115

Figure 12 shows the ratio of the average discharge period duration to the average no-discharge period. It was found that an increase of both the vibration frequency and amplitude led to a shorter discharge period. This was attributed to the relation of the vibration parameters and the electrode velocity, as detailed in Equation (1). The electrode velocity, v, varies with the time, t. The vibration frequency and amplitude are denoted by f and $\hat{a}$, while φ is the phase shift.

$$v(t) = \hat{a}\pi f \, \sin(2\pi f \, t + \varphi) \tag{1}$$

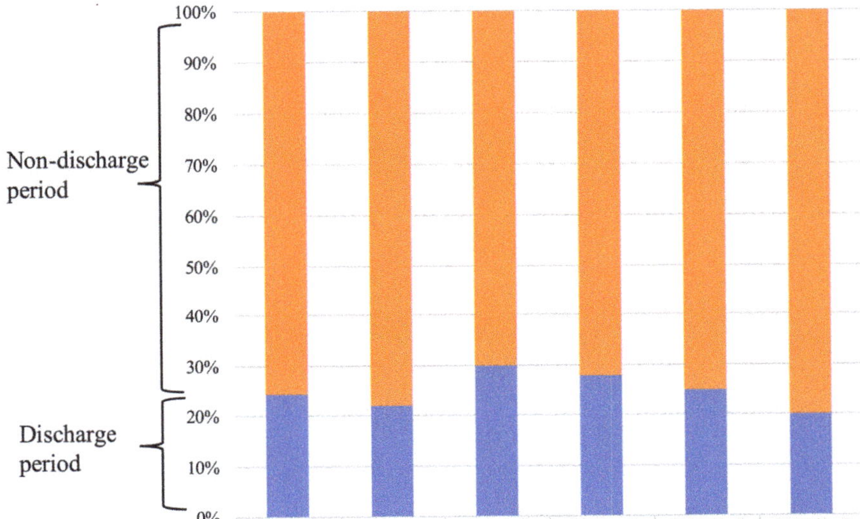

	50 Hz 3 µm	5 0 Hz 5 µm	400 Hz 3 µm	400 Hz 5 µm	1000 Hz 3 µm	1000 Hz 5 µm
Ratio of no-discharge period	76%	78%	70%	72%	75%	80%
Ratio of discharge period	24%	22%	30%	28%	25%	20%

Figure 12. Ratio of the average discharge period and no-discharge period duration in the vibration cycle.

The tool follows a sinusoidal trajectory, and increasing frequency and amplitude of the vibration leads to a higher electrode velocity. As shown in Figure 13, the residence time of the tool in the beneficial discharge range decreases when the amplitude is changed from 3 µm to 5 µm.

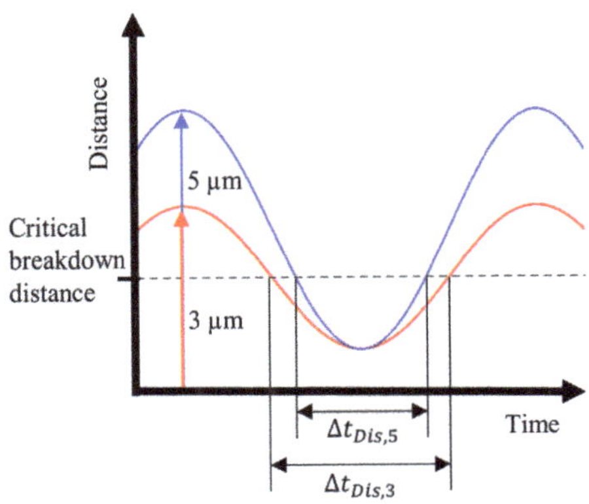

Figure 13. Distance of the tool electrode to the workpiece surface during one vibration cycle.

3.4. Effect of Vibration on the Machining Time

The progression of the tool was recorded, and the machining time to a depth of 80 μm is illustrated in Figure 14. The highest machining time was recorded without vibration. As seen from the tool feed in Figure 15, the tool progression followed the specified feed rate of 5 μm/second for only about 5 μm and then slowed down significantly. The machine control performed a reciprocation motion of the tool after abnormal discharge was detected to flush debris from the gap. This jump motion occurred frequently, indicating bad flushing conditions and regular short-circuiting.

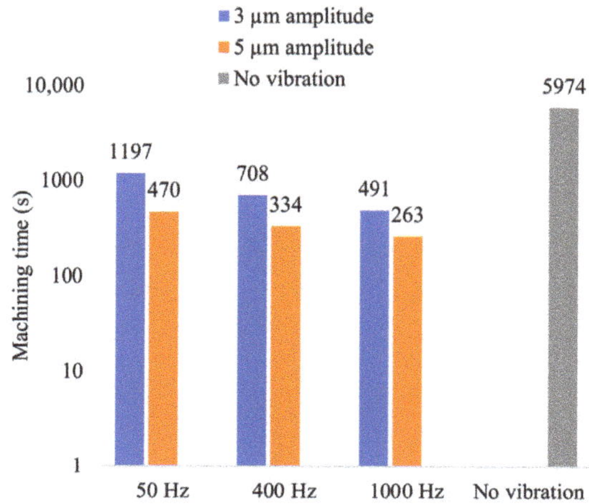

Figure 14. Machining time for a depth of 80 μm with the parameters specified in Table 2.

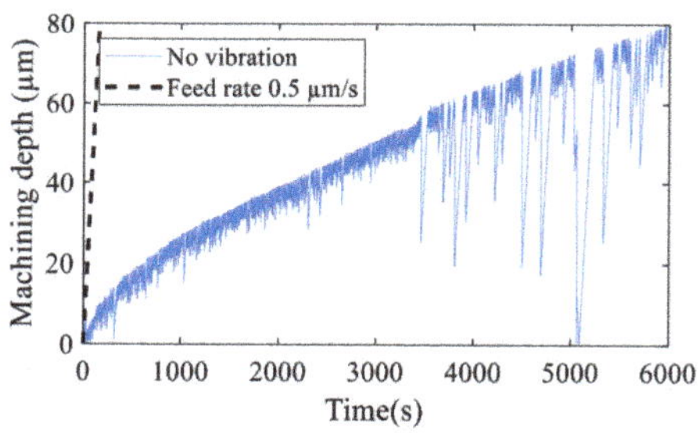

Figure 15. Tool feed without vibration and the parameters specified in Table 2.

The tool electrode feed with vibration over the machining time is illustrated in Figure 16. At a vibration with 3 μm amplitude, the machining rate followed the feed rate to a higher machining depth than without vibration. Tool jump motion did not occur up to a machining depth of 30 μm at 50 Hz and 400 Hz and up to 40 μm at 1000 Hz. The machining rate progressed slower from that point on. The frequency of tool reciprocation increased with a higher machining depth but remained less frequent than without vibration, as seen in Figure 15.

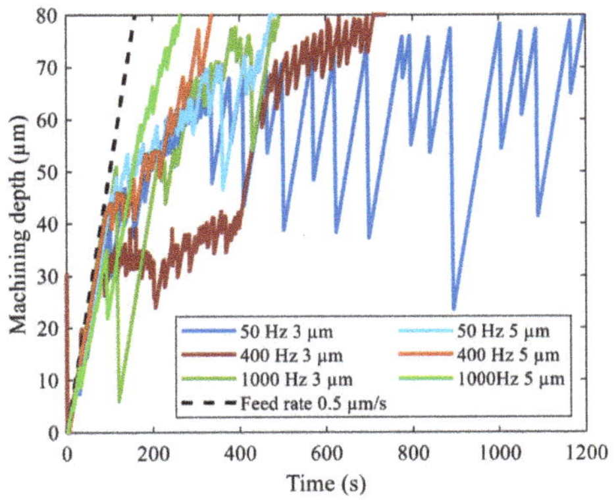

Figure 16. Tool feed with vibration and the parameters specified in Table 2.

The range of stable machining was further increased at a vibration amplitude of 5 μm. Increasing tool jump motion was observed after a machining depth of 40 μm at a vibration frequency of 50 Hz and 400 Hz. At 1000 Hz, tool reciprocation became more frequent after a machining depth of 55 μm was reached. The jump motion was less frequent than at 3 μm machining. The beneficial effect of tool vibration on the machining time is therefore attributed to better machining conditions that reduce the frequency of tool retraction.

3.5. *Effect of Vibration on Surface Roughness*

The arithmetic average roughness, Ra, of the array cavities was measured. Due to the small size of the cavities, a non-standardized measurement was performed. A scanning pitch of 0.5 μm, a sampling pitch of 0.5 μm, and a cutoff length of 0.25 mm were used. The mean value of the nine cavities of each array was calculated. As illustrated in Figure 17, machining with vibration at 50 Hz and 1000 Hz led to similar surface roughness to that without vibration. The surface roughness was slightly higher at 400 Hz vibration. In addition, as a general trend, a higher vibration amplitude decreased the surface roughness, except at 1000 Hz frequency.

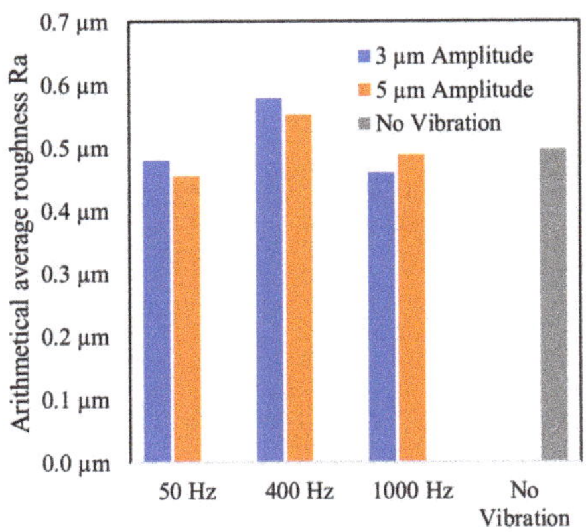

Figure 17. Average roughness of the array cavities.

Figure 18 shows an SEM image of the workpiece surface machined without vibration. The surface is characterized by large flat craters without distinctive crater walls. The density of the discharge craters was low, and adhesion of debris particles was visible.

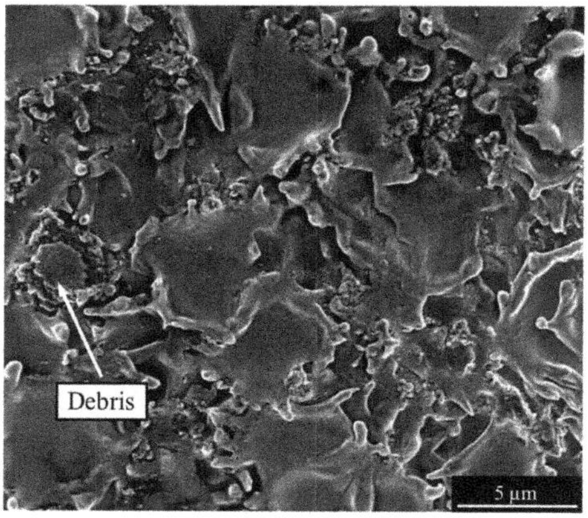

Figure 18. SEM image of the workpiece surface machined without vibration and the parameters specified in Table 2.

The SEM images in Figure 19 show the sample surfaces machined with tool vibration. Compared to Figure 18, the density of the discharge craters is higher, and the discharge craters are smaller but deeper. The SEM observation of the sample surface in Figure 19 revealed differences in the crater size and shape with different vibration parameters. At 50 Hz and 1000 Hz, the discharge craters appeared flat with thin walls. The surfaces machined at 400 Hz frequency in Figure 19c,d showed deeper craters. The thickness of the crater walls was higher, with more remolten material on the crater edges. The sample machined at 1000 Hz and 5 µm in Figure 19f also showed a few adhered debris particles and porosity of the crater walls. This might have caused the higher roughness despite smaller crater depths compared to machining with vibration at 1000 Hz and 3 µm.

The observation of the workpiece surface and measurement of surface roughness indicated that the discharge craters changed with different vibration settings. As Wang et al. [33] reported, a higher amplitude of the electrode vibration can lead to a shallower crater depth as a result of a shift in the discharge plasma origin. This might be a reason for the decrease in surface roughness from 3 µm to 5 µm vibration amplitudes at the frequencies of 50 Hz and 400 Hz. Another reason might be that the crater density at a larger amplitude is higher, which causes more overlaps among the craters, and in turn, lower roughness.

A higher discharge energy can melt more workpiece material, resulting in a correlation to the crater size [34]. The variance of the crater size indicates changing discharge energy at different vibration parameters. This could result from a difference in the gap width. A changing gap width is also demonstrated by the increasing intensity of the discharges at higher frequencies. With a smaller gap width and high discharge intensity, the risk of debris adhesion is increased, which was observed on the workpiece machined at 1000 Hz and 5 µm vibration. The workpiece machined without vibration provides evidence of a high concentration of debris in the gap, resulting in low discharge energy with small craters and a high probability of debris adhesion.

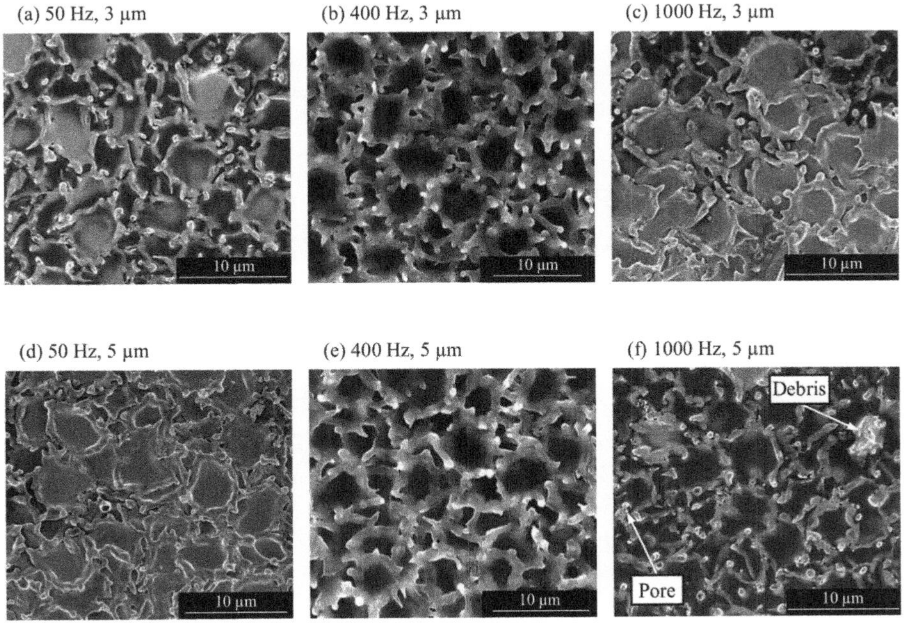

Figure 19. SEM image of the workpiece surface with the parameters specified in Table 2. The vibration frequencies and amplitudes are: (**a**) 50 Hz, 3 µm (**b**) 50 Hz, 5 µm (**c**) 400 Hz, 3 µm (**d**) 400 Hz, 5 µm (**e**) 1000 Hz, 3 µm (**f**) 1000 Hz, 5 µm.

3.6. *Effect of Vibration on Hole Diameter Uniformity*

One of the major challenges in micro-EDM of arrays is the non-uniformity of the machined features. Non-uniformity of the hole diameter was observed in experiments with and without tool vibration. So far, studies have focused on tool vibration in the ultrasonic range [12,20].

An experiment was performed with the same materials and parameters as the previous experiment specified in Table 2, and the hole diameter uniformity was studied. To prevent small deviations of the electrode diameter from influencing the results, the diametral overcut was used to assess the hole uniformity. The diametral overcut ratio is calculated according to Equation (2), where d_{hole} is the hole entrance diameter and $d_{electrode}$ is the electrode diameter. The hole entrance area, A_{hole}, and cross-sectional electrode area, $A_{electrode}$, were evaluated by 3D laser scanning microscope (LEXT OLS4100, Olympus Corp., Tokyo, Japan).

$$\text{Overcut Ratio} = \frac{\frac{\pi}{4}\left(d_{hole}{}^2 - d_{electrode}{}^2\right)}{\frac{\pi}{4}\, d_{hole}{}^2} = \frac{A_{hole} - A_{electrode}}{A_{hole}} \tag{2}$$

Figure 20 illustrates the overcut of the arrays. The evaluation was performed along the array's main diagonal, antidiagonal, horizontal, and vertical directions. It must be noted that the variance of hole number 3 is zero, as it represents the central hole along with all evaluation directions.

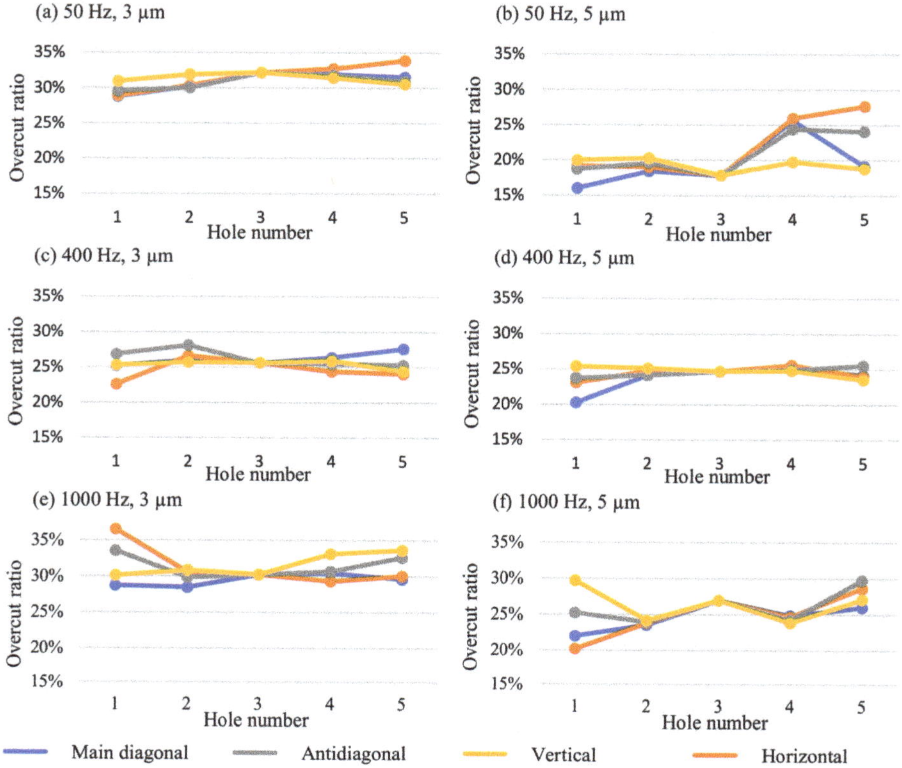

Figure 20. Ratio of the overcut to the hole entrance area with the parameters specified in Table 2. The vibration frequencies and amplitudes are: (**a**) 50 Hz, 3 μm (**b**) 50 Hz, 5 μm (**c**) 400 Hz, 3 μm (**d**) 400 Hz, 5 μm (**e**) 1000 Hz, 3 μm (**f**) 1000 Hz, 5 μm.

As illustrated in Figure 20a, the array machined at 50 Hz and 3 μm showed a higher overcut in the center of the array, which matched the results of previous studies. A standard deviation of 1.31% from the mean overcut ratio of 30.95% was obtained. The overcut ratio at 50 Hz and 5 μm was smaller, with a mean value of 21.31%. Linked to the decreased overcut ratio is a smaller gap width. This might explain the non-uniform overcut seen in Figure 20b as a cause of a more difficult debris removal through the smaller side gaps. The standard deviation of the overcut ratio was 3.47%. Machining at 400 Hz and 3 μm resulted in a mean overcut ratio of 25.48%. The values for the individual holes of the array machined at 400 Hz and 3 μm are illustrated in Figure 20c. The irregularity of the overcut ratio was smaller than at 50 Hz and 3 μm vibration, as indicated by the standard deviation of 1.2%. At 400 Hz and 5 μm, the best uniformity of the hole overcut ratio was achieved with a standard deviation of 1.09%. As demonstrated by Figure 20d, a uniform overcut in the center of the array and deviations with smaller overcut at the edges were observed. The mean overcut ratio was 24.23%. Figure 20e shows the overcut at 1000 Hz and 3 μm. There was a high variance of the overcut ratio at the edges of the array, while the uniformity increased towards the center. This was also represented by the standard deviation of 2.13%. The mean overcut ratio of 31.28% indicated a higher gap width. A similar trend was observed at 1000 Hz and 5 μm, as detailed in Figure 20f. A scattered overcut ratio on the edges was recorded that caused a high standard deviation of 2.74%. As seen by the higher overcut ratio, the gap width increased towards the center. The variance of the overcut decreased likewise. A mean overcut ratio of 25.98% was obtained.

The higher overcut around the edges of the array was unexpected. One possible explanation would be that debris from the electrodes in the center is transported to the outside by the vibrating movement. The debris is sucked back in the holes at the edges of the array and leads to the increased debris concentration in these areas. In earlier findings, a larger hole diameter was attributed to bad debris flushing from the center holes of the array. Hole arrays machined at 400 Hz and 1000 Hz with 5 μm vibration amplitude are compared in Figure 21. A higher variance of the hole diameter can be seen in Figure 21b.

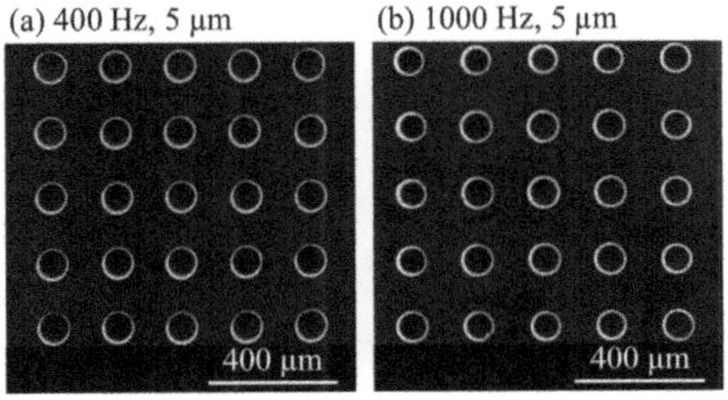

Figure 21. SEM images of 5 × 5 arrays (**a**) Machined with vibration at 400 Hz and 5 μm amplitude (**b**) Machined with vibration at 1000 Hz and 5 μm amplitude.

3.7. Effect of Vibration on Tool Wear

The tool wear was studied by SEM microscopy after a single use of the tool. Compared to the tool before use in Figure 4b, the rounded edges of the array used at 50 Hz and 3 μm vibration in Figure 22a indicate significant wear. The tool wear also caused the formation of craters on the top surface of the electrodes. Deposition of carbon and debris can be seen as black material in the upper part of the electrodes.

Figure 22. SEM images of the 5 × 5 array tools after use with the parameters specified in Table 2. The vibration frequencies and amplitudes are: (**a**) 50 Hz, 3 μm (**b**) 50 Hz, 5 μm (**c**) 400 Hz, 3 μm (**d**) 400 Hz, 5 μm (**e**) 1000 Hz, 3 μm (**f**) 1000 Hz, 5 μm.

The effects of wear were smaller on the array used at 400 Hz and 3 μm vibration. As shown in Figure 22b, craters were formed on some electrodes, mainly in the center of the array. A similar amount of carbon and debris was deposited on the electrode surface. The array used at 1000 Hz and 3 μm experienced less wear, as indicated by the sharper edges and flat electrode top faces in Figure 22c. Less material was deposited on the electrodes, leaving discharge craters visible. The tools used with a vibration amplitude of 5 μm showed a similar wear characteristic. High wear with craters in the electrode center was observed at 50 Hz vibration frequency in Figure 22d. Carbon and debris were deposited mainly on the electrode side faces, and the amount was smaller compared to the tool used at 50 Hz and 3 μm. The tool wear was significantly smaller after machining at 400 Hz and 5 μm, and no craters were formed on the electrodes, but a high build-up of material can be seen on the electrodes in Figure 22e. Vibration at 1000 Hz and 5 μm resulted in minimal tool wear. The edges of the electrodes were only slightly rounded, and the top surface remained flat, as shown in Figure 22f. Very little debris and carbon was deposited on the side faces, but a layer of material formed on the top surfaces of the electrodes.

3.8. Mechanism of Vibration-Induced Discharge Intensity Change

Tool vibration caused regular discharge intervals, which is in line with the concept of a beneficial discharge range. It was proposed that discharges only occur when the distance between tool and workpiece electrode is below a certain threshold distance [6]. Tong et al. [6] suggested a relationship between the optimum amplitude and the discharge distance between the tool and the workpiece. However, the model considered vibration to reduce the negative influence of a low response frequency of the machine control. It failed to address the effect of vibration on the discharge process.

It was observed in this study that the discharge intensity increased with higher amplitudes. Figure 23 compares the interaction of debris and discharges over one vibration cycle at low and high amplitude vibration. Due to the flushing of debris formed in the previous vibration cycle, the debris concentration in the gap is low when the electrode is in the top position, as shown in Figure 23(a1,a2). It is assumed that more effective flushing at higher amplitudes contributes to a lower debris concentration in Figure 23(a2). Therefore, the dielectric breakdown strength is higher. The first discharge of the cycle at the higher amplitude in Figure 23(b2) occurs at a smaller gap width, compared to the lower amplitude seen in Figure 23(b1). This discharge forms new debris particles in the gap, and the debris concentration increases. The rising debris concentration leads to a reduced breakdown strength and facilitates more discharges. Due to the smaller fluid volume at the higher amplitude, the debris concentration in Figure 23(c2) is significantly higher than in Figure 23(c1). More discharges are triggered as a result of the drastically reduced breakdown strength. As these discharges also release new debris particles, a self-reproduction effect of the discharges is achieved. This situation promotes the overlapping of craters, leading to a decrease in surface roughness, as seen in Figures 17 and 19. When the tool electrode moves upwards, new fluid enters the gap and reduces the debris concentration. The discharge process stops when the tool electrode surpasses the critical breakdown distance to the workpiece, as seen from Figure 23(d1,d2).

The concentrated discharge phenomenon might occur when the debris concentration becomes extremely high towards the end of the regular discharge period. As a consequence of the very low breakdown strength, the concentrated discharge can continue while the regular discharge is stopped. The gap observation showed that the discharge concentration was more likely to subside with tool vibration before arcing could occur. This provides evidence that the discharge period is linked to the material removal rate, as the material removal rate benefits from a high discharge rate and low frequency of short-circuiting. The tool vibration parameters must be carefully selected to achieve the optimum machining conditions. If the frequency is too low, the discharge period will continue too long, and short-circuiting will occur. If the frequency is too high, fewer discharges per cycle will occur, and the machining efficiency is reduced.

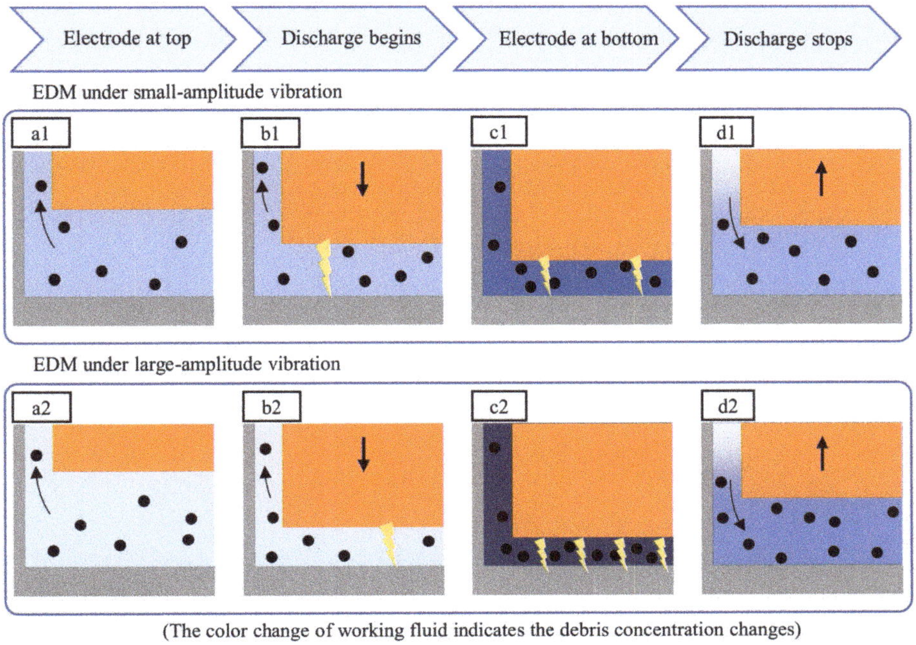

Figure 23. Mechanism of the increase of the discharge intensity with a higher vibration amplitude. (**a1**) Electrode at top, (**b1**) discharge begins, (**c1**) electrode at bottom, and (**d1**) discharge stops under small-amplitude vibration; (**a2**) Electrode at top, (**b2**) discharge begins, (**c2**) electrode at bottom, and (**d2**) discharge stops under large-amplitude vibration.

4. Conclusions

The effect of tool vibration on the electrical discharge machining process with micro array electrodes was studied by direct observation of the discharge gap. The influence of vibration on the microscopic surface topography and array uniformity was further evaluated. The following conclusions result from the study:

1. The uniformity of the discharge spot distribution on the individual electrodes of the array was improved by tool vibration. The higher uniformity resulted from a higher discharge rate and fewer irregular discharges, such as discharge concentration and short-circuiting.
2. Tool vibration led to a periodic discharge process. The discharges only occurred when the tool electrode was at a certain distance to the workpiece due to the electrical breakdown strength of the fluid.
3. Higher vibration frequency and amplitude reduced the ratio of the discharge period to the time without discharges in the vibration cycles. With increasing vibration amplitude, the intensity of the discharge period increased.
4. Tool vibration reduced the machining time. An increase in both vibration frequency and amplitude improved the machining time.
5. Vibration assistance led to a change in the surface roughness of the machined surfaces. This was primarily due to a difference in the discharge crater size and resolidified material. The crater size and density were reduced without vibration, and more debris adhered to the surface.
6. Arrays of 25 holes were machined, and the hole diameter uniformity was studied. The best uniformity was achieved at a vibration frequency of 400 Hz.

This work has demonstrated the potential of using tool vibration for assisting micro EDM with array electrodes and other complex geometries. The gap observation clarified the phenomena of the periodic discharge, which is necessary for obtaining the optimal vibration parameters. Vibration frequency and amplitude influenced the sample roughness and hole diameter uniformity. Further research will include the investigation of discharge energy and the direct observation of debris in the gap, among other things.

Author Contributions: Conceptualization, G.E. and J.Y.; methodology, G.E. and J.Y.; software, G.E.; validation, G.E.; formal analysis, G.E.; investigation, G.E.; resources, J.Y.; data curation, G.E.; writing—original draft preparation, G.E.; writing—review and editing, J.Y.; visualization, G.E.; supervision, J.Y.; project administration, J.Y.; funding acquisition, J.Y. All authors have read and agreed to the published version of the manuscript.

Funding: This research received no external funding.

Institutional Review Board Statement: Not applicable.

Informed Consent Statement: Not applicable.

Data Availability Statement: Not applicable.

Conflicts of Interest: The authors declare no conflict of interest.

References

1. Kibria, G.; Jahan, M.P.; Bhattacharyya, B. *Micro-Electrical Discharge Machining Processes*; Springer: Singapore, 2019; ISBN 978-981-13-3073-5.
2. Kitamura, T.; Kunieda, M.; Abe, K. High-Speed Imaging of EDM Gap Phenomena using Transparent Electrodes. *Procedia CIRP* **2013**, *6*, 314–319. [CrossRef]
3. Kunieda, M.; Masuzawa, T. A Fundamental Study on a Horizontal EDM. *CIRP Ann.* **1988**, *37*, 187–190. [CrossRef]
4. Endo, T.; Tsujimoto, T.; Mitsui, K. Study of vibration-assisted micro-EDM—The effect of vibration on machining time and stability of discharge. *Precis. Eng.* **2008**, *32*, 269–277. [CrossRef]
5. Schubert, A.; Zeidler, H.; Oschätzchen, M.H.; Schneider, J.; Hahn, M. Enhancing Micro-EDM using Ultrasonic Vibration and Approaches for Machining of Nonconducting Ceramics. *SV-JME* **2013**, *59*, 156–164. [CrossRef]
6. Tong, H.; Li, Y.; Wang, Y. Experimental research on vibration assisted EDM of micro-structures with non-circular cross-section. *J. Mater. Processing Technol.* **2008**, *208*, 289–298. [CrossRef]
7. Li, Z.; Bai, J.; Cao, Y.; Wang, Y.; Zhu, G. Fabrication of microelectrode with large aspect ratio and precision machining of micro-hole array by micro-EDM. *J. Mater. Processing Technol.* **2019**, *268*, 70–79. [CrossRef]
8. Li, K.; Xu, G.; Huang, X.; Xie, Z.; Gong, F. Manufacturing of Micro-Lens Array Using Contactless Micro-Embossing with an EDM-Mold. *Appl. Sci.* **2019**, *9*, 85. [CrossRef]
9. Huan, L.; Jicheng, B.; Yan, C.; Guozheng, Z.; Shaojie, H. Micro-electrode wear and compensation to ensure the dimensional consistency accuracy of micro-hole array in micro-EDM drilling. *Int. J. Adv. Manuf. Technol.* **2020**, *111*, 2653–2665. [CrossRef]
10. Zhang, L.; Tong, H.; Li, Y. Precision machining of micro tool electrodes in micro EDM for drilling array micro holes. *Precis. Eng.* **2015**, *39*, 100–106. [CrossRef]
11. Debnath, T.; Patowari, P.K. Fabrication of an array of micro-fins using Wire-EDM and its parametric analysis. *Mater. Manuf. Processes* **2019**, *34*, 580–589. [CrossRef]
12. Takahata, K.; Gianchandani, Y.B. Batch mode micro-EDM for high-density and high-throughput micromachining. In *Technical Digest, MEMS 2001, Proceedings of the 14th IEEE International Conference on Micro Electro Mechanical Systems (Cat. No.01CH37090), Interlaken, Switzerland, 25 January 2001*; IEEE: Piscataway, NJ, USA, 2001; pp. 72–75.
13. Yi, S.M.; Park, M.S.; Lee, Y.S.; Chu, C.N. Fabrication of a stainless steel shadow mask using batch mode micro-EDM. *Microsyst Technol* **2008**, *14*, 411–417. [CrossRef]
14. Zeng, W.L.; Gong, Y.P.; Liu, Y.; Wang, Z.L. Experimental Study of Microelectrode Array and Micro-Hole Array Fabricated by Ultrasonic Enhanced Micro-EDM. *KEM* **2007**, *364–366*, 482–487. [CrossRef]
15. Kim, B.H.; Park, B.J.; Chu, C.N. Fabrication of multiple electrodes by reverse EDM and their application in micro ECM. *J. Micromech. Microeng.* **2006**, *16*, 843–850. [CrossRef]
16. Chen, S.-T. Fabrication of high-density micro holes by upward batch micro EDM. *J. Micromech. Microeng.* **2008**, *18*, 85002. [CrossRef]
17. Takahata, K.; Shibaike, N.; Guckel, H. High-aspect-ratio WC-Co microstructure produced by the combination of LIGA and micro-EDM. *Microsyst Technol* **2000**, *6*, 175–178. [CrossRef]
18. Wang, H.; Li, Y.; Chen, X.; Jia, Y.; Chi, G.; Wang, Y.; Wang, Z. Micro-milling/micro-EDM combined processing technology for complex microarray cavity fabrication. *Int. J. Adv. Manuf. Technol.* **2021**, *113*, 1057–1071. [CrossRef]
19. Sun, Y.; Gong, Y.; Wen, X.; Xin, B.; Yin, G.; Meng, F.; Tang, B. Evaluation of dimensional accuracy and surface integrity of cylindrical array microelectrodes and cylindrical array microholes machined by EDM. *Archiv. Civ. Mech. Eng.* **2022**, *22*, 46. [CrossRef]

20. Zhang, Y.; Xie, B. Investigation on hole diameter non-uniformity of hole arrays by ultrasonic vibration-assisted EDM. *Int. J. Adv. Manuf. Technol.* **2021**, *112*, 3083–3091. [CrossRef]
21. Lee, P.A.; Kim, Y.; Kim, B.H. Effect of low frequency vibration on micro EDM drilling. *Int. J. Precis. Eng. Manuf.* **2015**, *16*, 2617–2622. [CrossRef]
22. Mastud, S.A.; Kothari, N.S.; Singh, R.K.; Joshi, S.S. Modeling Debris Motion in Vibration Assisted Reverse Micro Electrical Discharge Machining Process (R-MEDM). *J. Microelectromech. Syst.* **2015**, *24*, 661–676. [CrossRef]
23. Uhlmann, E.; Domingos, D.C. Investigations on Vibration-assisted EDM-machining of Seal Slots in High-Temperature Resistant Materials for Turbine Components—Part II. *Procedia CIRP* **2016**, *42*, 334–339. [CrossRef]
24. Shitara, T.; Fujita, K.; Yan, J. Direct observation of discharging phenomena in vibration-assisted micro-electrical discharge machining. *Int. J. Adv. Manuf. Technol.* **2020**, *108*, 1125–1138. [CrossRef]
25. Prihandana, G.S.; Mahardika, M.; Hamdi, M.; Mitsui, K. Effect of low-frequency vibration on workpiece in EDM processes. *J Mech Sci Technol* **2011**, *25*, 1231–1234. [CrossRef]
26. Hinduja, S.; Kunieda, M. Modelling of ECM and EDM processes. *CIRP Ann.* **2013**, *62*, 775–797. [CrossRef]
27. Mori, A.; Kunieda, M.; Abe, K. Clarification of Gap Phenomena in Wire EDM Using Transparent Electrodes. *Procedia CIRP* **2016**, *42*, 601–605. [CrossRef]
28. Yue, X.; Yang, X.; Li, Q.; Li, X. Novel methods for high-speed observation of material removal and molten pool movement in EDM. *Precis. Eng.* **2020**, *66*, 295–305. [CrossRef]
29. Uhlmann, E.; Domingos, D.C. Investigations on Vibration-Assisted EDM-Machining of Seal Slots in High-Temperature Resistant Materials for Turbine Components. *Procedia CIRP* **2013**, *6*, 71–76. [CrossRef]
30. Kitamura, T.; Kunieda, M. Clarification of EDM gap phenomena using transparent electrodes. *CIRP Ann.* **2014**, *63*, 213–216. [CrossRef]
31. Järrendahl, K.; Davis, R.F. Chapter 1 Materials Properties and Characterization of SiC. In *SiC Materials and Devices*; Park, Y.S., Ed.; Academic Press: London, UK, 2010; pp. 1–20. ISBN 9780127521602.
32. Jernigan, G.G.; VanMil, B.L.; Tedesco, J.L.; Tischler, J.G.; Glaser, E.R.; Davidson, A.; Campbell, P.M.; Gaskill, D.K. Comparison of epitaxial graphene on Si-face and C-face 4H SiC formed by ultrahigh vacuum and RF furnace production. *Nano Lett.* **2009**, *9*, 2605–2609. [CrossRef]
33. Wang, Y.; Liu, Z.; Shi, J.; Dong, Y.; Yang, S.; Zhang, X.; Lin, B. Analysis of material removal and surface generation mechanism of ultrasonic vibration–assisted EDM. *Int. J. Adv. Manuf. Technol.* **2020**, *110*, 177–189. [CrossRef]
34. Hou, S.; Bai, J. A geometric prediction model of surface morphology in micro-EDM considering stochastic characteristics of discharge crater size. *Int. J. Adv. Manuf. Technol.* **2021**, *117*, 1147–1162. [CrossRef]

micromachines

Article

Laser Micromachining in Fabrication of Reverse-µEDM Tools for Producing Arrayed Protrusions

Hreetabh Kishore, Chandrakant Kumar Nirala * and Anupam Agrawal

Department of Mechanical Engineering, Indian Institute of Technology Ropar, Rupnagar 140001, Punjab, India; 2017mez0007@iitrpr.ac.in (H.K.); anupam@iitrpr.ac.in (A.A.)
* Correspondence: nirala@iitrpr.ac.in

Abstract: This paper focuses on the fabrication of high-quality novel products using a µEDM process variant called Reverse-µEDM. The tool plate required for the Reverse-µEDM is fabricated using Nd: YAG-based laser beam micromachining (LBµM) at the optimized process parameters. The Grey relation analysis technique is used for optimizing LBµM parameters for producing tool plates with arrayed micro-holes in elliptical and droplet profiles. Titanium sheets of 0.5 mm thickness were used for such micro-holes, which can be used as a Reverse-µEDM tool. The duty cycle (a combination of pulse width and frequency) and current percentage are considered as significant input process parameters for the LBµM affecting the quality of the micro-holes. A duty cycle of 1.25% and a current of 20% were found to be an optimal setting for the fabrication of burr-free shallow striation micro-holes with a minimal dimensional error. Thereafter, analogous protrusions of high dimensional accuracy and minimum deterioration were produced by Reverse-µEDM using the LBµM fabricated tool plates.

Keywords: Reverse-µEDM; Nd: YAG LBµM; micro-holes; burrs; protrusions

Citation: Kishore, H.; Nirala, C.K.; Agrawal, A. Laser Micromachining in Fabrication of Reverse-µEDM Tools for Producing Arrayed Protrusions. *Micromachines* **2022**, *13*, 306. https://doi.org/10.3390/mi13020306

Academic Editor: Francesco Modica

Received: 10 January 2022
Accepted: 8 February 2022
Published: 17 February 2022

1. Introduction

The technology for fabricating micro-scale engineering components and features is of interest to the biomedical, optical, and electronical industries [1]. One of the essential components is an arrayed protrusion (or micro pin-fins) of complex cross-sectional profiles (i.e., elliptical, circular, and diamond shapes), which is used for high heat dissipation from high-performance microelectronics [2]. The fabrication of such protrusions requires dedicated technology to achieve high dimensional accuracy and cost-effectiveness in the products. Amongst the recently developed non-conventional micro-fabrication technologies, the Reverse-µEDM process has emerged as a promising technique for fabricating micro-scale protrusions of high aspect ratio [3] and complex cross-sections [4]. Reverse-µEDM works on reversing the conventional µEDM polarity in which material removal takes place due to electron work function and electrical resistivity associated with the discharge energy ratio [5]. These fabricated protrusions have cross-sectional profiles similar to the shape of the micro-holes on the tool plate fabricated by LBµM. Notably, while fabricating a dense array of unconventional protrusions using the above-said process, issues of high machining time and damaged tips of the arrayed structures, particularly in the central zone, are found [6]. Of these two significant issues, the former is related to the process capability. The latter, which is of particular interest to the present research, is due to the poor quality of those micro-holes in the tool plate. Kishore et al. [7] explored LBµM as an auxiliary process used for fabricating tool plates with the desired shape micro-holes in different arrangements. Both the processes were facilitated on a single CNC machine tool with an axial resolution of 0.1 µm. Nevertheless, there remains a presence of cleavage burrs on the side walls of each micro-hole that lead to damaging the subsequently fabricated protrusions. The reason was anticipated to be a poor selection of process parameters.

However, LBμM can machine a wide range of difficult-to-cut materials, such as ceramics [8,9], polymers [10,11], and various metals such as aluminum alloys [12], stainless steel [13], and titanium alloys [14]. The inherent burrs and recast layer formation, especially in metallic samples, results in inaccurate dimensions and poor surface quality of fabricated microcavities. It may be due to the high heat input of an intense laser beam, which directly vaporizes the molten metal from the localized zone, resulting in the vapor and plasma pressure generation from the microcavity [15]. Appropriate heat input parameters can control the excessive burrs and recast layer formation. This includes wavelength [16,17], pulse duration [18,19], and laser power [20], which are the primary LBμM parameters that affect the machining characteristics. Several works were carried out to analyze the effect of these process parameters on microcavity fabrication. Tunna et al. [21] investigated the impact of varying wavelengths (355 nm, 532 nm, and 1064 nm) and laser intensity (0.5–57.9 GWm^{-2}) in pulsed Nd: YAG LBμM over the copper foil. They observed the maximum etch per pulse at 532 nm wavelength while the minimum was at 1064 nm due to the higher reflectivity of copper. Leitz et al. [22] conducted a detailed comparative study of pulse durations in micro, nano, pico and femtosecond LBμM. They are followed by Liu et al. [23], who reported that ultra-short laser pulses result in comparatively better precision in micromachining in terms of surface quality but with poor machining responses.

The enhanced machinability of the Nd: YAG-pulsed laser has attracted researchers to investigate other associated process parameters, such as laser intensity, frequency, scanning speed, and line spacing on the *MRR* and surface roughness of thin sheet [24]. Demir et al. [19] investigated the pulse width (12 ns and 200 ns) effect of nanosecond pulsed LBμM on the higher productivity of TiN coatings.

Considering the importance and capability of LBμM, the present work optimizes Nd: YAG-based pulsed fibre LBμM parameters for fabricating high-quality tool plates for the Reverse-μEDM process. The quality is considered in terms of burrs, striation marks, and dimensional accuracy of the machined arrayed micro-holes. The tool plate is then demonstrated for producing damage-free and dimensionally accurate arrayed elliptical and droplet protrusions using the Reverse-μEDM process and thus can be considered the main contribution of the present work.

2. Materials and Methods

2.1. Complete Process Configuration

Schematic for integration of Reverse-μEDM and LBμM for fabrication of the arrayed protrusions as the final product is shown in Figure 1. The Reverse-μEDM process consists of an RC discharge circuit with multiple options for capacitance in parallel connections and discharge voltages, a tool electrode, a workpiece electrode, and a dielectric medium. The stored energy from the capacitor is released instantaneously, due to which electro-thermal erosion occurs from the tool and the workpiece leading to material removal.

Here, the Reverse-μEDM is mainly used to fabricate single or arrayed 3D protruded structures of varying aspect ratios and cross-sectional profiles [25]. The basic tool-work configuration and polarity alteration of μEDM to achieve Reverse-μEDM is depicted in Figure 1a. In Reverse-μEDM, generally, a work material (anode) with a flat face is attached to the non-rotating spindle fed towards the tool plate (cathode). Figure 1b shows that the tool plate has an array of micro-holes in a pattern similar to the required array of protruded structures. These micro-holes are fabricated by progressive LBμM head, customized to gain an identical positional accuracy. Thus, it is possible to fabricate precise micro-holes using LBμM in any profile, which is almost impossible through any other mechanical micromachining.

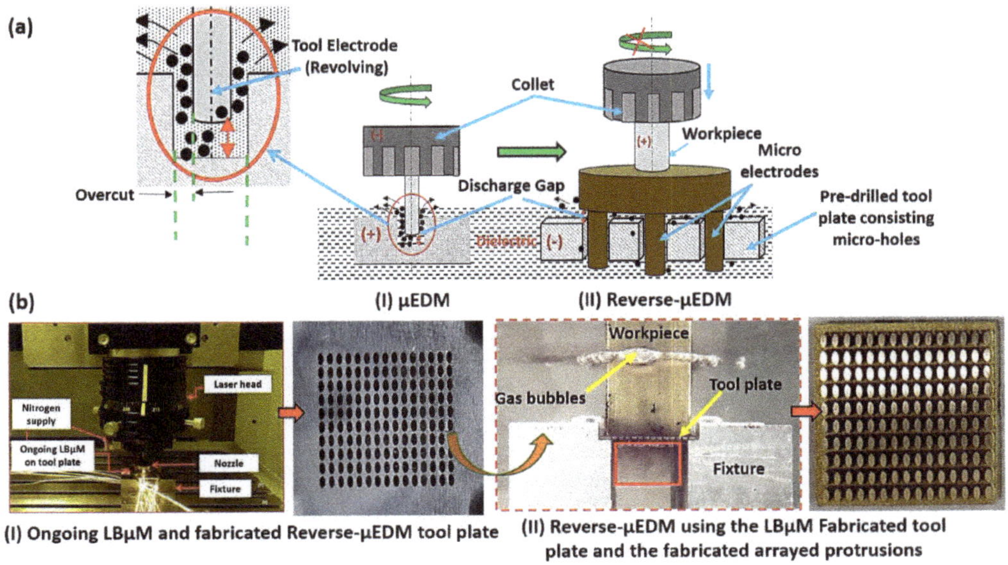

Figure 1. (**a**) Schematic representation of the Reverse-μEDM achieved through μEDM drilling, and (**b**) (I) the LBμM for producing tool plate, and (II) the Reverse-μEDM for fabricating arrayed protrusions.

2.2. LBμM Experiments for Fabrication of Reverse-μEDM Tool Plate

The experiments are performed on a commercially available pure titanium (ASTM Grade 2) sheet of 0.5 mm thickness. The sheet thickness is chosen considering enough margin after the possible erosion while used as a tool plate in the Reverse-μEDM. At the same time, it is also ensured that the available LBμM can cut such a difficult-to-cut material of 0.5 mm thickness. Titanium was chosen for its outstanding high strength, low weight, and corrosion resistance properties. It has led to a diversified range of fruitful applications in MEMS devices fabrication [14]. The nanosecond pulsed Nd: YAG fibre laser (Class 4, iPG Photonics, Yokohama, Japan) is used to fabricate micro-holes of desired shapes (refer to Figure 2a). Pulse width, pulse frequency, and average laser power expressed as the percentage of the applied current to the diode lasers are the essential process parameters considered for optimization study.

Average peak power, resulting in material removal, is an essential parameter for achieving dimensionally accurate high-quality micro-holes is presented through Equations (1)–(3) [20];

$$T_p = Duty\ cycle/f \tag{1}$$

$$P_{peak} = P_{avg}/f \times T_p \tag{2}$$

$$E_{pulse} = P_{peak} \times T_p \tag{3}$$

Here ("T_p") is pulse width and ("f") is the pulse frequency. It can be observed from Equation (2) that the peak power of the laser is a function of average power ("P_{avg}") developed for a given pulse frequency and pulse width. The present work utilizes a smaller spot size (with a minimum spot diameter of 50 μm) high-intensity laser beam profile, usually based on the fluence profile of the Gaussian distribution function with two thresholds beam diameter, as shown in Figure 2d [26]. The irradiated laser beam follows the trepanning scanning tool path (refer to Figure 2c) in the x-axis on the sheet.

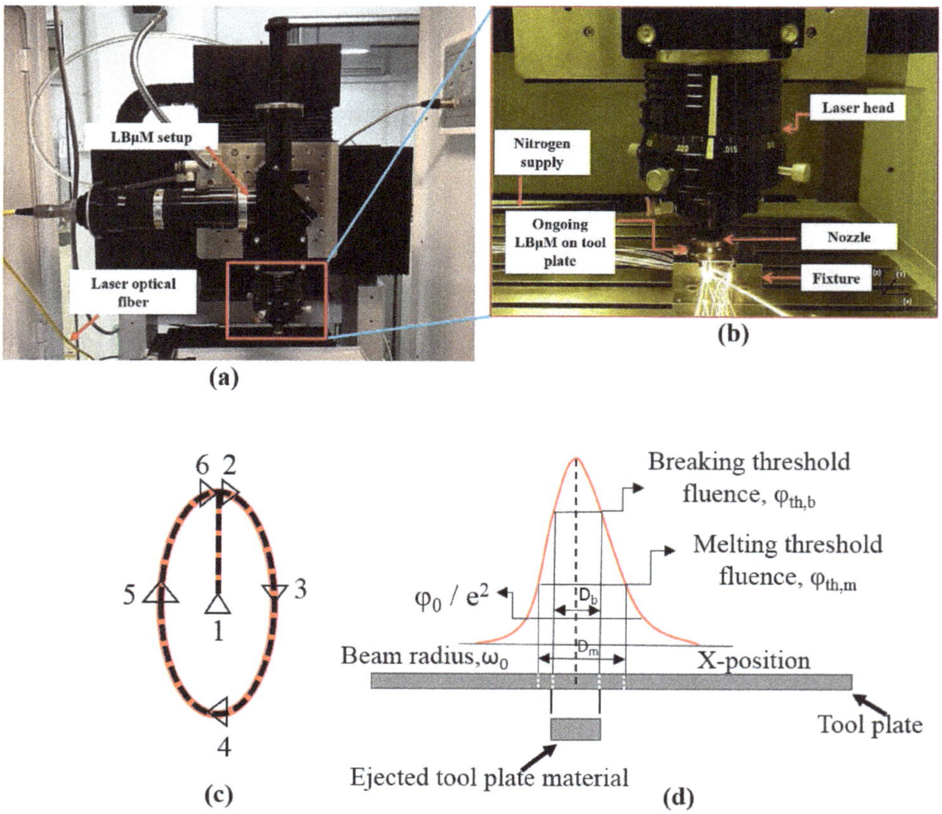

Figure 2. (**a**) The LBμM setup, (**b**) Ongoing LBμM for fabrication of micro-holes, (**c**) trepanning scanning pattern, and (**d**) Gaussian beam profile distribution.

A dedicated fixture is used to hold the sheet (tool plate for Reverse-μEDM) in such a way that there remains a gap for the free flow of ejected melted material (refer to Figure 3I). The flow of the molten metal is assisted by a shielding gas (nitrogen), which is coaxially supplied to the laser head. An illustration of ongoing LBμM for micro-hole fabrication on the sheet is shown in Figure 2b. Elliptical micro-holes, with the major and minor axes of 950 μm and 500 μm are fabricated at different LBμM parameters. The different LBμM process parameters and their levels considered to perform the experimental runs are summarized in Table 1. A design of experiments (DOE) approach based on Taguchi *L16* orthogonal array is applied to identify the best possible parametric combination with minimum experimental runs required, as shown in Table 2. The auxiliary parameters viz. stand-off distance, assisting gas pressure, and scanning speed is kept constant at 300 μm, 9 bar, and 150 mm/min. The quality criterion of LBμM includes the measured machining responses of each micro-hole in terms of minimum recast layer height, Ra_T, taper, and maximum MRR_T, along with HAZ microhardness. It is noteworthy that each experiment is repeated thrice, and the average response values are tabulated.

The optimal LBμM machining parameters are obtained based on the interaction of multiple responses using Grey relational analysis. The optimal parametric combination is obtained from the highest grey relational grades obtained by calculating the mean of all the coefficients associated with the recorded experimental responses [27]. Moreover, a confirmatory experimental run is conducted to analyze the deviation in the recorded responses, and the discussion in detail on the recorded responses is presented in Section 3.1.

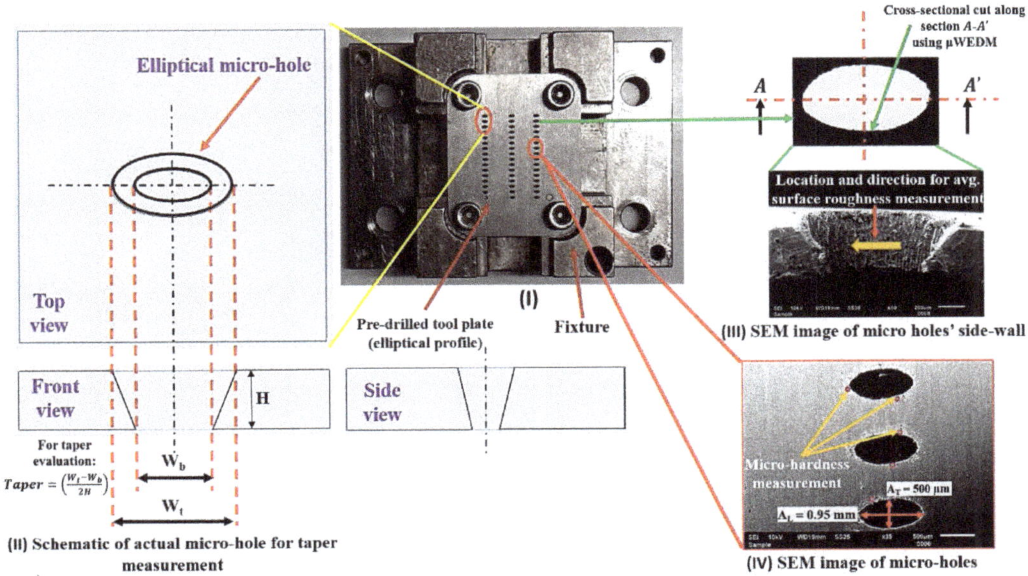

Figure 3. Illustration for estimating (**I**), (**II**) taper, (**III**) Ra_T, and (**IV**) micro-hardness of tool plate.

Table 1. Factors and their levels.

Input Factors	Units	Level			
		1	2	3	4
Pulse width	ms	0.25	0.5	0.75	1
Pulse frequency	Hz	50	65	80	95
Current percentage	DC (%)	20	40	60	80

The machined micro-holes in the elliptical cross-section are evaluated for detailed dimensional analysis, chemical composition, and surface characterization alteration. A 2D image acquisition of the fabricated holes is conducted using an optical microscope (ZEISS Axio Vert. A1, Carl Zeiss Microscopy GmbH, Jena, Germany). It is followed by microstructural and EDS analysis on the micro-holes surface using SEM (JSM6610LV, Jeol Ltd., Freising, Germany) equipped with an XFlash 6130 QUANTAX (Bruker Ltd., Bremen, Germany) EDS system. Besides these, various essential LBμM responses were evaluated, such as MRR_T, Ra_T (using roughness tester, Mitutoyo, 178–923E SJ210 Series (Mitutoyo South Asia Pvt. Ltd., New Delhi, India) (refer to Figure 3III). The top and bottom kerf width ("W_t") and ("W_b"), and taper (refer to Figure 3II) are also evaluated for each micro-hole. For the precise calculation of mass loss, an electronic micro-weighing balance (MYA 21.4Y Microbalance, RADWAG Balances and Scales Ltd., Radom, Poland) with a repeatability of 0.1 μg is used. Another response, i.e., HAZ micro-hardness measurements, is performed using a Vickers hardness testing machine (Wilson Instruments, 402 MVD, Esslingen, Germany) near the cut edge of each micro-hole (refer to Figure 3IV).

2.3. Reverse-μEDM Experiments for Fabricated Arrayed Protrusions

The obtained optimal LBμM parametric combination is used further to fabricate tool plate consisting of a single or an array of micro-holes as an essential component in Reverse-μEDM for producing arrayed protrusions. The machining conditions for Reverse-μEDM and optimal LBμM are given in Table 3. Three essential output responses, i.e., MRR_P, Ra_P, *TWR*, were rigorously monitored and recorded (approaches are depicted in Figure 4) as they are significantly affected by the nature of the machining environment. In addition to geometrical analysis, surface characterization, "Ra_P" and micro-hardness of tool plates were also evaluated.

Table 2. Experimental runs and the recorded LBμM responses.

	Factors				Responses				
	A	B	C	D = A × B × 10^{-1}					
Exp No.	Pulse Width (ms)	Frequency (Hz)	Current (%)	Duty Cycle	MRR_T (mm^3/min)	Height of Recast Layer (mm)	Ra_T (μm)	Taper (Rad)	Hv, HAZ (Micro-Holes)
1	**0.25**	**50**	**20**	**1.250**	**0.738**	**0.034**	**1.46**	**0.01**	**123**
2	0.25	65	40	1.625	0.753	0.042	1.78	0.029	128
3	0.25	80	60	2.000	0.782	0.049	1.91	0.043	130
4	0.25	95	80	2.375	0.807	0.056	2.02	0.051	134
5	0.50	50	40	2.500	0.831	0.062	2.14	0.068	139
6	0.50	65	20	3.250	0.878	0.078	2.58	0.077	145
7	0.50	80	80	4.000	0.911	0.083	2.87	0.112	151
8	0.50	95	60	4.750	0.963	0.107	3.17	0.124	162
9	0.75	50	60	3.750	0.876	0.071	2.47	0.094	156
10	0.75	65	80	4.875	0.942	0.093	2.98	0.132	163
11	0.75	80	20	6.000	1.015	0.111	3.58	0.159	171
12	0.75	95	40	7.125	1.062	0.123	3.95	0.181	179
13	1	50	80	5.000	0.952	0.096	3.02	0.146	165
14	1	65	60	6.500	1.029	0.119	3.70	0.168	173
15	1	80	40	8.000	1.094	0.128	4.35	0.211	183
16	1	95	20	9.500	1.108	0.132	4.68	0.237	188

Table 3. Process conditions of Reverse-μEDM and LBμM (Machine tool: Model: DT110i; Mikrotools Pte Ltd., Singapore).

Reverse-μEDM Parameters (Based on Expertise and Availability)		**LBμM Parameters (Based on GRA Optimization)**	
Setup	RC based	LASER type	Nd-YAG YLR-150/1500-QCW-MM-AC-Y11
Resolution (X, Y, Z)	0.1 μm	Wavelength	1070 nm
Tool plate	Titanium	Power	150 W
Workpiece	Brass	Frequency	50 Hz
Gap voltage	110 V	Pulse width	0.25 ms
Capacitance	10 nF	Spot diameter	55 μm
Electrode Feed rate	5 μm/s	current (%)	20
Dielectric oil (type)	NICUT LL21 E		

Measured Responses after Reverse-μEDM

Reverse-μEDM Using Tool Plate Fabricated (Droplet Protrusions)	**Approximately Time (h)**	***MRR_P* (mm^3/min)**	***TWR* (mm^3/min)**	***Ra_P* (μm)**	**Micro-Hardness of Tool Plate (Hv)**	
					Before M/cing	**After M/cing**
(I) with random parametric set	32	0.119	0.0053	1.63	118	125
(II) with optimal parametric set	28	0.142	0.0042	1.46	118	129

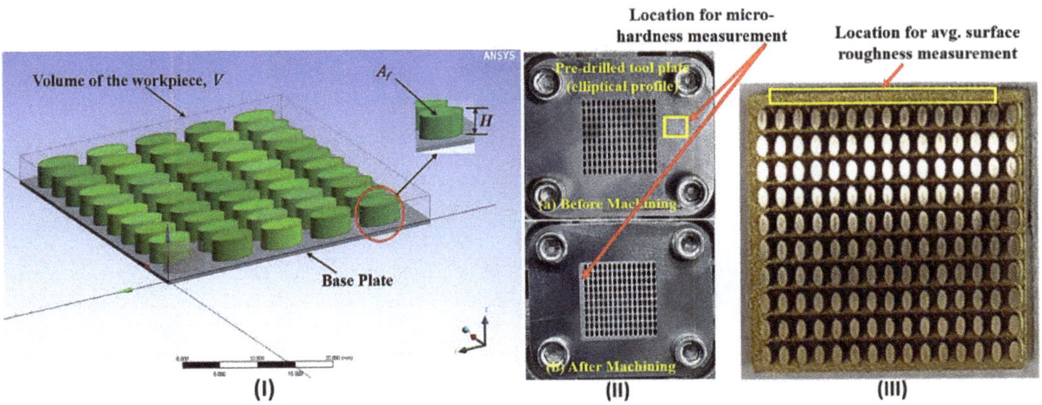

$$MRR_P = \frac{V_R}{T} = \frac{[V - nv]}{T}$$

$'V_R'$ = volume of the material removed

$'T'$ = time taken in the fabrication of a complete array

$'V'$ = workpiece volume undergone Reverse-μEDM

$'v'$ = volume of a single structure

$'n'$ = number of protruded structures in a complete array

$'A_f'$ and $'H'$ = cross-sectional area and height of a single protruded structure

$$TWR = \frac{1000\,(W_0 - W_1)}{D_t T_m}$$

$'W_0'$ =weight of the tool plate before Reverse-μEDM

$'W_1'$= weight of the tool plate after Reverse-μEDM

$'T_m'$ is total machining time (mins)

Figure 4. Illustration for **(I)** *MRR*, **(II)** *TWR* estimation, and **(III)** location for Ra_P measurement.

3. Results and Discussion

This section analyzes the recorded responses mentioned in Section 2 and the optimal parametric combination of LBμM parameters. It is followed by a detailed discussion regarding the dimensional and surface quality evaluations of Reverse-μEDM fabricated arrayed protrusions using tool plates fabricated at optimal LBμM parameters in Section 3.2.

3.1. LBμM Experimental Results

The optical images of all the fabricated micro-holes at different duty cycles are shown in Figure 5. It is observed that the micro-holes fabricated at lower duty cycles have shown good dimensional accuracy with minimal burr formation than the micro-holes fabricated at higher duty cycles. It may be due to the better efficiency of molten metal removal from the cut kerf at lower duty cycles. Post examination of fabricated micro-holes reveals the striation patterns on the holes' side wall at lower moderate and higher duty cycles, as shown in Figure 6. It is a well-accepted fact in LBμM that the formation of either uniform or non-uniform striation patterns on the holes' side walls is solely associated with the molten metal viscosity and purging of assisting gas. The purging gas enhances the cooling effect and generates the drag force on the molten metal through the cut kerf, which may stick to the micro-holes' side walls. The rate of flow of molten metal (fluid strain) also changes due to the movement of the nozzle head along with the desired holes' profile, which alters the amount of molten metal purging outside the cut kerf of the micro-holes. The lesser molten metal's viscosity and surface tension along the tool plate thickness results in the retardation of the molten metal streamlines to accumulate at the bottom of the micro-hole. This results in better dimensional accuracy and a shallow striation pattern at lower duty cycles than the ejected melted metal at higher duty cycles, shown in Figure 6 (recorded data are tabulated in Table 2).

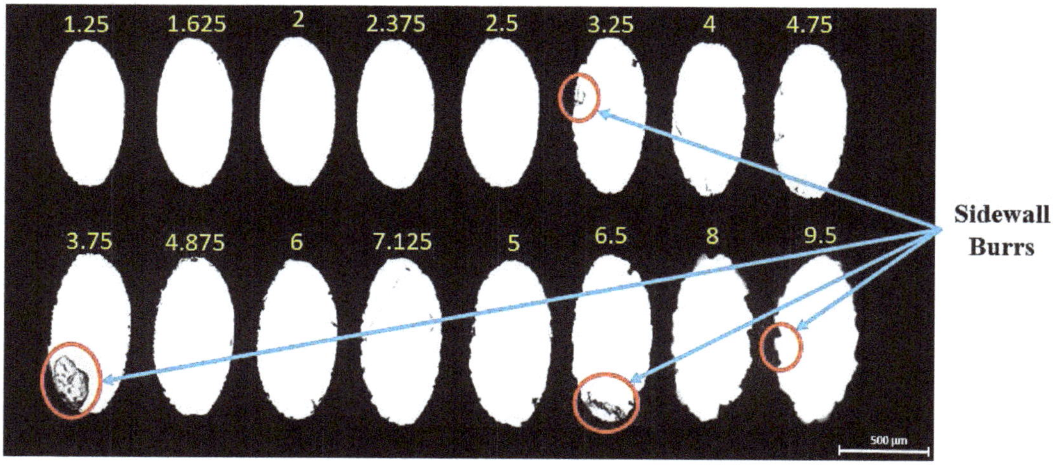

Figure 5. Micro-holes at different parametric combinations (indicated by corresponding duty cycle values).

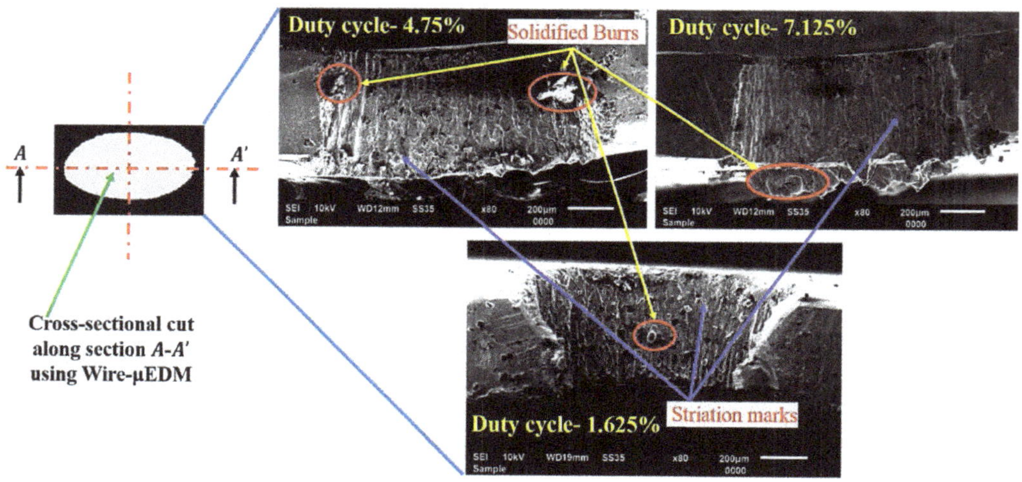

Figure 6. SEM images of holes' cut-section depicting striation marks.

3.1.1. Microstructure and EDS

During the LBμM process, the thermal gradient is generated near the cut edge as the machining progresses. The developed thermal gradient alters the microstructure of the base metal, as observed in a small zone near the cut edge as HAZ. Figure 7b(I,II) shows the base metal's microstructure and HAZ developed at a duty cycle of 1.25% and a current percentage of 20%. Careful observation of the HAZ microstructure also shows the difference in the microstructural details compared to the base metal. The grains are uniformly distributed in the HAZ region compared to the random distribution on the bulk metal surface at lower duty cycles. Additionally, the amount may be altered due to the wide range of process parameters and thermal strains generated at high-intensity laser beam irradiation.

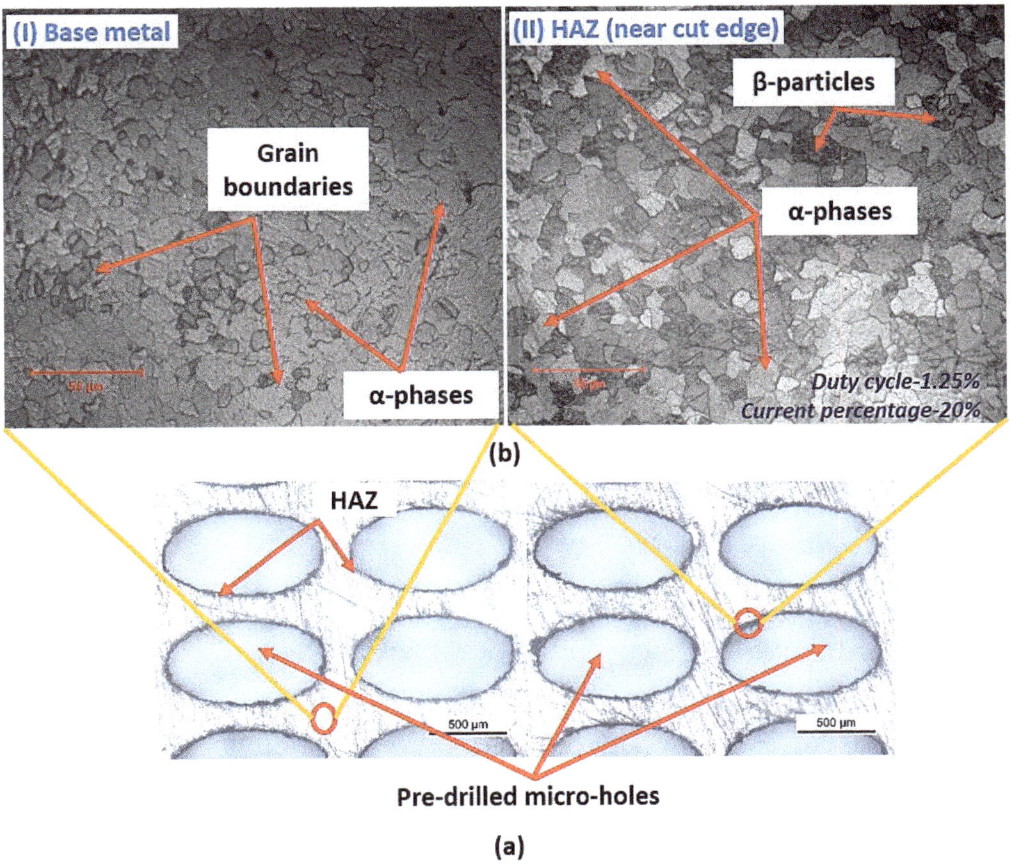

Figure 7. (**a**) Tool plate with micro-holes, and (**b**) microstructure at (**I**) base metal surface, and ((**II**) at HAZ location.

Additionally, the extent of the alteration in the HAZ region is also analyzed through the elemental distribution of the titanium tool plate. Quantitatively, the elemental data were presented to better compare the base metal and the HAZ region of the micro-hole shown in Figure 8a(i,ii). To illustrate both regions on the tool plate surface, the spectrum is highlighted with a red box in the corresponding SEM image of the micro-holes shown in Figure 8a.

A significant alteration is found in the elemental composition near the cut edge, presenting the elements' details from the EDS plot. The presence of smaller oxidation zones and carbide formation may be due to the reaction of the molten metal to the environmental gases present in the machining chamber or surface oxidation. However, the EDS of the base metal shows only the titanium content. Some micro-sized split boundaries, micropores, and globular solidification impressions are found at the cut edges, as shown in Figure 8b. These features might be occurring due to sudden vaporization and bubble formation in the melted metal layer. A dense recast layer adjacent to the micro-holes cut edge is evidenced, of which sufficient explanation is provided in the following sub-section.

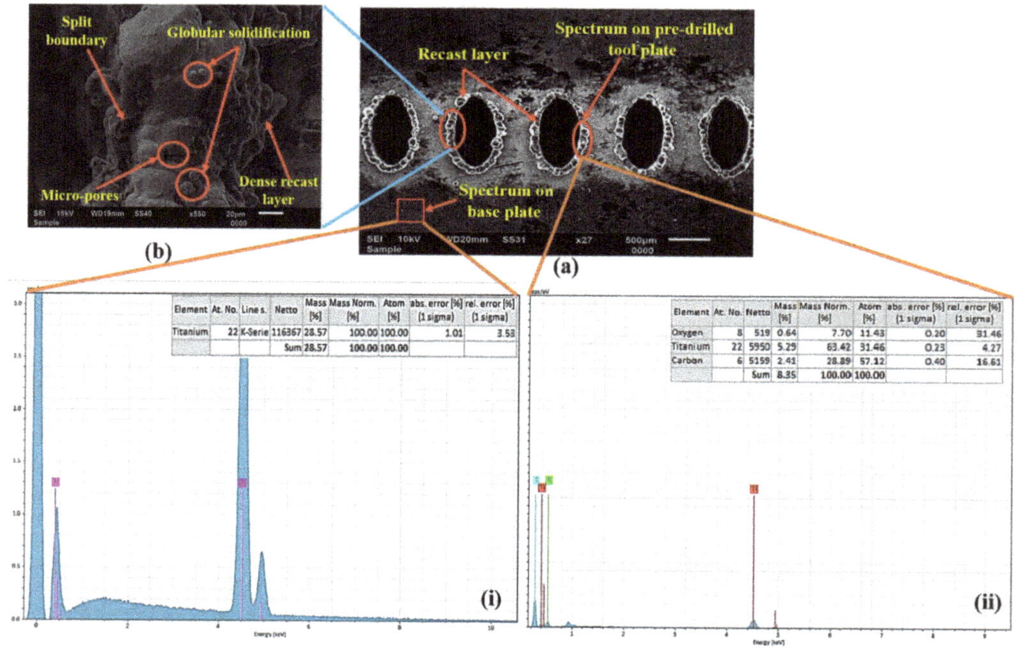

Element	At. No.	Line s.	Netto	Mass [%]	Mass Norm. [%]	Atom [%]	abs. error [%] (1 sigma)	rel. error [%] (1 sigma)
Titanium	22	K-Serie	116367	28.57	100.00	100.00	1.01	3.53
			Sum	28.57	100.00	100.00		

Element	At. No.	Netto	Mass [%]	Mass Norm. [%]	Atom [%]	abs. error [%] (1 sigma)	rel. error [%] (1 sigma)
Oxygen	8	519	0.64	7.70	11.43	0.20	31.46
Titanium	22	5950	5.29	63.42	31.46	0.23	4.27
Carbon	6	5159	2.41	28.89	57.12	0.40	16.61
		Sum	8.35	100.00	100.00		

Figure 8. (**a**) EDS plots on tool plate (i) base metal, (ii) cut edge of micro-hole, and (**b**) enlarge view of the cut edge.

3.1.2. Recast Layer and *MRR*

The top surface of the workpiece is subjected to focused laser beam irradiation results in sequential heating, melting, and vaporization, leading to the metal's removal through micro-holes. While some portion of the resolidified molten metal gets interlocked at the cut edge, known as the recast layer. The SEM image of micro-holes confirms the recast layer formation, shown in Figure 8a,b.

Figure 9a shows the line graph of the recast layer height against the increasing duty cycle at a constant current of 20% (% of avg. peak power) for all micro-holes. It is observed that the trend for recast layer height is growing with an increase when the duty cycle and current percentage lie from 20–80% and between 0.034 and 0.132 mm. This is because of the longer interaction time between the tool plate and the intense laser beam at a particular parametric combination. The increasing laser beam energy per unit area, leads to more melting, causing more recast layer formation at the cut edges. Additionally, the high vapor pressure generation on the molten metal surface causes a thermal gradient effect from the adjacent base metal. It generates hot plasma at the top of the cut edge, absorbs laser energy coming towards the base metal surface, for which the intensity of the laser beam is somewhat reduced. At higher duty cycles, high thermal gradients occur between the base metal and cut edge of the tool plate surfaces at a particularly assisting gas pressure resulting in the thicker recast layer height.

The influence of LBμM parameters on MRR_T is also evaluated for the productivity of the micro-holes. The corresponding MRR_T data for each experimental run shown in Table 2 varies from 0.738 to 1.108 mm^3/min for lower to higher duty cycles. Figure 9b shows the variation for MRR_T at a constant current of 20% (% of avg. peak power). A similar trend was observed for the increased MRR_T with an increasing current percentage from 20% to 80%. With the increasing current percentage (at the higher duty cycle), the tool plate's top surface is susceptible to intense laser irradiation and faster melting and vaporization, which leads to an improved MRR_T.

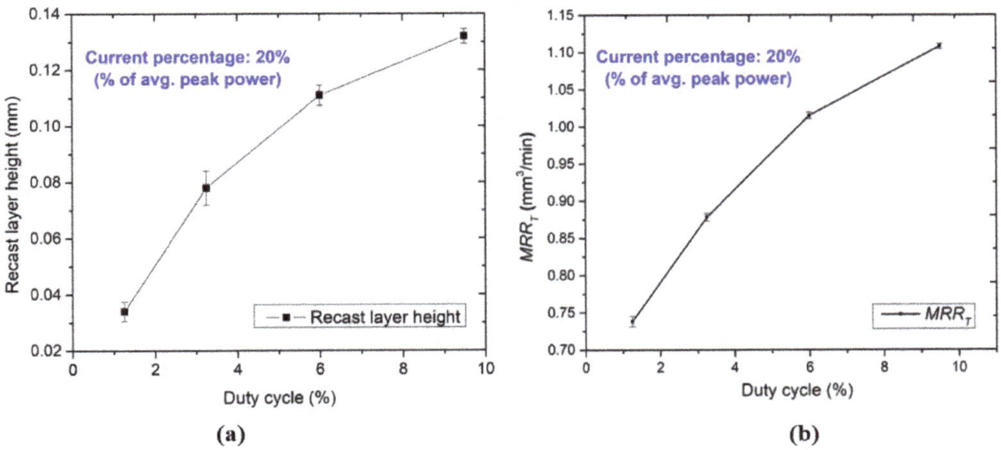

Figure 9. Variation of (**a**) recast layer heights (**b**) MRR_T with duty cycles.

3.1.3. Taper

The micro-holes cut edge and the side wall surface gets slightly modified, causing the non-identical entrance and exit, producing a taper due to the varying heat input and assisting gas pressure. Figure 10 exhibits the effect of increasing the duty cycle at a constant current percentage of 20% (% of avg. peak power) on the taper of the micro-holes. A similar trend of the increasing taper is observed for every consistent current percentage from 20% to 80%, showing the dominant contribution among other input parameters. As the power of the laser beam increases, the thermal energy of the incident beams directly transfers to the top surface of the tool plate. As a result, the top surface of the tool plate gets severely melted and vaporized and the strength of the beam decreases at the bottom edge, causing metal removal through a narrow gap. In addition, assisting gas pressure causes an increment in the taper by simultaneously cooling and ejecting the melted metal from the top surface of the tool plate at higher duty cycles.

3.1.4. Avg. Surface Roughness, Ra_T

An increase in laser intensity irradiation on the tool plate surface results in higher amounts of molten metal removal. Additionally, the drag force generated by assisting gas pressure purging out the molten metal faster causes unevenness on the side-wall surface of micro-holes. The effect of LBμM parameters on varying Ra_T of the micro-holes' side wall can be analyzed through the line plot in Figure 11a. The variation of Ra_T value with duty cycle at different current percentages ranging from 20% to 80% is observed as 1.46 to 4.68 μm (refer to Table 2).

It is observed that the Ra_T values are minimal at the lower duty cycle in LBμM. A lower duty cycle ensures the availability of great timing for the laser beam to melt and vaporize the molten metal. Additionally, the melted metal removal from the cut kerf is supported by suitable assisting gas pressure, which drags the molten metal uniformly. However, with an increase in duty cycle, the Ra_T values show an increasing trend. Excessive melt pool formation caused by partial melting of base metal near the cut edge creates non-uniformity on the side wall surface as non-uniform striation marks and high Ra_T values.

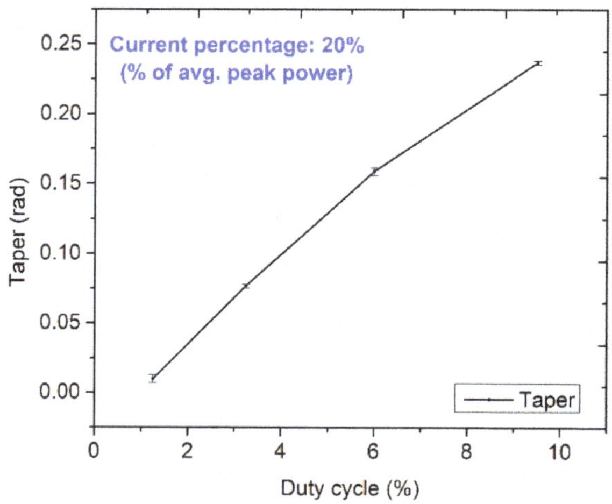

Figure 10. Variation of taper with duty cycles.

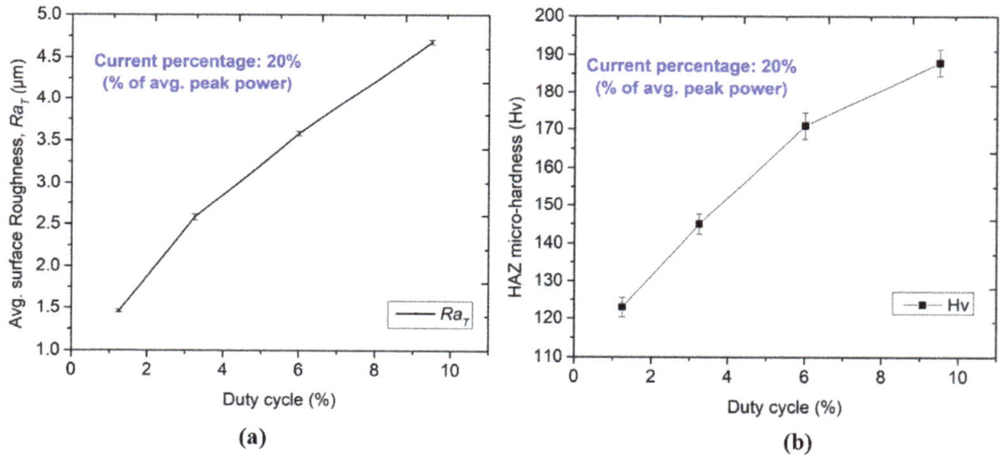

Figure 11. Variation of (**a**) Ra_T, and (**b**) HAZ micro-hardness with duty cycles.

3.1.5. Micro-Hardness of HAZ

The variation of HAZ micro-hardness with increasing duty cycle at constant current percentage is illustrated in Figure 11b. A similar trend was observed for increased micro-hardness with an increasing current percentage from 20% to 80%. It is observed that the hardness near the cut edge of each micro-hole is significantly higher than the hardness of the base metal. The formation of intense layers of titanium carbides near the cut edges and varying thermal loading leads to a surface modification at the edges. The formation of these layers may be due to the chosen assisting gas pressure and grain refinement. The traces of carbide formation at the cut edge of the micro-holes are confirmed by EDS analysis, as shown in Figure 8a(ii).

From the optimization, the parametric combination for the first experimental run (refer to Table 2) has shown the best performance for the multi-objective optimization (maximizing MRR_T, HAZ micro-hardness, and minimizing recast layer height, taper, Ra_T). The reproduced machined micro-hole fabricated at the optimal parametric combination (refer to Table 2) has presented satisfactory machining responses with less than 1% error.

This is confirmed through an experimental validation run shown in the inset (refer to Figure 12) that has no evidence of side wall burrs and no significant deviation in the taper. This parametric combination is further utilized in the micromachining of single and arrayed micro-holes for producing arrayed protrusions.

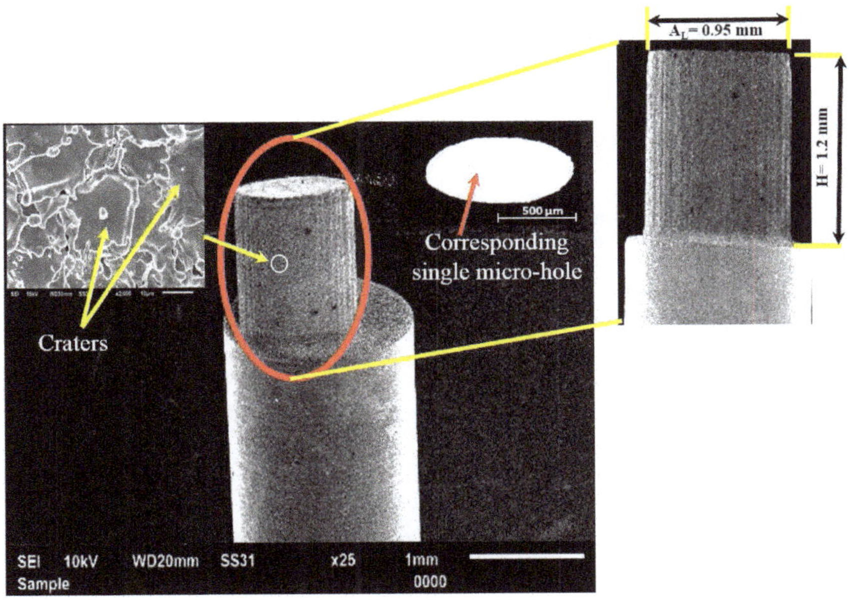

Figure 12. SEM image of single elliptical protrusion produced using micro-hole shown in the inset.

3.2. Reverse-µEDM Using Optimal Tool Plate

The micromachined single micro-hole at the optimized LBµM parameters is used to fabricate a single protrusion using the Reverse-µEDM. The SEM images of the single elliptical protrusion and the fabricated micro-hole are shown in Figure 12. The fabricated protrusion is almost free from tip damage around its periphery and perpendicular along the orthogonal length.

The potentiality of the obtained optimal LBµM parametric combination is further used for fabricating an array of elliptical and droplet micro-holes and the subsequent arrayed protrusions. For this, a fabricated tool plate with a collection of 10 × 15 micro-holes is fabricated at two different LBµM conditions: (I) a randomly selected combination (includes a pulse width of 0.75 ms, pulse frequency of 75 Hz, and a current of 20% (% of avg. peak power)), and (II) an optimal parametric combination, as highlighted in Table 2. SEM images of fabricated protruded structures at both mentioned conditions are shown in Figure 13a,b. The fabricated arrayed protrusions, with the tool plate micromachined at the random parametric set, encounter the issue of damaged tips for a few of the structures, as seen in Figure 13a. However, the presence of burrs causes longer machining time due to non-contributing discharge pulses resulting in non-uniform material removal from the workpiece electrode until it gets removed from the entire micro-holes. In contrast, it was not significant while fabricated at optimized LBµM parameters, as shown in Figure 13b. The reason could be well understood by looking back at the micro-hole fabricated at the optimized LBµM parameters, as shown in Figure 12 (inset). Since it has no apparent cleavage or burrs at the micro-hole cut edges, it allows faster machining by restricting high-order discharges and short-circuiting during Reverse-µEDM.

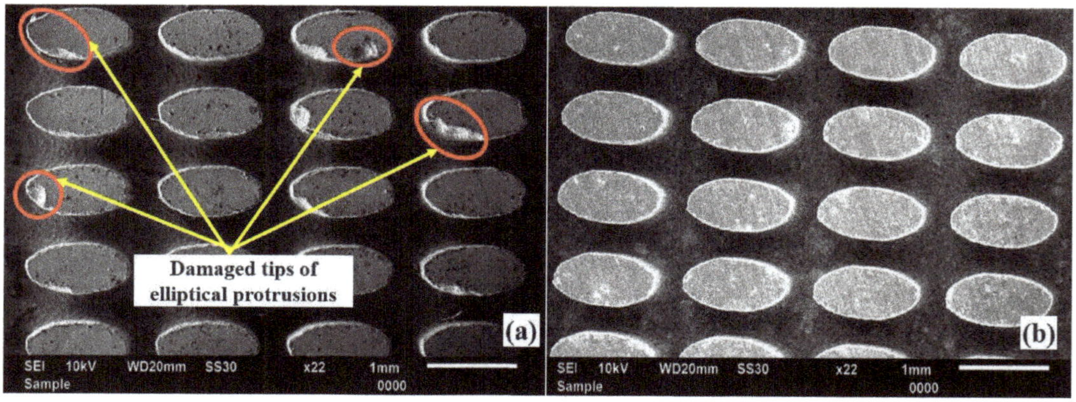

Figure 13. SEM images of fabricated elliptical arrayed protrusions with (**a**) randomly selected and (**b**) optimal LBμM parametric combinations.

Moreover, a droplet cross-section profile with a minimum inter-electrode gap of 100 μm in a staggered configuration is also fabricated. Figure 14a,b show partial magnified images of a few protrusions from an array taken at the center.

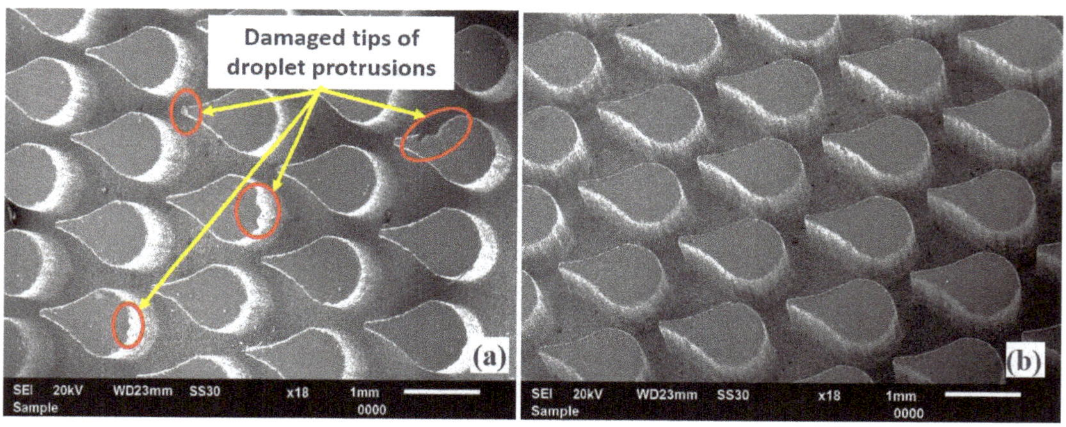

Figure 14. SEM images of fabricated droplet arrayed protrusions with (**a**) randomly selected and (**b**) optimal LBμM parametric combinations.

The recorded dimensions of fabricated protrusions are depicted in the same Figure in which an almost negligible taper is evidenced. The reason for high-quality droplet protrusions adheres to a similar explanation, as reported in the case of arrayed elliptical protrusions.

In Reverse-μEDM, the workpiece being as anode and tool plate as cathode, are subjected to generating a new surface due to frequent electrical discharging between them. The alteration in the modified surface of the tool plate is confirmed through elemental analysis captured at the zone where the machining takes place after Reverse-μEDM. Figure 15a depicts the spectrum, whereas Figure 15b shows the various elemental composition percentages on the machined tool plate.

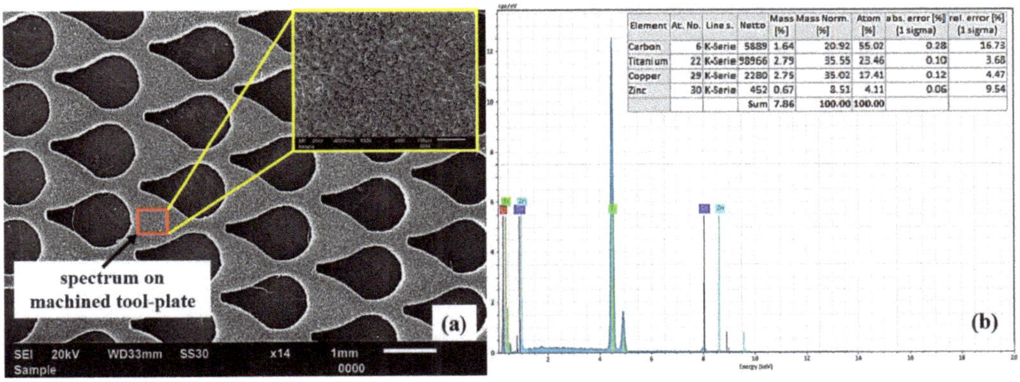

Figure 15. Energy spectrum on (**a**) tool plate after Reverse-μEDM, and (**b**) corresponding EDS plot.

Responses recorded from the Reverse-μEDM are presented in Table 3. MRR_P, TWR and Ra_P recorded improvements of more than 16%, 20%, and 10%, respectively, while using the tool plate fabricated at optimized LBμM parameters. The burr-free tool plate at the optimized LBμM parameter leads to uninterrupted machining, hence, improved MRR_P, TWR, and Ra_P. Additionally, it leads to the freezing of abnormal discharges by proper debris evacuation from the discharge gap. However, abnormal discharges due to debris accumulation are significant for increased tool wear in μEDM [28]. As a result, there is less possibility for debris re-solidification, due to its rapid cooling, on the machined tool plate surface in Reverse-μEDM. In contrast, the tool plate fabricated at the optimized LBμM parameters, comparatively, provides more scope for its rapid cooling during Reverse-μEDM. Hence, it leads to slightly better micro-hardness of the pre-drilled tool plate.

4. Conclusions

High-quality protrusions are fabricated in the shape of elliptical and droplet cross-sections using Reverse-μEDM. The tool plate required for the Reverse-μEDM is fabricated using Nd: YAG-based LBμM at the optimized process parameters. The Nd: YAG LBμM parameters are analyzed to achieve burr-free, minimum taper, and shallow striation marks of micro-holes fabricated on a 0.5 mm thick titanium sheet. The micro-holes fabricated at optimal LBμM parameters are used as a tool plate in Reverse-μEDM for producing high-quality protrusions.

The following conclusions may be drawn upon the analysis of the fabricated products:

- The LBμM at the lowest duty cycle and current percentage, as the optimized LBμM parameters, resulted in minimum recast layer height, minimum taper, and average surface roughness ("Ra_T") with almost negligible burrs with shallow side wall striation marks of micro-holes.
- The pulse width of 0.25 ms, pulse frequency of 50 Hz, and a current percentage of 20% (% of avg. peak power) were the optimal parametric combinations for LBμM obtained by Grey relation analysis.
- The optimized LBμM parameters have demonstrated high-quality arrayed micro-holes and are further used to produce arrayed elliptical and droplet protrusions through Reverse-μEDM.
- Damage-free protrusions with an improved MRR_P, TWR, and Ra_P by more than 16%, 20% and 10%, respectively, are achieved by Reverse-μEDM upon using the optimized tool plate.

Author Contributions: Writing—original draft, H.K.; Writing—review & editing, C.K.N. and A.A. All authors have read and agreed to the published version of the manuscript.

Funding: This research was funded by Indian Institute of Technology Ropar for providing financial support under the ISIRD grant (F. No. 9-282/2017IITRPR/705).

Institutional Review Board Statement: Not applicable.

Informed Consent Statement: Not applicable.

Conflicts of Interest: The authors declare no conflict of interest.

Nomenclature

µEDM	Micro-electro-discharge-machining
Reverse-µEDM	Reverse-micro-electro-discharge-machining
LBµM	Laser beam micromachining
Nd: YAG	Neodymium-doped yttrium aluminum garnet
HAZ	Heat affected zone
RC	Resistance-capacitance
SEM	Scanning electron microscope
EDS	Energy Dispersive Spectroscopy
TWR	Tool wear rate (mm^3/min)
GRA	Grey relational analysis
Hv	Vicker micro-hardness
Ra_T	Average surface roughness of micro-holes side-wall surface (µm)
Ra_P	Average surface roughness of protrusions surface (µm)
MRR_T	Material removal rate (mm^3/min) of tool plate
MRR_P	Material removal rate (mm^3/min) of protrusions

References

1. Lei, S.; Zhao, X.; Yu, X.; Hu, A.; Vukelic, S.; Jun, M.B.G.; Joe, H.E.; Lawrence Yao, Y.; Shin, Y.C. Ultrafast Laser Applications in Manufacturing Processes: A State-of-the-Art Review. *J. Manuf. Sci. Eng. Trans. ASME* **2020**, *142*, 031005.
2. Qin, L.; Hua, J.; Zhao, X.; Zhu, Y.; Li, D.; Liu, Z. Micro-PIV and Numerical Study on Influence of Vortex on Flow and Heat Transfer Performance in Micro Arrays. *Appl. Therm. Eng.* **2019**, *161*, 114186. [CrossRef]
3. Nirala, C.K.; Saha, P. A New Approach of Tool Wear Monitoring and Compensation in RµEDM Process. *Mater. Manuf. Process.* **2016**, *31*, 483–494. [CrossRef]
4. Kishore, H.; Nadda, R.; Nirala, C.K.; Agrawal, A. Modelling and Simulation Based Surface Characterization of Reverse-MEDM Fabricated Micro Pin-Fins. *Procedia CIRP* **2019**, *81*, 1230–1235. [CrossRef]
5. Ivanov, A.; Lahiri, A.; Baldzhiev, V.; Trych-Wildner, A. Suggested Research Trends in the Area of Micro-EDM—Study of Some Parameters Affecting Micro-EDM. *Micromachines* **2021**, *12*, 1184. [CrossRef] [PubMed]
6. Kishore, H.; Nirala, C.K.; Agrawal, A. Feasibility Demonstration of MEDM for Fabrication of Arrayed Micro Pin-Fins of Complex Cross-Sections. *Manuf. Lett.* **2020**, *23*, 14–18. [CrossRef]
7. Kishore, H.; Nirala, C.K.; Agrawal, A. Assessment of Process Parameters and Performance Enhancement through a Novel Suction Flushing Technology in RµEDM. *Mater. Manuf. Process.* **2021**, *36*, 1476–1488. [CrossRef]
8. Hanon, M.M.; Akman, E.; Genc Oztoprak, B.; Gunes, M.; Taha, Z.A.; Hajim, K.I.; Kacar, E.; Gundogdu, O.; Demir, A. Experimental and Theoretical Investigation of the Drilling of Alumina Ceramic Using Nd:YAG Pulsed Laser. *Opt. Laser Technol.* **2012**, *44*, 913–922.
9. Wang, X.; Huang, Y.; Xing, Y.; Fu, X.; Zhang, Z.; Ma, C. Fabrication of Micro-Channels on Al_2O_3/TiC Ceramics Using Picosecond Laser Induced Plasma Micromachining. *J. Manuf. Process.* **2019**, *44*, 102–112. [CrossRef]
10. Roth, G.L.; Haubner, J.; Kefer, S.; Esen, C.; Hellmann, R. Fs-Laser Based Hybrid Micromachining for Polymer Micro-Opto Electrical Systems. *Opt. Lasers Eng.* **2021**, *137*, 106362.
11. Pecholt, B.; Vendan, M.; Dong, Y.; Molian, P. Ultrafast Laser Micromachining of 3C-SiC Thin Films for MEMS Device Fabrication. *Int. J. Adv. Manuf. Technol.* **2008**, *39*, 239–250.
12. Mandal, K.K.; Kuar, A.S.; Mitra, S. Experimental investigation on laser micro-machining of Al 7075 Alloy. *Opt. Laser Technol.* **2018**, *107*, 260–267. [CrossRef]
13. Mishra, S.; Yadava, V. Prediction of hole characteristics and hole productivity during pulsed Nd:YAG laser beam percussion drilling. *Proc. Inst. Mech. Eng. Part B J. Eng. Manuf.* **2013**, *227*, 494–507. [CrossRef]
14. Chatterjee, S.; Mahapatra, S.S.; Bharadwaj, V.; Choubey, A.; Upadhyay, B.N.; Bindra, K.S. Drilling of micro-holes on titanium alloy using pulsed Nd:YAG laser: Parametric appraisal and prediction of performance characteristics. *Proc. Inst. Mech. Eng. Part B J. Eng. Manuf.* **2019**, *233*, 1872–1889. [CrossRef]
15. Singh, S.S.; Baruah, P.K.; Khare, A.; Joshi, S.N. Effect of Laser Beam Conditioning on Fabrication of Clean Micro-Channel on Stainless Steel 316L Using Second Harmonic of Q-Switched Nd:YAG Laser. *Opt. Laser Technol.* **2018**, *99*, 107–117. [CrossRef]

16. Lee, S.W.; Shin, H.S.; Chu, C.N. Fabrication of Micro-Pin Array with High Aspect Ratio on Stainless Steel Using Nanosecond Laser Beam Machining. *Appl. Surf. Sci.* **2013**, *264*, 653–663. [CrossRef]
17. Martan, J.; Prokešová, L.; Moskal, D.; Ferreira de Faria, B.C.; Honner, M.; Lang, V. Heat Accumulation Temperature Measurement in Ultrashort Pulse Laser Micromachining. *Int. J. Heat Mass Transf.* **2021**, *168*, 120866.
18. Schille, J.; Schneider, L.; Loeschner, U. Process Optimization in High-Average-Power Ultrashort Pulse Laser Microfabrication: How Laser Process Parameters Influence Efficiency, Throughput and Quality. *Appl. Phys. A Mater. Sci. Process.* **2015**, *120*, 847–855. [CrossRef]
19. Demir, A.G.; Pangovski, K.; O'Neill, W.; Previtali, B. Laser Micromachining of TiN Coatings with Variable Pulse Durations and Shapes in Ns Regime. *Surf. Coat. Technol.* **2014**, *258*, 240–248.
20. Giorleo, L.; Ceretti, E.; Giardini, C. Optimization of Laser Micromachining Process for Biomedical Device Fabrication. *Int. J. Adv. Manuf. Technol.* **2016**, *82*, 901–907. [CrossRef]
21. Tunna, L.; O'Neill, W.; Khan, A.; Sutcliffe, C. Analysis of Laser Micro Drilled Holes through Aluminium for Micro-Manufacturing Applications. *Opt. Lasers Eng.* **2005**, *43*, 937–950. [CrossRef]
22. Leitz, K.H.; Redlingshöer, B.; Reg, Y.; Otto, A.; Schmidt, M. Metal ablation with short and ultrashort laser pulses. *Phys. Procedia* **2011**, *12*, 230–238. [CrossRef]
23. Liu, D.X.; Sun, Y.L.; Dong, W.F.; Yang, R.Z.; Chen, Q.D.; Sun, H.B. Dynamic Laser Prototyping for Biomimetic Nanofabrication. *Laser Photonics Rev.* **2014**, *8*, 882–888. [CrossRef]
24. Cicală, E.; Soveja, A.; Sallamand, P.; Grevey, D.; Jouvard, J.M. The Application of the Random Balance Method in Laser Machining of Metals. *J. Mater. Process. Technol.* **2008**, *196*, 393–401. [CrossRef]
25. Hreetabh; Nirala, C.K.; Agrawal, A. A New Approach for Fabrication of Complex-Shaped Arrayed Micro Electrodes. In *Advances in Micro and Nano Manufacturing and Surface Engineering*; Lecture Notes on Multidisciplinary Industrial Engineering; Shunmugam, M., Kanthababu, M., Eds.; Springer: Singapore, 2019. [CrossRef]
26. Benton, M.; Hossan, M.R.; Konari, P.R.; Gamagedara, S. Effect of Process Parameters and Material Properties on Laser Micromachining of Microchannels. Micromachines. *Micromachines* **2019**, *10*, 123. [CrossRef] [PubMed]
27. Padhee, S.; Pani, S.; Mahapatra, S.S. A parametric study on laser drilling of Al/SiCp metal-matrix composite. *Proc. Inst. Mech. Eng. Part B J. Eng. Manuf.* **2012**, *226*, 76–91. [CrossRef]
28. Nirala, C.K.; Saha, P. Evaluation of MEDM-Drilling and MEDM-Dressing Performances Based on Online Monitoring of Discharge Gap Conditions. *Int. J. Adv. Manuf. Technol.* **2016**, *85*, 1995–2012. [CrossRef]

micromachines

MDPI

Article

Implementation of Micro-EDM Monitoring System to Fabricate Antimicrobial Nanosilver Colloid

Kuo-Hsiung Tseng [1], Meng-Yun Chung [1,*] and Juei-Long Chiu [2]

1 Department of Electrical Engineering, National Taipei University of Technology, Taipei 10608, Taiwan; f10473@mail.ntut.edu.tw
2 Business Planning Development Department, Panasonic Eco Solution Sales Taiwan Co., Ltd., Taipei 10608, Taiwan; chiu.paul@tw.panasonic.com
* Correspondence: alexmychung@gmail.com

Abstract: This study implemented a discharge energy and success-rate monitoring system to replace the traditional oscillograph observation method and conducted a microbial control test for a nanosilver colloid prepared by an Electrical Discharge Machine (EDM). The advantage of this system is that the discharge conditions can be instantly and continuously observed, and the optimized discharge parameter settings can be recorded. The monitoring system can use the arcing rate to control the energy consumption of the electrodes to standardize the nanosilver colloid. The results show that the arcing rate, electrode weight loss, and absorption peak wavelength are very accurate. The nanosilver colloid prepared by EDM is free of any chemical additive, and in comparison to other preparation methods, it is more applicable to biotechnology, even to the human body. The microbial control test for the nanosilver colloid included a Bathroom sample, Penicillium, Aspergillus niger, and Aspergillus flavus. In test solution NO.1 (prepared by micro-EDM), the effects of all four samples were inhibited at 14mm in a metal ring experiment, and in the cotton pad experiment, Penicillium was inhibited at 17 mm. In the metal ring experiment, test solution NO. 2 (prepared by EDM) had an effect at 20 mm on the bathroom samples, but at only 15 mm on flavus. In the cotton pad experiment, the inhibited effect was more effective in Penicillium and Aspergillus Niger; both inhibited effects occurred at 25 mm. Test solutions NO.3 (prepared by micro-EDM) and NO.4 (32 ppm Ag+) had a 14–15 mm effect on all samples in the metal ring experiment. In the cotton pad experiment, NO.3 had an effect on Penicillium at 19 mm while the effect on the others occurred at 14 mm, and NO.4 had an effect at 25 mm in Penicillium and Aspergillus Niger, and only at 14 mm in the bathroom and Aspergillus flavus samples.

Keywords: electrical discharge machine; energy consumption; nanosilver colloid; antimicrobial

Citation: Tseng, K.-H.; Chung, M.-Y.; Chiu, J.-L. Implementation of Micro-EDM Monitoring System to Fabricate Antimicrobial Nanosilver Colloid. *Micromachines* **2022**, *13*, 790. https://doi.org/10.3390/mi13050790

Academic Editors: Irene Fassi and Francesco Modica

Received: 12 April 2022
Accepted: 17 May 2022
Published: 18 May 2022

1. Introduction

Nanotechnology, biotechnology, and information technology are three basic technologies in the 21st century [1,2]. In Germany in 1984, H. Gleiter developed the first nanoparticle, initiating the application of nanotechnology. Scientists refer to powder particle aggregates in material particles with a diameter smaller than 100 nm as nanoparticles. When silver is made at a nanoscale, the particles are minified while the surface is enlarged so that the functions of the silver are greatly increased. With the appearance of new functions, many new applications of nanosilver have been developed [3]. Nanotechnology discusses new physical and chemical characteristics of technology at a scale under 100 nm. In recent years, the development of nanotechnology and nanomaterial has gradually entered the application phase. In terms of the current conditions and trends of the development of the nanosilver market, according to a Future Markets investigation, the global nanosilver application market accounted for USD 17.8 billion in 2011. In total, 95% of nanosilver market applications are based on the antibacterial property of nanosilver; the majority of those applications are in the cosmetic and medical markets, with an output value of about USD 8.4 billion, accounting for 47% of the nanosilver application market. The market applications of nanosilver in textiles and the packing industry account for 17% and 8% of nanosilver market applications, respectively. Other nanosilver market applications, mostly for high-level electronic materials, account for 5%. Since the effect of nanomaterial on the human body is yet uncertain, nanosilver has been conservatively applied in the cosmetic and medical markets. Consumers' concerns regarding the safety of nanoproducts influence the market output value of nanosilver in cosmetics, medical treatments, textiles, and packing [4–6].

In addition, as the nanosilver technique gradually becomes mature, the application of nanosilver to the market of paints, pigments, and coating is estimated to increase in the future [7,8]. In terms of the biomedical industry, in order to guarantee the advantage of talent in biotechnology competition in the 21st century, the Ministry of Education added six original, major resource centers in the rising field of "biomedical nano, stem cell, and tissue engineering" in 2004, becoming the instructional resource centers for seven major fields. The general objective is to cultivate top biotechnical talents with foresight; cross-domain, enhanced industrial experience; and an international view. According to statistics from the Industrial Development Bureau, Ministry of Economic Affairs, the nanomaterials commonly used in Taiwan are TiO_2, SiO_2, and nanosilver, and the annual input of labor, financial resources, material resources, and turnover has increased in recent years. Nanotechnology development has been an important technical means for adding industrial value [9]. The main goals of this study are the following: (1) to implement a real-time monitoring system (discharge energy and success rate) for a micro-Electrical Discharge Machine (micro-EDM) with instant and continuous observation of the results, and to record the optimized discharge parameter settings; (2) to use ESDM to fabricate an environmental-friendly nanosilver colloid to inhibit widespread fungi that are harmful to the human body, including those from a bathroom sample, Penicillium, Aspergillus niger, and Aspergillus flavus.

2. Materials and Methods

2.1. Preparation of Nanosilver Colloid by Electrical Spark Discharge Method (ESDM)

There are two ways to fabricate nanoparticles: one is physical, and the other is a chemical method. The chemical way uses the "bottom-up" method, which is similar to the "self-assembly" technology of substances and functional substances produced by nature. These methods are all constructed from the level of the molecule to the nanoscale. The physical preparation method is generally a "top-down" etching technique. Mechanical pulverization and grinding use higher-hardness materials as a medium to grind particles by shear force, friction force, and impact force to break down the particles to a small size.

The electrical discharge machine (EDM) is traditionally used for processing metal materials with high toughness and hardness [10,11]. Its principle is that, when the switch is turned on, a spark is generated to consume the metallic contact surface. However, contact will become loose if the consumption occurs over time [12–14]. Therefore, the EDM takes advantage of the principle to put the processing electrode and the object in an insulation liquid [15]. When the two electrodes are placed at a very small distance (μm), the electrons will flow from the cathode to the anode and damage the insulation state of the insulation liquid. The electrons continuously impact on the surface of the object [16]. This will generate a high spark with a high temperature. The spark temperature is around 6000 °C to 10,000 °C, and it will melt the surface into micro- or nanoparticles [17,18]. Using the principle of an EDM to fabricate the nanoparticle is called the Electrical Spark Discharge Method (ESDM).

The proposed EDM system mainly consists of the following sub-systems:

1. Power system: 100 DC volts and above;
2. Servo control system: controls the z-axis motor to maintain the two electrodes at a distance of micrometers;
3. Parameter control panel: adjusts the discharge cycle or z-axis speed, etc.

The EDM monitoring system has traditionally used an oscillator to observe the voltage and current wavelengths between electrodes, as shown in Figure 1 [19–21]. T_{on} is the discharging time, when the electrodes will successfully carry out spark discharge; T_{off} is the stopping time. The period (T_{on} + T_{off}) will continuously repeat during the processing time, for example, 20 min and 40 min in this study. When the discharge is successful, the period T_{on} when the pulse voltage appears is composed of a period of spark discharge ignition delay time and a period of spark discharge time. When the discharge is successful, the initial stage of the pulse voltage is the ignition delay time, and the electric field strength provided by the pulse voltage during this period is not enough to break down the dielectric of the electrode gap. Therefore, the voltage between the gap (V_{IEG}) will be maintained at the open circuit voltage, and the current between the gap (I_{IEG}) will be 0A, as shown in Figure 1a. The dielectric strength of the electrode gap gradually decreases during the ignition delay of the spark discharge. The period after the dielectric in the electrode gap is broken down by the electric field strength is the spark discharge period. During the spark discharge, the electrode gap presents a low-resistance state, so I_{IEG} rapidly rises to a maximum value, and V_{IEG} drops to a very low voltage value, during which spark discharge occurs between the tool and the workpiece, as shown in Figure 1b. During a successful spark discharge time, the spark will be generated, as shown in Figure 1c. When the power pulse voltage is turned off during T_{off}, the electrode gap will immediately end the spark discharge state because no pulse voltage provides the energy required for discharge. During the T_{off} period, I_{IEG} and V_{IEG} both drop to zero, as shown in Figure 1d. During T_{off}, the electrode gap will gradually recover the insulation, and the insulation degree of the electrode gap at the end of T_{off} will affect the ignition delay time of the next cycle of discharge. In addition, after a successful electrode discharge, the deionized water (DW) will be restored to a state of insulation in order to facilitate the next cycle of discharge and eliminate the metallic particles between the electrodes. Therefore, the T_{on}-T_{off} settings affect processing efficiency and quality [22]. There are three scenarios of discharge [23–25] as shown in the following:

1. Discharge failure (situation 1): There is voltage without a current. This is the same as with an open circuit.
2. Discharge failure (situation 2): As the two electrodes are very close, it is likely to cause a short circuit, similar to a current without voltage.
3. Discharge success: The spark damages the DW insulation and successfully generates a spark discharge. In addition, there are voltage and a current at the same time.

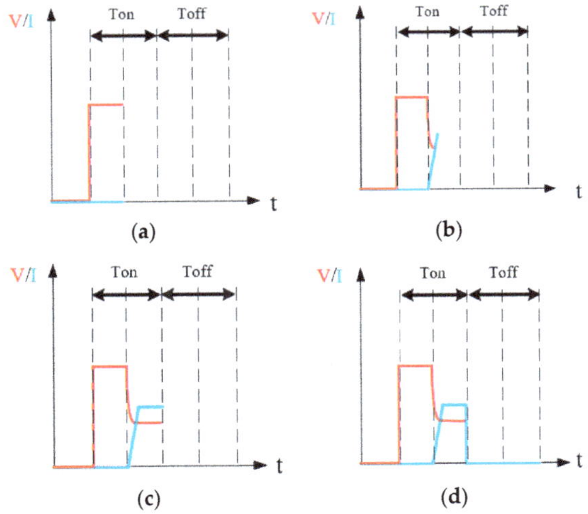

Figure 1. Electrical Spark Discharge Method (discharge success): (**a**) voltage without current; (**b**) discharge start; (**c**) spark discharge occurs; (**d**) discharge off.

Nanometal colloids, such as nanogold, nanosilver, nano-copper, nano-aluminum, and nano-titanium, can be made with any method able to conduct electricity using an EDM [26–30]. Nanosilver colloids have been applied in biomedical research, and silver ions can resist bacteria growth. For example, an enzyme with silver ion cannot ferment [31,32].

The nanometal colloids can be analyzed using a spectrophotometer, which can measure two indicative parameters:

1. Absorption peak: The instrument will emit UV and visible light to radiate the object under analysis. Through the sensor receiving the light, it can measure the absorption peak of each wavelength of the colloid. Higher absorption peaks suggest higher concentrations of colloid at the wavelength. On the contrary, a lower absorption rate suggests a lower concentration of colloid [33–35].
2. Wavelength of the highest absorption peak: Nanoparticles will radiate under the stimulation of certain wavelengths. For example, nanosilver has the highest absorption value around 400 nm [36], while nano-copper is around 280 nm [37].

As an EDM during discharge has no cyclic changes, and the frequency is very high, the discharge feature cannot be observed using an oscillator [38]; however, it can be observed by capturing an instant image of it. The disadvantage of the traditional method is that it cannot accumulate the successful discharge times. In other words, using an oscillator to observe the efficiency of the EDM is very inconvenient. This study proposed a monitoring system that can realize real-time discharge monitoring and obtain the statistics for successful discharge times, electrode energy consumption, and discharge success rate.

The EDM vaporizes the silver wire into nano-sized silver particles. The working electrodes and workpiece in this study are distinguished as electrodes. The electrode connected to positive electricity is the anode, while that connected to negative electricity is the cathode. The nanosilver colloid is prepared by ESDM in this study. The silver wire (with a diameter of 1 mm) is ground into nano-sized (1~100 nm) silver nanoparticles by the high temperature of the ESDM [39]. The discharge parameter setting panel is used to adjust the process parameters of the EDM, and the setting panel provides a voltmeter and ammeter for measuring the mean values of the voltage and current between the electrodes. Figure 2a is a diagram of the EDM and its discharge parameter setting panel. The anode and cathode connection connects the two electrodes in the beaker on the platform. The electromagnetic heating stirrer is used to stir the deionized water inside the beaker. The reason is that the electrode gap will gradually recover the insulation, and the insulation degree of the electrode gap at the end of T_{off} will affect the ignition delay time of the next cycle of discharge, so the stirring moves the nanoparticles away from the gap between the electrodes to make the gap recover the insulation more quickly. The oscilloscope is used to observe the voltage and current of the gap between the electrodes. In Figure 2b, Z-Axis is used to control the rise/fall of the Z-axis. CAPACITOR is used to set the capacitance value and is proportional to the transient current of discharge pulse at T_{on}. T_{on} and T_{off} control the discharge and off time of the electrodes. Polarity adjusts the polarity of the upper and lower electrodes. SERVO controls the servo motor speed and is proportional to the motor speed. Stabilizing controls the sensitivity of the feedback circuit and is also proportional to the motor speed. HV switches the DC power supply from 140 V to 240 V. In addition, Ip is proportional to the discharge current and processing rate, and inversely proportional to precision.

This study integrated multi-field technology research and successfully developed a micro-EDM set as shown in Figure 3. This micro-EDM was composed of an EDM jig mechanism, a hardware circuit system, and a software monitoring interface. In terms of design, the application of 3D printing and PLA materials can greatly reduce the manufacturing cost and the size of the equipment. The 3D-printed fix jig can hold the electrodes in the beaker. The hardware circuit part is designed from the circuit design of a PCB board combined with electronic components to achieve a large electric discharge machine. The PCB boards are the following: 1. motor control circuit and discharge feedback circuit (for controlling the motor to move forward and backward according to the feedback of the voltage); 2. discharge control circuit (for giving the output of the 100V DC pulse wave at the electrodes); and 3. discharge success rate circuit (for calculating the discharge success times, rate, and energy consumption with the software) [40–42]. The electrical spark discharge function and signal feedback are realized by VisSim software 6.0 (Visual Solution Inc., Pleasant Prairie, the USA) and an RT-DAC4/PCI interface card. Through this interface card, the software can be used to replace local electronic circuits to achieve the goals of a smaller circuit size and lower cost. Unlike the EDM, which needs an oscilloscope to observe the voltage and current of the gap between the electrodes and is not likely to count the discharge success times, the micro-EDM can accomplish these jobs with VisSim software. Because the micro-EDM is self-made, any repair is clear and easy. Although both an EDM and a micro-EDM can fabricate a nanosilver colloid by ESDM, the micro-EDM can make smaller-sized particles and better suspension colloids than the EDM does.

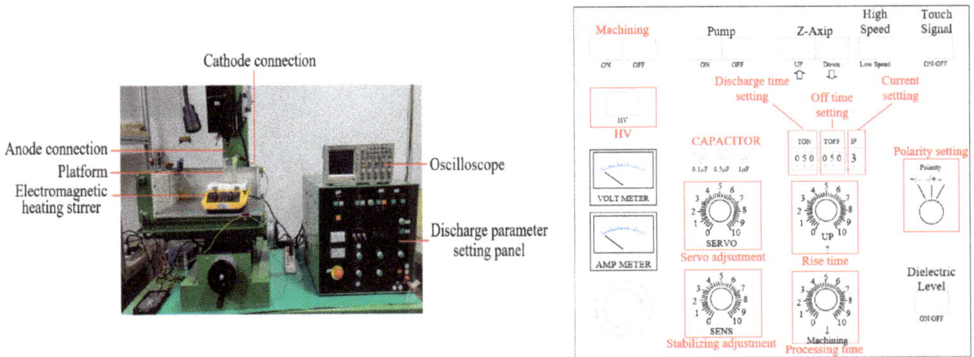

(a) (b)

Figure 2. (**a**) Picture of EDM. (**b**) Discharge parameter setting panel.

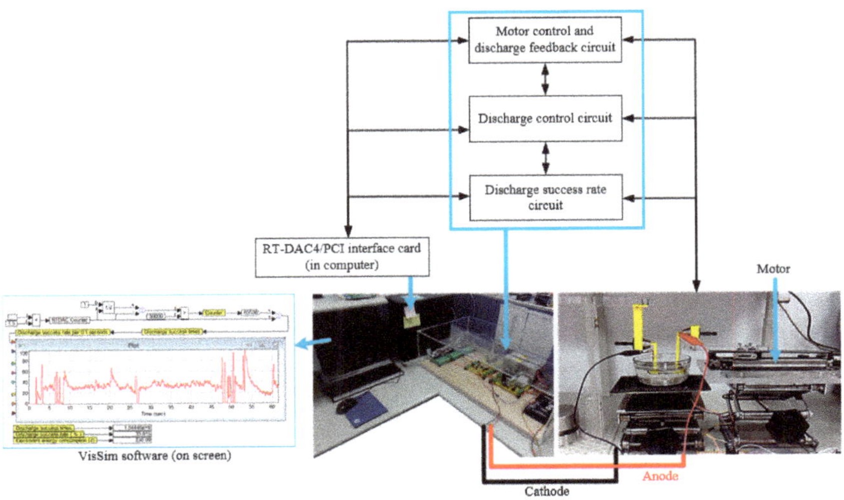

Figure 3. Picture of micro-EDM.

2.2. *Activation and Cultivation of Strains*

2.2.1. Inhibition Zone Test

The other main purpose of the study was to use the traditional EDM and the micro-EDM that produced the nanosilver colloid to evaluate its effectiveness in the microbial control of fungi; as the particle size, zeta potential, and concentration were different, so its physical properties and chemical properties were also different. The study procedures were divided into following:

1. Test solution preparation: NO.1 was the nanosilver colloid made using the micro-EDM, with a silver ion concentration of 5 ppm. NO.2 was the nanosilver colloid produced by the traditional EDM; the silver ion concentration was 32 ppm, and the silver particle concentration was about 300 ppm. NO.3 was the nanosilver colloid produced by the micro-EDM; while the silver ion concentration was 5 ppm, the nanoparticle was the smallest. NO.4 was a 32 ppm silver ion titer.
2. Fungus sample preparation: There are four kinds of fungus samples: (a) the strain was randomly collected from the dim and damp parts of a bathroom; the strain was smeared on the Potato Dextrose Agar (PDA) and cultured; (b) penicillium: the strain was collected from moldy oranges; the strain was smeared on PDA and cultured; (c) Aspergillus Niger: pure strain was bought from BCRC (Bioresource Collection and Research Center, Hsinchu, Taiwan); the strain was smeared on Malt Extract Agar (MEA) and cultured; (d) Aspergillus Flavus: pure strain was bought from BCRC; the strain was smeared on PDA and cultured. The bathroom sample, Penicillium, Aspergillus niger, and Aspergillus flavus are the four targets in this study. These are widespread both indoors and outdoors, and those exposed to them will sometimes have allergies, get sick, or suffer an infection or other conditions that may harm the body. These four targets are easily obtained, simple to cultivate, and most importantly, they are often sources of trouble in normal life in Taiwan (due to the wet weather). Therefore, this study used the micro-EDM to fabricate a nanosilver colloid that can inhibit the four targets [43,44].
3. Experimental method design: Inhibition zone test: the size of the inhibition zones formed by the NO.1~NO.4 test solutions in four fungus culture media were observed in order to judge the microbial control effect of each test solution. The test process is shown in Figure 4. The test configuration is shown in Table 1. The NO.1~NO.4 test solutions were injected into the inhibition zones in the culture media cultivating different strains, respectively. The corresponding color labels are: NO.1 (Red), NO.2 (Yellow), NO.3 (Green), NO.4 (Blue).

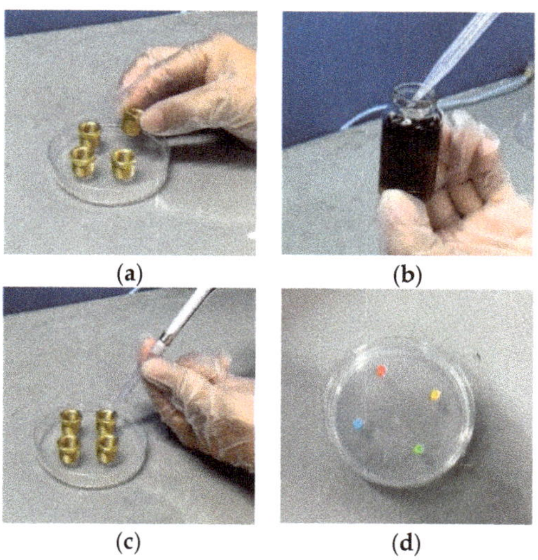

Figure 4. Pictures of inhibition zone test process: (**a**) metal rings placed on culture medium; (**b**) pipette extracts 0.15 mL test solution; (**c**) NO.1~NO.4 test solutions injected into metal rings, respectively; (**d**) metal rings carefully removed 30 min later and signs placed.

Table 1. Experimental design and configuration.

Strain		Culture Medium
Name	Code	
Bathroom sample	A	MEA
Penicillium	P	MEA
Aspergillus niger	N	MEA
Aspergillus flavus	F	PDA

2.2.2. Microbial Control Cotton Pad Test

Cotton pads with a diameter of 140 mm were labeled 1, 2, 3, and 4, soaked in the NO.1~NO.4 test solutions, respectively, and placed in the culture media cultivating different strains; the inhibition zone sizes of the cotton pads were observed to judge the microbial control effects [45]. The test process is shown in Figure 5. The corresponding symbols are: NO.1 (1), NO.2 (2), NO.3 (3), NO.4 (4).

(a) (b) (c)

Figure 5. Pictures of cotton pad test process: (**a**) 140 mm labeled cotton pad soaked in NO.1~NO.4 test solutions; (**b**) cotton pad carefully placed in culture medium with tweezers; (**c**) the aforesaid steps repeated four times.

3. Results

3.1. The Results of the Nanosilver Colloid

The micro-EDM was used to prepare the nanosilver colloid for test solution NO.1, and the related parameter settings are described in Tables 2 and 3. The weight loss of the experiment was only 0.38 mg, which means only 0.38 mg nanoparticles were in the DW, as shown in Table 4. In addition, the silver ion concentration was about 5 ppm. In the discharge success rate circuit, the voltage signals must first be processed using two series resistors to divide the interelectrode gap voltage V_{gap}. The V_{gap} was 0.0526 times that of the interelectrode gap voltage. The I_{gap} signals of the circuit were equal to the current that flows through the electrode gap. The conditions for the circuit to judge the successful discharge were $0.5\ V \leqq V_{gap} \leqq 2.4\ V$ and $I_{gap} \geq 2.4\ A$. If the discharge was successful (met the criteria above), the signal output from the circuit to the counting pin of the RT-DAC4/PCI card of the VisSim was 1 (counting pin = 1); on the contrary, if the discharge was not successful, the counting pin = 0. The discharge success rate was monitored through the signal transmission output by the discharge success rate circuit to the CNT0 counter of the RT-DAC4/PCI card, and the signal was processed by VisSim to obtain the discharge success rate and the cumulative number of successful discharges (summing all the counting pins during the processing time). The flow chart of the discharge success rate counting is shown in Figure 6. The calculation of discharge success rate is shown in Equations (1) and (2), where N_D is the cumulative number of successful sampling and discharging events in time T_{on}; N_{total} is the total number of discharge samplings in time T_{on}; D_{cycle} is the duty cycle ($T_{on} + T_{off}$); t_{proc} is the production process time; and f_{sample} is the sampling frequency. This study was able to obtain the total discharge success times and determine the signals through this digital input/output interface card. It can replace the complex electrical circuit, which can count more than 1 million times per minute. The VisSim software used in this study is PC-based. By using the PC-based VisSim, this study was able to observe the real-time discharge times. In addition, it immediately computed the successful discharge times, electrode energy consumption, and discharge success rate, as shown in Figure 7. The success rate was 27.4861%, and the energy consumption was 236.09J. With these settings, this study was able to obtain the nanosilver characteristics. The UV wavelength of the nanosilver was 398 nm, and the absorbance was 0.56, as shown in Figure 8a. The zeta potential was −28.2 mV, as shown in Figure 8b. The size was around 5~8 nm, as shown in Figure 8c.

$$\text{Discharge success rate} = \frac{N_D}{N_{total}} \quad (1)$$

$$N_{total} = D_{cycle} \times t_{proc} \times f_{sample} \quad (2)$$

Table 2. Parameter settings of the micro-EDM.

Voltage & Current	**V = 100 V, I = 4.2 A**	**SENS.**	**Kp = 0.75, Ki = 0.045, Kd = 0.035**
T_{on}-T_{off}	10–10	Z-Axis	off
Capacitor	off	Machining	off
Servo	—	HV	off

Table 3. Materials and testing conditions of the micro-EDM.

Diameter of Ag	**Anode: 1 mm; Cathode: 2 mm**	**Beaker**	**200 mL**
Processing time	10 min	Filter paper	Advantec
Dielectric fluid	DW	ATM	1 atm

Table 4. Weight loss calculations of the micro-EDM.

Items \ Weight	**Before W_0 (mg)**	**After W_1 (mg)**	**W_0–W_1 (mg)**
E_1(anode)	5081.25	5080.96	0.29
E_2(cathode)	3186.96	3186.87	0.09
$E_1 + E_2$	8268.21	8267.83	0.38
Temperature(°C)	25 °C	28 °C	—

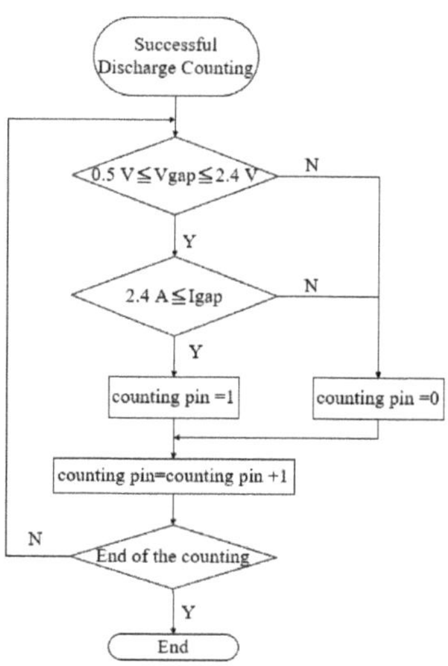

Figure 6. Flow chart of discharge success rate counting.

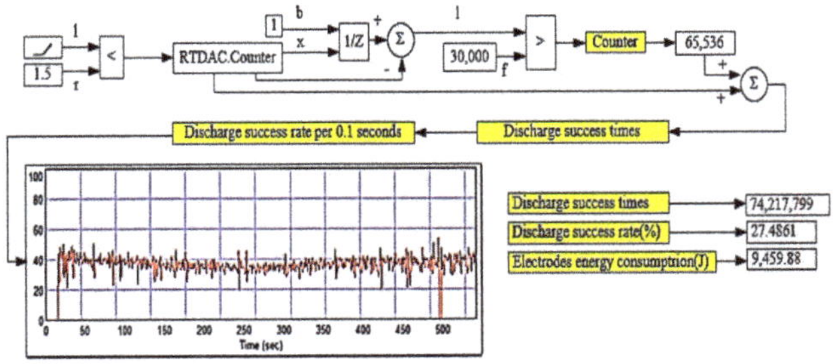

Figure 7. VisSim monitoring of the wave pattern.

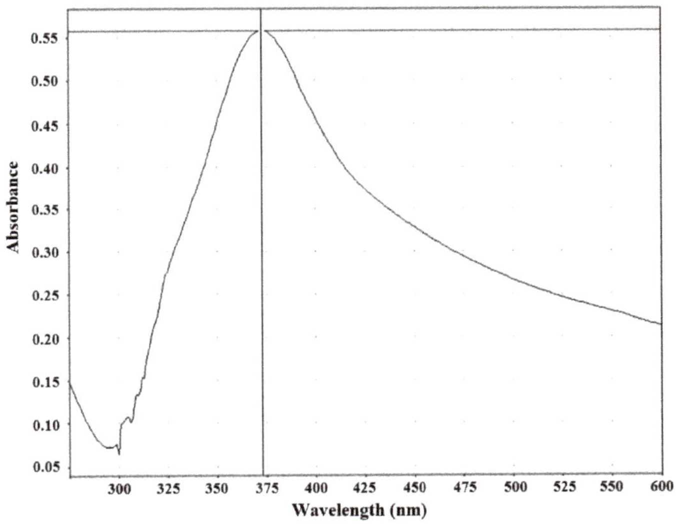

(**a**)

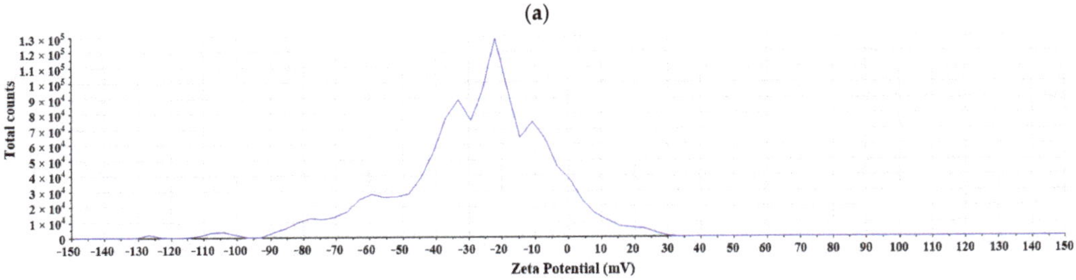

(**b**)

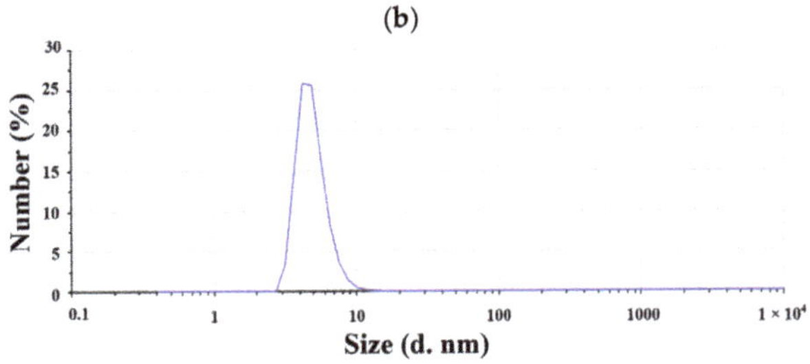

(**c**)

Figure 8. Characteristics of nanosilver colloid prepared using micro-EDM: (**a**) UV absorption wavelength; (**b**) zeta potential; (**c**) number PSD.

The traditional EDM was used to prepare the nanosilver colloid for test solution NO.2, and the related parameter settings are described in Tables 5 and 6. The weight loss of the experiment was only 124.43 mg, which means only 124.43 mg nanoparticles were in the DW, as shown in Table 7. The prepared nanosilver colloid was kept still for 2~3 weeks and then filtered through filter paper, which removed about 77.28 mg of precipitates and big silver particles; the estimated concentration was (124.43–77.28)/0.25 = 308 ppm. The UV wavelength of the nanosilver was 398 nm, and the absorbance was 1.1. The zeta potential was −33 mV, as shown in Figure 9a. The size was around 5~8 nm, as shown in Figure 9b.

Table 5. Parameter settings of the EDM.

Voltage & Current	V = 140 V, I = 11.8 A	SENS	1/2
T_{on}–T_{off}	50–50	Z-Axis	off
Capacitor	off	Machining	off
Servo	1/2	HV	off

Table 6. Materials and testing conditions of the EDM.

Diameter of Ag	Anode: 1 mm; Cathode: 2 mm	Beaker	250 mL
Processing time	40 min	Filter paper	Advantec
Dielectric fluid	DW	ATM	1 atm

Table 7. Weight loss calculations of the EDM.

Items \ Weight	Before W_0 (mg)	After W_1 (mg)	W_0–W_1 (mg)
E_1(anode)	640.70	518.49	122.21
E_2(cathode)	5775.78	5773.56	2.22
$E_1 + E_2$	6416.48	6292.05	124.43
Temperature(°C)	21 °C	32.6 °C	—

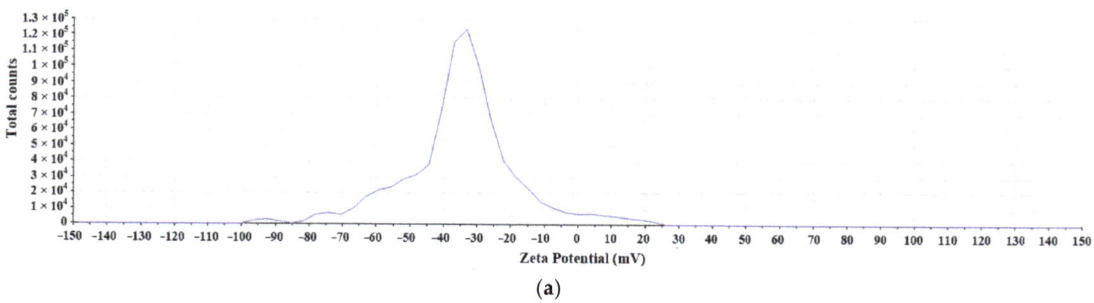

(a)

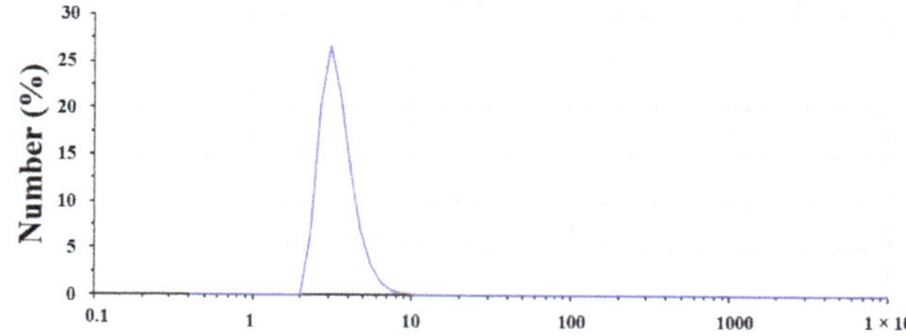

(b)

Figure 9. Characteristics of nanosilver colloid prepared using EDM: (**a**) zeta potential distribution; (**b**) number PSD.

The micro-EDM was used to prepare the nanosilver colloid for testing solution NO.3. In addition, the zeta potential was −40 mV, as shown in Figure 10a. The size was around 0.5~1 nm, as shown in Figure 10b. Testing solution NO.4 was a 1000 ppm Ag^+ standard solution.

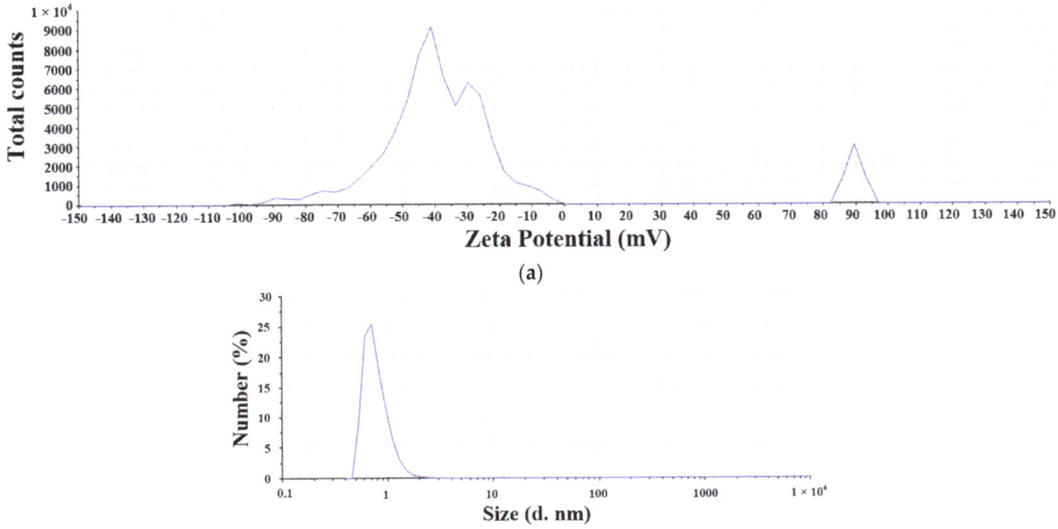

Figure 10. Characteristics of nanosilver colloid with a smaller size: (**a**) zeta potential distribution; (**b**) number PSD.

3.2. *Microbial Control Cotton Pad Test Results and Discussion*

The experimental results of the inhibition zone test (metal ring) are shown in Figure 11. The experimental results of the inhibition zone test (cotton pad) are shown in Figure 12.

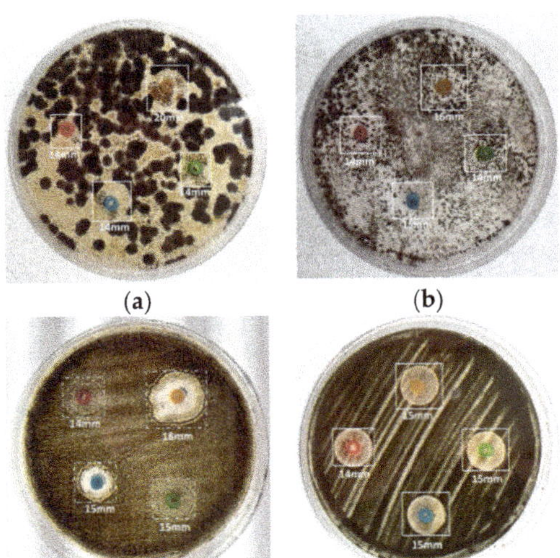

Figure 11. Pictures of inhibition zone test: (**a**) bathroom sample; (**b**) Penicillium; (**c**) Aspergillus Niger; (**d**) Aspergillus Flavus.

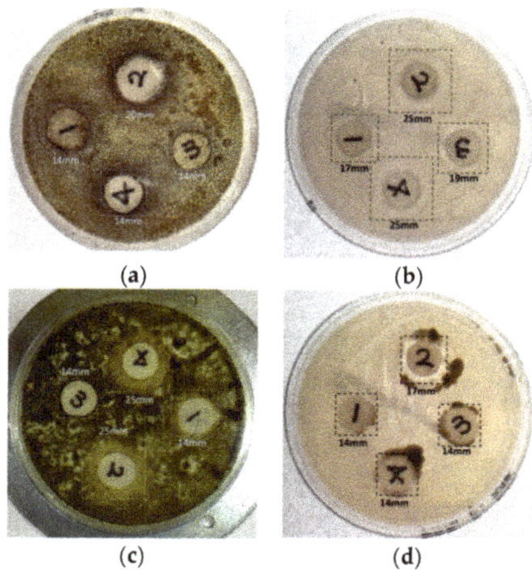

Figure 12. Pictures of cotton pad test: (**a**) bathroom sample; (**b**) Penicillium; (**c**) Aspergillus Niger; (**d**) Aspergillus Flavus.

3.3. *The Results and Discussion of Inhibition Zone Tests*

The results of the inhibition zone tests are described as follows. In addition, the results are shown in Tables 8 and 9 and Figure 13 [46,47].

Table 8. Results of microbial test (metal ring), diameter (mm).

Test Liquids / Samples	NO.1	NO.2	NO.3	NO.4
Bathroom (A)	14	20	14	14
Penicillium (P)	14	16	14	15
Aspergillus niger (N)	14	18	15	15
Aspergillus flavus (F)	14	15	15	15

Table 9. Results of microbial test (cotton pad), diameter (mm).

Test Solution / Samples	NO.1	NO.2	NO.3	NO.4
Bathroom (A)	14	20	14	14
Penicillium (P)	17	25	19	25
Aspergillus Niger (N)	14	25	14	25
Aspergillus Flavus (F)	14	17	14	14

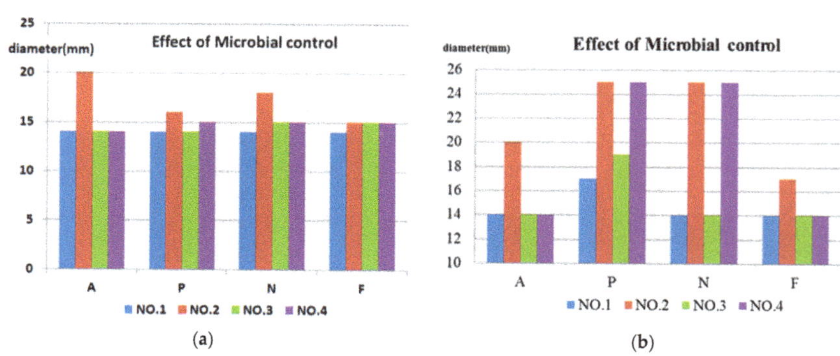

Figure 13. Effects of microbial control: (**a**) metal ring and (**b**) cotton pad.

1. In the inhibition zone test, the metal ring was placed on the culture medium, and the test solution was then injected into the metal ring [48]. The test solution was slowly absorbed by the culture medium before the metal ring was removed. If the metal ring is removed before the test solution is dried up, the flow of the test solution may influence the experimental results.
2. The microbial control cotton pad test was an improvement proposed for inhibition zone testing [49,50]. Compared to the experiment with the metal ring, in which the test solution needs to be injected in the middle of the ring, the cotton pad holds water, so the test solution will not flow on the culture medium, and the waiting time for drying during the experiment can be shortened.
3. The microbial control cotton pad test results were relatively objective and observable. It was observed that the NO.2 and NO.4 test solutions had the best microbial control effects.
4. The inhibition zone test was implemented for four kinds of fungi in the NO.1~NO.3 nanosilver colloids with different concentrations and characteristics and the NO.4 Ag^+ standard solution. It was observed that the NO.2 32 ppm nanosilver colloid had the best microbial control effect, then the NO.4 Ag^+ standard solution at the same concentration. It may be that, because the silver nanoparticles release Ag^+ continuously at the same Ag^+ concentration, the microbial control effect is better than the standard Ag^+ solution. While the concentration of the NO.1 and NO.3 nanosilver colloids was about 5ppm, NO.3 had the smallest silver nanoparticle, meaning a larger contact surface area and the highest zeta potential; thus, its microbial control effectiveness was better than NO.4.

Comparing the two inhibition zone tests, the metal ring limited the expansion of the inhibition zone when the fungus grew, so it is difficult to observe the actual microbial control effect. Figure 14 shows the experimental results of using the metal ring and the cotton pad. The figure "(C)" represents the use of a cotton pad, and "(R)" represents the use of a metal ring. In the NO.1 test liquid, the effect on all four samples was the same in the metal ring experiment, all inhibited to 14mm. In addition, in the cotton pad experiment, the nanosilver had more effect on Penicillium at 17 mm. The NO.2 test liquid had more effect on the bathroom samples at 20 mm but had only a 15 mm inhibited effect on flavus. In the cotton pad experiment, the inhibited effect was more effective in Penicillium and Aspergillus Niger, both with inhibiting effects at 25 mm. The NO.3 and NO.4 test liquids had little difference in effect on the four samples and were all about 14–15 mm in the metal ring experiment. In the cotton pad experiment, NO.3 had more effect on Penicillium at 19 mm, while the effect on the others was 14 mm; and NO.4 had an effect of 25 mm on Penicillium and Aspergillus Niger, and the effect on the others (bathroom and Aspergillus flavus samples) was only 14 mm. The order of the microbial test effect is Penicillium (P) > Aspergillus niger (N) > bathroom sample (A) > Aspergillus flavus (F).

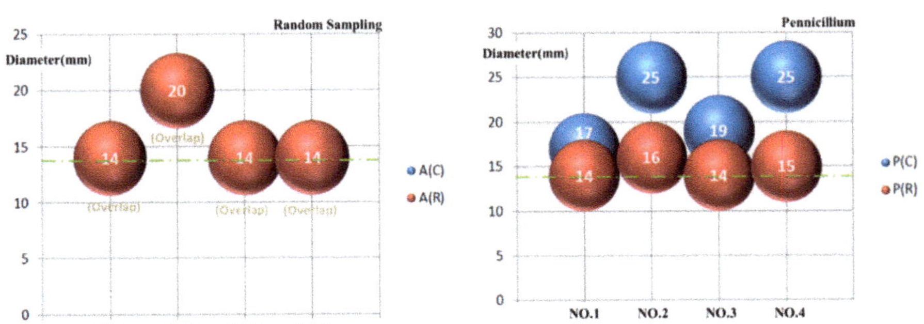

(a) (b)

Figure 14. *Cont.*

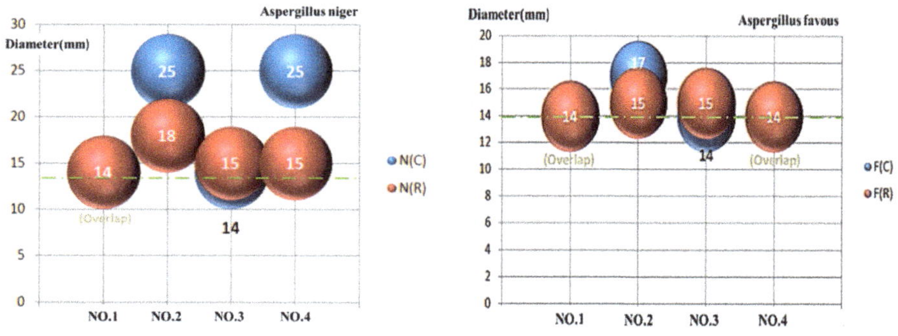

(c) (d)

Figure 14. Comparison of microbial control tests: (**a**) bathroom sample; (**b**) Penicillium; (**c**) Aspergillus niger; (**d**) Aspergillus Flavus.

4. Conclusions

There are two ways to fabricate nanoparticles: one is physical (the "bottom-up"), and the other is a chemical (the "top-down") technique. The chemical method uses chemical agents, which is not environmentally friendly. In addition, other physical methods consume energy/time and cannot instantly monitor the process. This study developed a real-time monitoring system for the Electrical Spark Discharge Method (ESDM) using a micro-Electrical Discharge Machine. This study has two limitations. One is that the ESDM is a physical method and, unlike the chemical way, cannot obtain the precise size and concentration of the nanoparticle. The other is that the inhibition of Niger cannot obtain a precise concentration but a relative one. However, all the experiments in this study were conducted with a standardized process in order to reduce the errors made by the human hand. With this system, the micro-EDM can obtain the successful discharge times, electrode energy consumption, and discharge success rate. The conclusions of this study are as follows:

1. The user may adjust the successful discharge times, electrode energy consumption, and discharge success rate at any time to obtain the optimal parameters to fabricate the nanosilver colloid using the Electrical Spark Discharge Method.
2. In the inhibition zone experiment, this study proposes use of the microbial control cotton pad. The microbial control cotton pad is characterized by its low cost, quick experimentation, and easy observation. It is unlikely to fail, and there is no metal ring; thus, the presentation of the inhibition zone will match the actual result better.
3. In the NO.1 liquid, the effects were inhibited at 14 mm in all four samples in the metal ring experiment and at 17 mm in Penicillium in the cotton pad experiment. In the metal ring experiment, the NO.2 liquid had an effect at 20 mm in the bathroom samples and a 15 mm inhibited effect on flavus. In the cotton pad experiment, the inhibited effect was more effective in Penicillium and Aspergillus Niger, both with an inhibited effect at 25 mm. The NO.3 and NO.4 liquids had a14–15 mm effect on all samples in the metal ring experiment. In the cotton pad experiment, NO.3 had an effect on Penicillium at 19 mm, while its effect on the others was only 14 mm. NO.4 had an effect of 25 mm on Penicillium and Aspergillus Niger and only 14mm on the bathroom and Aspergillus flavus samples.
4. The inhibition zone experiment found that the Ag^+ and nanosilver colloids have a better microbial control effect on penicillium and Aspergillus niger and a worse effect on Aspergillus flavus.
5. Although this study has verified that a nanosilver colloid made using an EDM has microbial control effects, it is only a small step. The issue of nanomaterials is a big subject in terms of antibacterial, biomedical, and environmental aspects. All of the

research topics are only just beginning. It is worth investing in more research to benefit more people in the near future.

Author Contributions: Project administration, K.-H.T.; Resources, M.-Y.C.; Supervision, K.-H.T.; Funding acquisition, M.-Y.C.; Data curation, J.-L.C. and M.-Y.C.; Formal analysis, J.-L.C.; Methodology, J.-L.C. and M.-Y.C.; Validation, M.-Y.C.; Writing—original draft, J.-L.C.; Writing—review and editing, M.-Y.C. All authors have read and agreed to the published version of the manuscript.

Funding: The Ministry of Science and Technology (MOST 109-2222-E-027-005-).

Data Availability Statement: Not applicable.

Acknowledgments: The authors would like to thank the Precision Research and Analysis Center, National Taipei University of Technology for technical support of this research.

Conflicts of Interest: The authors declare no conflict of interest.

Abbreviations

Electrical Discharge Machine (EDM)	Direct Current (DC)
Micro-Electrical Discharge Machine (micro-EDM)	Potato Dextrose Agar (PDA)
Electrical Spark Discharge Method (ESDM)	Bioresource Collection and Research Center (BCRC)
Deionized Water (DW)	Malt Extract Agar (MEA)

References

1. Roco, M.C.; Bainbridge, W.S. *Converging Technologies for Improving Human Performance: Nanotechnology, Biotechnology, Information Technology and Cognitive Science*; Springer Science & Business Media: Berlin, Germany, 2013.
2. Aithal, S.; Aithal, P.S. Green and Eco-friendly Nanotechnology—Concepts and Industrial Prospects. *Int. J. Manag. Technol. Soc. Sci.* **2021**, *6*, 1–31. [CrossRef]
3. Roukes, M. Plenty of Room, Indeed. *Sci. Am.* **2007**, *9*, 48–57. [CrossRef]
4. Gajbhiye, S.; Satish, S. Silver nanoparticles in cosmetics. *J. Cosmet. Dermatol. Sci. Appl.* **2016**, *6.1*, 48–53. [CrossRef]
5. Yang, L.; Liu, F.; Chen, Y.; Liu, Z.; Zhang, G. Research on the Treatment of Diabetic Foot with Ulcer Based on Nano-Silver Antibacterial Dressing. *J. Nanosci. Nanotechnol.* **2021**, *21*, 1220–1229. [CrossRef]
6. Abu-Qdais, H.; Abu-Dalo, M.; Hajeer, Y. Impacts of Nanosilver-Based Textile Products Using a Life Cycle Assessment. *Sustainability* **2021**, *13*, 3436. [CrossRef]
7. Temizel-Sekeryan, S.; Hicks, A.L. Global environmental impacts of silver nanoparticle production methods supported by life cycle assessment. *Resour. Conserv. Recycl.* **2020**, *156*, 104676. [CrossRef]
8. Cui, J.; Shao, Y.; Zhang, H.; Zhang, H.; Zhu, J. Development of a novel silver ions-nanosilver complementary composite as antimicrobial additive for powder coating. *Chem. Eng. J.* **2021**, *420*, 127633. [CrossRef]
9. Kidd, J.; Westerhoff, P.; Maynard, A. Survey of industrial perceptions for the use of nanomaterials for in-home drinking water purification devices. *NanoImpact* **2021**, *22*, 100320. [CrossRef]
10. Gouda, D.; Panda, A.; Nanda, B.K.; Kumar, R.; Sahoo, A.K.; Routara, B.C. Recently evaluated Electrical Discharge Machining (EDM) process performances: A research perspective. *Mater. Today Proc.* **2021**, *44*, 2087–2092. [CrossRef]
11. Joshi, A.Y.; Joshi, A.Y. A systematic review on powder mixed electrical discharge machining. *Heliyon* **2019**, *5*, e02963. [CrossRef]
12. Reece, M.P. The vacuum switch. Part 1: Properties of the vacuum arc. *Proc. Inst. Electr. Eng.* **1963**, *110*, 793–802. [CrossRef]
13. Sheth, M.; Gajjar, K.; Jain, A.; Shah, V.; Patel, H.; Chaudhari, R.; Vora, J. Multi-objective Optimization of Inconel 718 Using Combined Approach of Taguchi—Grey Relational Analysis. In *Advances in Mechanical Engineering*; Springer: Singapore, 2021; pp. 229–235.
14. Chaudhari, R.; Vora, J.J.; Pramanik, A.; Parikh, D. Optimization of Parameters of Spark Erosion Based Processes. In *Spark Erosion Machining*; CRC Press: Boca Raton, FL, USA, 2020; pp. 190–216.
15. Sarker, B.; Shankar, C. Discriminant Analysis-based Parametric Study of an Electrical Discharge Machining Process. *Sci. Iran.* **2021**. [CrossRef]
16. Kumar, S.; Singh, R.; Singh, T.P.; Sethi, B. Surface modification by electrical discharge machining: A review. *J. Mater. Process. Technol.* **2009**, *209*, 3675–3687. [CrossRef]
17. Schnick, M.; Füssel, U.; Hertel, M.; Spille-Kohoff, A.; Murphy, A. Metal vapour causes a central minimum in arc temperature in gas–metal arc welding through increased radiative emission. *J. Phys. D Appl. Phys.* **2009**, *43*, 022001. [CrossRef]
18. Zhang, Z.; Ming, W.; Zhang, G.; Huang, Y.; Wen, X.; Huang, H. A new method for on-line monitoring discharge pulse in WEDM-MS process. *Int. J. Adv. Manuf. Technol.* **2015**, *81*, 1403–1418. [CrossRef]

19. Tseng, K.-H.; Yeh, C.-T.; Chung, M.-Y.; Lin, Y.-S.; Qui, N. A study of preparing silver iodide nanocolloid by electrical spark discharge method and its properties. *Sci. Rep.* **2021**, *11*, 20457. [CrossRef]
20. Li, Z.-L.; Xi, X.-C.; Chu, H.-Y.; Xu, L.-Y.; Gao, Q.; Zhao, W.-S. Improving RT-WEDM performance with a radio frequency signal monitoring system. *Int. J. Adv. Manuf. Technol.* **2022**, *118*, 391–404. [CrossRef]
21. Elyass, W.M.; Sharma, V.; Kantak, G.; Alam, M.M. Performance Investigation of Wire EDM Process Parameters. *IOP Conf. Ser. Mater. Sci. Eng.* **2022**, *1224*, 012001. [CrossRef]
22. Yu, S.F.; Lee, B.Y.; Lin, W.S. Waveform Monitoring of Electric Discharge Machining by Wavelet Transform. *Int. J. Adv. Manuf. Technol.* **2001**, *17*, 339–343. [CrossRef]
23. Chen, L. An On-Line Monitoring System for Discharge Condition in LS-WEDM. *Machin. Design Manufact.* **2017**, *2*, 128–131.
24. Liu, K.; Bert, L.; Dominiek, R. Process capabilities of Micro-EDM and its applications. *Int. J. Adv. Manufact. Technol.* **2010**, *47*, 11–19. [CrossRef]
25. Abhilash, P.; Chakradhar, D. Failure detection and control for wire EDM process using multiple sensors. *CIRP J. Manuf. Sci. Technol.* **2021**, *33*, 315–326. [CrossRef]
26. Tseng, K.-H.; Chung, M.-Y.; Chang, C.-Y. Parameters for Fabricating Nano-Au Colloids through the Electric Spark Discharge Method with Micro-Electrical Discharge Machining. *Nanomaterials* **2017**, *7*, 133. [CrossRef] [PubMed]
27. Tien, D.C.; Liao, C.-Y.; Huang, J.-C.; Tseng, K.-H.; Lung, J.-K.; Tsung, T.-T.; Kao, W.-S.; Tsai, T.-H.; Cheng, T.W.; Yu, B.-S.; et al. Novel technique for preparing a nano-silver water suspension by the arc-discharge method. *Rev. Adv. Mater. Sci.* **2008**, *18*, 750–756.
28. Tseng, K.-H.; Ke, H.-K.; Ku, H.-C. Parameters and properties for the preparation of Cu nanocolloids containing polyvinyl alcohol using the electrical spark discharge method. *Nanomat. Nanotechnol.* **2021**, *11*, 18479804211035190. [CrossRef]
29. Liao, C.-Y.; Tseng, K.-H.; Lin, H.-S. Preparation of Metallic Aluminum Compound Particles by Submerged Arc Discharge Method in Aqueous Media. *Met. Mater. Trans. A* **2013**, *44*, 91–97. [CrossRef]
30. Chang, H.; Tsung, T.-T.; Lo, C.-H.; Lin, H.-M.; Lin, C.-K.; Chen, L.-C.; Jwo, C.-S. A study of nanoparticle manufacturing process using vacuum submerged arc machining with aid of enhanced ultrasonic vibration. *J. Mater. Sci.* **2005**, *40*, 1005–1010. [CrossRef]
31. Tien, D.-C.; Tseng, K.-H.; Liao, C.-H.; Tsung, T.-T. Identification and quantification of ionic silver from colloidal silver prepared by electric spark discharge system and its antimicrobial potency study. *J. Alloys Comp.* **2009**, *473*, 298–302. [CrossRef]
32. Tseng, K.-H.; Lee, H.-L.; Tien, D.-C.; Tang, Y.-L.; Kao, Y.-S. A Study of Antibioactivity of Nanosilver Colloid and Silver Ion Solution. *Adv. Mater. Sci. Eng.* **2014**, *2014*, 371483. [CrossRef]
33. Katta, V.K.M.; Dubey, R.S. Green synthesis of silver nanoparticles using Tagetes erecta plant and investigation of their structural, optical, chemical and morphological properties. *Mater. Today Proc.* **2021**, *45*, 794–798. [CrossRef]
34. He, W.-M.; Zhou, Z.; Han, Z.; Li, S.; Zhou, Z.; Ma, L.-F.; Zhang, S.-Q. Ultrafast Size Expansion and Turn-On Luminescence of Atomically Precise Silver Clusters by Hydrogen Sulfide. *Angewandte Chem. Int. Ed.* **2021**, *60*, 8505–8509. [CrossRef]
35. Liaw, J.-W.; Tsai, S.-W.; Lin, H.-H.; Yen, T.-C.; Chen, B.-R. Wavelength-dependent Faraday–Tyndall effect on laser-induced microbubble in gold colloid. *J. Quant. Spectrosc. Radiat. Transf.* **2012**, *113*, 2234–2242. [CrossRef]
36. Tseng, K.-H.; Chen, Y.-C.; Shyue, J.-J. Continuous synthesis of colloidal silver nanoparticles by electrochemical discharge in aqueous solutions. *J. Nanopart. Res.* **2011**, *13*, 1865–1872. [CrossRef]
37. Mann, V.; Hugger, F.; Roth, S.; Schmidt, M. Influence of Temperature and Wavelength on Optical Behavior of Copper Alloys. *Appl. Mech. Mater.* **2014**, *655*, 89–94. [CrossRef]
38. Yang, F.; Liu, Y.; Liu, W.; Yao, K. One-Cycle Control for Pulse Power Generator in Electrical-Discharge-Machining. *IEEE Trans. Ind. Electron.* **2021**, 1. [CrossRef]
39. Tseng, K.-H.; Chiu, J.-L.; Lee, H.-L.; Liao, C.-Y.; Lin, H.-S.; Kao, Y.-S. Preparation of Ag/Cu/Ti Nanofluids by Spark Discharge System and Its Control Parameters Study. *Adv. Mater. Sci. Eng.* **2015**, *2015*, 694672. [CrossRef]
40. Tseng, K.-H.; Chang, C.-Y.; Cahyadi, Y.; Chung, M.-Y.; Hsieh, C.-L. Development of Proportional–Integrative–Derivative (PID) Optimized for the MicroElectric Discharge Machine Fabrication of Nano-Bismuth Colloid. *Micromachines* **2020**, *11*, 1065. [CrossRef]
41. Tseng, K.-H.; Chen, K.-H.; Chang, C.-Y.; Cahyadi, Y.; Chung, M.-Y. Implementation of a Micro-Electrical Discharge Machine System to Fabricate TiO_2 Nanocolloid. *Mechatronics* **2021**, *79*, 102649. [CrossRef]
42. Tseng, K.-H.; Chang, K.-Y.; Chen, M.-J.; Tseng, Y.-K. Novel electrical discharge machining system with real-time control and monitoring for preparing nanoiron colloid. *Adv. Mechan. Eng.* **2018**, *10*, 1687814018791705. [CrossRef]
43. Adedayo, M.R.; Mohammed, M.T.; Ajiboye, A.E.; Abdulmumini, S.A. Pectinolytic activity of Aspergillus niger and Aspergillus flavus grown on grapefruit (citrus Parasidis) peel in solid state fermentation. *Glob. J. Pure Appl. Sci.* **2021**, *27*, 93–105. [CrossRef]
44. Sampson, T. Mycological Survey of Unused Tissue Papers in Public Toilets within a University Campus in Port Harcourt, Rivers State, Nigeria. *Saudi J. Pathol. Microbiol.* **2021**, *6*, 447–450.
45. Mia, R.; Sk, S.; Oli, Z.B.S.; Ahmed, T.; Kabir, S.; Waqar, A. Functionalizing cotton fabrics through herbally synthesized nanosilver. *Clean. Eng. Technol.* **2021**, *4*, 100227. [CrossRef]
46. Tseng, K.-H.; Chang, C.-Y.; Chung, M.-Y.; Tang, Y.-L. Bacteriostatic Substrate by Conductivity Method and Electric Spark Discharge Method Combined with Electrospinning for Silver Dressing. *Int. J. Polym. Sci.* **2016**, *2016*, 9425358. [CrossRef]
47. Xu, L.; Wang, Y.-Y.; Huang, J.; Chen, C.-Y.; Wang, Z.-X.; Xie, H. Silver nanoparticles: Synthesis, medical applications and biosafety. *Theranostics* **2020**, *10*, 8996–9031. [CrossRef] [PubMed]

48. Oladipo, I.C.; Lateef, A.; Azeez, M.A.; Asafa, T.B.; Yekeen, T.A.; Akinboro, A.; Akinwale, A.S.; Gueguim-Kana, E.B.; Beukes, L.S. Green Synthesis and Antimicrobial Activities of Silver Nanoparticles using Cell Free-Extracts of Enterococcus species. *Not. Sci. Biol.* **2017**, *9*, 196–203. [CrossRef]
49. Wasif, A.I.; Laga, S.-K. Use of nano silver as an antimicrobial agent for cotton. *AUTEX Res. J.* **2009**, *9*, 5–13.
50. Fouda, A.; Shaheen, T.I. Silver Nanoparticles: Biosynthesis, Characterization and Application on Cotton Fabrics. *Microbiol. Res. J. Int.* **2017**, *20*, 1–14. [CrossRef]

MDPI
St. Alban-Anlage 66
4052 Basel
Switzerland
Tel. +41 61 683 77 34
Fax +41 61 302 89 18
www.mdpi.com

Micromachines Editorial Office
E-mail: micromachines@mdpi.com
www.mdpi.com/journal/micromachines

www.ingramcontent.com/pod-product-compliance
Lightning Source LLC
Chambersburg PA
CBHW061500210926
51850CB00008B/1778
* 9 7 8 3 0 3 6 5 6 9 8 7 1 *